The Role of Dietary Bioactive Compounds in Human Health

The Role of Dietary Bioactive Compounds in Human Health

Guest Editors

Baojun Xu
Bin Du

MDPI

Basel • Beijing • Wuhan • Barcelona • Belgrade • Novi Sad • Cluj • Manchester

Guest Editors
Baojun Xu
Life Sciences
Beijing Normal-Hong Kong
Baptist University
Zhuhai
China

Bin Du
Analysis and Testing Center
Hebei Normal University of
Science and Technology
Qinhuangdao
China

Editorial Office
MDPI AG
Grosspeteranlage 5
4052 Basel, Switzerland

This is a reprint of the Special Issue, published open access by the journal *Molecules* (ISSN 1420-3049), freely accessible at: https://www.mdpi.com/journal/molecules/special_issues/Dietary_Health.

For citation purposes, cite each article independently as indicated on the article page online and as indicated below:

Lastname, A.A.; Lastname, B.B. Article Title. *Journal Name* **Year**, *Volume Number*, Page Range.

ISBN 978-3-7258-4370-1 (Hbk)
ISBN 978-3-7258-4369-5 (PDF)
https://doi.org/10.3390/books978-3-7258-4369-5

Contents

About the Editors

Baojun Xu

Dr. Xu is a Chair Professor in Beijing Normal-Hong Kong Baptist University (BNBU, a full English teaching university in China), Fellow of the Royal Society of Chemistry, Zhuhai Scholar Distinguished Professor, Department Head of the Department of Life Sciences, Program Director of the Food Science and Technology Program, and author of over 400 peer-reviewed papers. Dr. Xu received his Ph.D. in Food Science from Chungnam National University, South Korea. He conducted postdoctoral research work at North Dakota State University (NDSU), Purdue University, and Gerald P. Murphy Cancer Foundation in USA during 2005-2009. He was a short-term visiting researcher at NDSU in 2012 and University Georgia in 2014, followed by being a visiting researcher during his sabbatical leave (7 months) at Pennsylvania State University in the USA in 2016. Dr. Xu is serving as an Associate Editor-in-Chief of Food Science and Human Wellness, Associate Editor of Food Research International, Associate Editor of Food Frontiers, and Editorial Board Member of around 10 international journals. He received the inaugural President's Award for Outstanding Research of UIC in 2016 and the President's Award for Outstanding Service of UIC in 2020. Dr. Xu has been listed in the world's top 2% scientists by Stanford University for the past five consecutive years and has been included in the best scientists in the world list in the fields of biology and biochemistry at Research.com in 2023 and 2024. Prof. Xu was named a Highly Ranked Scholar (top 0.05%) by ScholarGPSTM, ranked at #8 in food science and technology in the world and #14 in agricultural and natural sciences.

Bin Du

Dr. Du is a Professor at Hebei Normal University of Science and Technology, Director of the Analysis and Testing Center, and Vice Dean of Hebei Key Laboratory of Natural Products Activity Components and Function. Dr. Du has served as a principal investigator or co-investigator on several extramural research grants from funding agencies. His research interests primarily focus on the isolation and purification of phytochemicals (such as polysaccharide, phenolics, and flavones) in fruits, vegetables, and mushrooms. He is also interested in the structure and physicochemical and biological properties (such as anti-inflammatory and antioxidant activities) of these phytochemicals. He has authored over 30 scientific peer-reviewed papers in different prestigious journals and served as an invited scientific reviewer for many highly respected journals. Dr. Du has frequently been invited to give oral presentations at different academic occasions and present his studies at several national and international conferences. During his doctoral training, Dr. Du received Travel Grants from different agencies. His work aims to understand how nutrition affects the functioning of the human body, since a better understanding of this is key to developing strategies to improve human health and well-being, lower disease risk, and treat nutrition-related illnesses.

molecules

MDPI

Editorial

Editorial for the Special Issue "The Role of Dietary Bioactive Compounds in Human Health" in *Molecules*

Bin Du [1] and Baojun Xu [2,*]

1 Hebei Key Laboratory of Natural Products Activity Components and Function, Hebei Normal University of Science and Technology, Qinhuangdao 066004, China; bindufood@aliyun.com
2 Food Science and Technology Program, Department of Life Sciences, Beijing Normal-Hong Kong Baptist University, Zhuhai 519087, China
* Correspondence: baojunxu@uic.edu.cn

The prevalence of chronic diseases, such as cardiovascular disease, diabetes, chronic obstructive pulmonary disease and severe mental health disorders, has been constantly increasing over the last two decades [1]. It has been recognized that greater adherence to the dietary advice of specialists is a critical component in preventing and managing chronic diseases. Bioactive compounds are ingredients obtained from natural raw materials (e.g., plants, fruits, vegetables, grains, meat, or seafood). Dietary bioactive compounds play an important role in the daily lives of consumers as they are shown to improve health [2]. However, the molecular mechanisms behind the anti-tumor, anti-inflammatory and antioxidant activities of these compounds are not quite clear yet.

This Special Edition is aimed at collecting and summarizing the existing knowledge on the disease prevention effects (including antioxidative, anti-tumor, anti-obesity, anti-diabetes, anti-inflammatory, cardiovascular-protective, skin-protective, and neuroprotective activities, etc.) of dietary bioactive compounds. This Special Issue includes eleven papers covering the above-mentioned aspects, including six reviews and five research papers.

Saponins are a diverse group of compounds that are widely distributed throughout the plant kingdom, and are characterized by their structures containing a triterpene or steroid aglycone and one or more sugar chains [3]. Saponins have various pharmacological activities. Alzheimer's disease (AD), a neurodegenerative disorder, presents a significant global health burden. Saponins have displayed therapeutic potential in treating Alzheimer's disease. Moussa et al. investigated the binding potential of saponins from *Platycodon grandiflorum* with six AD-related proteins using computational and quantum chemistry simulations (Contribution 1). Molecular docking revealed interactions between saponins and key AD enzymes, including GSK-3β and synapsins I–III. MD simulations highlighted three compounds—polygalacin D2, polygalacin D, and platycodin D—showing binding profiles comparable to standard drugs (ifenprodil and donepezil). Further DFT analysis elucidated their electronic properties and anti-AD potential.

Phenolic compounds are a large group of plant metabolites that have long attracted interest from researchers worldwide due to their functions in human health and well-being [4]. In Contribution 2 of this Special Issue, polyphenol and caffeine levels in popular types of coffee from six Warsaw-based franchises (120 samples total) were analyzed. A strong positive correlation was found between polyphenol and caffeine contents (Contribution 2).

In another contribution to this Special Issue, the authors explored whether esculetin can alleviate H_2O_2-induced apoptosis and pyroptosis in HepG2 cells. The results showed that esculetin pretreatment significantly improved cell viability, reduced ROS levels, and suppressed apoptosis by modulating the expression of Bax, Bcl-2, cleaved Caspase-3, cleaved PARP, and Cyt-c. It also attenuated pyroptosis by lowering LDH release and the levels of NLRP3, cleaved Caspase-1, IL-1β, and GSDMD-N. Mechanistically, esculetin inhibited the JNK pathway (p-JNK, p-c-Fos, p-c-Jun), and its protective effects were reversed by the JNK activator anisomycin (Contribution 3).

Furthermore, Wu et al. developed an extraction method combining hot water extraction with three-phase partitioning, followed by ultrasonic degradation to produce ultrasonic *Laminaria japonica* polysaccharides (ULJPs). A comparative analysis revealed that ULJPs had a lower molecular weight (153 kDa) and a more porous structure than LJPs, while maintaining similar primary structures (α-hexopyranoses with 1,4-glycosidic bonds). Notably, ULJP exhibited enhanced bioactivity (Contribution 4).

While sodium butyrate (NaB) is recognized for its therapeutic potential, its effects on macrophage-mediated inflammation remain unclear. One study investigated NaB's protective mechanisms against AGEs-induced damage in THP-1 macrophages through PI3K-dependent autophagy pathways (Contribution 5).

There is one review related to functional foods in this Special Issue. Flaxseeds, valued for millennia, are gaining modern recognition for their rich nutritional profile, including omega-3 fatty acids, lignans, protein, and fiber. These components contribute to diverse health benefits, such as cardiovascular protection, cancer prevention, blood sugar regulation, digestive health, and weight management. As a functional food, flaxseeds offer significant health advantages but continued research and technological advancements are required to optimize their benefits while minimizing risks (Contribution 6). Another contribution examines the unique characteristics and applications of Bactrian camel (BC) milk, an emerging functional food with distinct advantages over other mammalian milk. As interest grows in BC milk's functional food applications, this synthesis provides essential guidance for research and commercial development in this promising sector. The contribution highlights both the scientific basis and economic opportunities surrounding BC milk utilization (Contribution 7).

UV-induced skin photoaging disrupts skin metabolism and impacts well-being. While conventional treatments often cause side effects, natural compounds like epigallocatechin gallate (EGCG), the primary bioactive in tea polyphenols, offer safer alternatives. Another contribution to this Special Issue examines EGCG's mechanisms and benefits and innovations regarding its delivery for photoaging repair (Contribution 8).

Phytochemicals in plant-based foods, particularly polyphenols, demonstrate significant medicinal value despite being non-essential nutrients. Contribution 9 examines their cardioprotective effects through three key mechanisms: oxidative stress regulation, vascular protection, and anti-inflammatory action (Contribution 9).

Sambucus nigra L. (elderberry) has been traditionally valued for its medicinal properties, attributed to its rich antioxidant content. Contribution 10 evaluates the anti-inflammatory and anticancer effects of extracts and components of *S. nigra* L. flowers and fruits (Contribution 10).

There is one systematic review of fruits and vegetables included in this Special Issue. Fruits and vegetables not only serve as nutritional sources but also as natural therapeutics, containing bioactive compounds like hirsutine (HSN), an indole alkaloid found in Uncaria species and various foods (e.g., seafood, grains, and fruits such as bananas and citrus). This contribution synthesizes the current evidence (up to July 2023) from PubMed, Scopus, and other databases on HSN's pharmacological mechanisms and biopharmaceutical properties (Contribution 11).

In summary, the results of the above-mentioned studies will fill the gap between our knowledge of natural bioactive compounds and their underlying cellular signaling and molecular mechanisms. Further studies are needed to discover the cross-connections between oxidative stress and disease prevention in signaling pathways networks.

Author Contributions: B.D. wrote the original manuscript. B.X. wrote the central part and improved this manuscript. All authors contributed equally to the article. All authors have read and agreed to the published version of the manuscript.

Conflicts of Interest: The authors declare no conflicts of interest.

List of Contributions:

1. Moussa, A.Y.; Alanzi, A.R.; Luo, J.; Wang, J.; Cheang, W.S.; Xu, B. Role of saponins from *Platycodon grandiflorum* in Alzheimer's disease: DFT, molecular docking, and simulation studies in key enzymes. *Molecules* **2025**, *30*, 1812. https://doi.org/10.3390/molecules30081812
2. Wierzejska, R.E.; Gielecińska, I.; Hallmann, E.; Wojda, B. Polyphenols vs. caffeine in coffee from Franchise coffee shops: Which serving of coffee provides the optimal amount of this compounds to the body. *Molecules* **2024**, *29*, 2231. https://doi.org/10.3390/molecules29102231
3. Luo, Y.; Chang, T.; Huang, S.; Xiang, J.; Tang, S.; Shen, H. Protective effects and mechanisms of esculetin against H_2O_2-induced oxidative stress, apoptosis, and pyroptosis in human hepatoma HepG2 cells. *Molecules* **2024**, *29*, 1415. https://doi.org/10.3390/molecules29071415
4. Wu, J.; Wang, H.; Liu, Y.; Xu, B.; Du, B.; Yang, Y. Effect of ultrasonic irradiation on the physicochemical and structural properties of *Laminaria japonica* polysaccharides and their performance in biological activities. *Molecules* **2023**, *28*, 8. https://doi.org/10.3390/molecules28010008
5. Yan, M.; Li, X.; Sun, C.; Tan, J.; Liu, Y.; Li, M.; Qi, Z.; He, J.; Wang, D.; Wu, L. Sodium butyrate attenuates AGEs-induced oxidative stress and inflammation by inhibiting autophagy and affecting cellular metabolism in THP-1 cells. *Molecules* **2022**, *27*, 8715. https://doi.org/10.3390/molecules27248715
6. Duarte, S.; Shah, M.A.; Sanches Silva, A. Flaxseed in diet: A comprehensive look at pros and cons. *Molecules* **2025**, *30*, 1335. https://doi.org/10.3390/molecules30061335
7. Seyiti, S.; Kelimu, A.; Yusufu, G. Bactrian camel milk: Chemical composition, bioactivities, processing techniques, and economic potential in China. *Molecules* **2024**, *29*, 4680. https://doi.org/10.3390/molecules29194680
8. Sun, J.; Jiang, Y.; Fu, J.; He, L.; Guo, X.; Ye, H.; Yin, C.; Li, H.; Jiang, H. Beneficial effects of epigallocatechin gallate in preventing skin photoaging: A review. *Molecules* **2024**, *29*, 5226. https://doi.org/10.3390/molecules29225226
9. Iqbal, I.; Wilairatana, P.; Saqib, F.; Nasir, B.; Wahid, M.; Latif, M.F.; Iqbal, A.; Naz, R.; Mubarak, M.S. Plant polyphenols and their potential benefits on cardiovascular health: A review. *Molecules* **2023**, *28*, 6403. https://doi.org/10.3390/molecules28176403
10. Stępień, A.E.; Trojniak, J.; Tabarkiewicz, J. Health-promoting properties: Anti-inflammatory and anticancer properties of *Sambucus nigra* L. flowers and fruits. *Molecules* **2023**, *28*, 6235. https://doi.org/10.3390/molecules28176235
11. Bhuia, M.S.; Wilairatana, P.; Ferdous, J.; Chowdhury, R.; Bappi, M.H.; Rahman, M.A.; Mubarak, M.S.; Islam, M.T. Hirsutine, an emerging natural product with promising therapeutic benefits: A systematic review. *Molecules* **2023**, *28*, 6141. https://doi.org/10.3390/molecules28166141

References

1. Airoldi, C.; Pagnoni, F.; Cena, T.; Ceriotti, D.; De Ambrosi, D.; De Vito, M.; Faggiano, F. Estimate of the prevalence of subjects with chronic diseases in a province of Northern Italy: A retrospective study based on administrative databases. *BMJ Open* **2023**, *13*, e070820. [CrossRef] [PubMed]
2. Vieira, I.R.S.; Conte-Junior, C.A. Dietary bioactive compounds and human health: The role of bioavailability. *Nutrients* **2024**, *17*, 48. [CrossRef] [PubMed]

3. Güçlü-Ustündağ, O.; Mazza, G. Saponins: Properties, applications and processing. *Crit. Rev. Food Sci. Nutr.* **2007**, *47*, 231–258. [CrossRef] [PubMed]
4. Ferreira, T.; Gomes, S.M.; Santos, L. Elevating cereal-based nutrition: *Moringa oleifera* supplemented bread and biscuits. *Antioxidants* **2023**, *12*, 2069. [CrossRef] [PubMed]

molecules

MDPI

Article

Role of Saponins from *Platycodon grandiflorum* in Alzheimer's Disease: DFT, Molecular Docking, and Simulation Studies in Key Enzymes

Ashaimaa Y. Moussa [1], Abdulah R. Alanzi [2], Jinhai Luo [3], Jingwen Wang [3], Wai San Cheang [4,*] and Baojun Xu [3,*]

[1] Department of Pharmacognosy, Faculty of Pharmacy, Ain Shams University, Abbassia, Cairo 11566, Egypt; ashaimaa_yehia@pharma.asu.edu.eg

[2] Department of Pharmacognosy, College of Pharmacy, King Saud University, Riyadh 11451, Saudi Arabia; aralonazi@ksu.edu.sa

[3] Food Science and Technology Program, Department of Life Sciences, Beijing Normal-Hong Kong Baptist University, Zhuhai 519087, China

[4] State Key Laboratory of Quality Research in Chinese Medicine, Institute of Chinese Medical Sciences, University of Macau, Macao SAR, China

* Correspondence: annacheang@um.edu.mo (W.S.C.); baojunxu@uic.edu.cn (B.X.); Tel.: +86-756-3620636 (B.X.)

Abstract: Alzheimer's disease (AD), one of the neurodegenerative disorders, afflicts negatively across the whole world. Due to its complex etiology, no available treatments are disease-altering. This study aimed to explore isolated saponins profiles from *Platycodon grandiflorum* in the binding pockets of six target proteins of AD using computational and quantum chemistry simulations. Initially, saponin compounds were docked to AD enzymes, such as GSK-3β and synapsin I, II, and III. The subsequent research from MD simulations of the best three docked compounds (polygalacin D2, polygalacin D, and platycodin D) suggested that their profiles match with the binding of standard active drugs like ifenprodil and donepezil to the six enzymes. Moreover, analyzing DFT quantum calculations of top-scoring compounds fully unravels their electronic and quantum properties and potential in anti-AD. The subtle differences between polygalacin D and D2, and platycodin D, were studied at the level of theory DFT/B3LYP, showing that the electron-donating effect of the hydroxy ethyl group in platycodin D rendering this compound of moderate electrophilicity and reactivity. Polygalacin D2 diglucoside substituent in position-2 contributed to its best binding and intermolecular interactions more than polygalacin D and prosapogenin D, which acted as the negative decoy drug.

Keywords: Alzheimer's; density functional theory; molecular dynamics simulations; saponins; Platycodon

1. Introduction

Neurodegenerative disorders are more than ever before afflicting populations worldwide. Around one billion people suffer from neurological and age-related ailments [1]; particularly, Alzheimer's disease (AD), which is considered the fifth cause of death globally according to recent WHO reports [2]. By 2030, the number of AD American patients is prospected to rise by more than 35%, and the total AD cases may reach 150 million worldwide in 2050 [3].

The key symptoms of AD patients are cognitive and memory impairment, loss of self-care capacity, apathy, and agitation [4]. The pathological basis of these disorders is complicated and, so far, not totally understood, yet tau aggregation and phosphorylation as well as activation of the amyloidogenic pathway and oxidative stress seemed to be implicated in the progression of AD. The aforementioned disease pathologies pointed out the possible involvement of enzymes like NMDA, GSK-3β, BACE-1, AChE, BuChE, MAO-A, MOA-B, and ROCK2, whose alteration might occur concomitantly or separately [5,6]. In the meantime, effective treatments are missing that can completely cure AD, and even drugs that could partially manage the symptoms are limited to two classes only [7,8]. With the limited efficacy of previously discovered drugs like memantine, donepezil, galantamine, and rivastigmine, and their side effects, few options are left for patients. Currently, only aducanumab is used, which was introduced into the market in 2003 [9].

Based on several clinical studies, the multitargeted approach was proved to be the best choice for management of AD due to the complex nature of the disease and the need to induce or inhibit multiple genes and proteins. Patients who suffer from AD mainly have pathological hallmarks like neurofibrillary tangles and amyloid plaques, which are caused by the NMDA receptor (NMDA), BACE1, and GSK-3β hyperactivity. While β-secretase acts to cleave the β-amyloid protein to amyloid beta peptides 42 (Aβ-42) and subsequently to amyloid plaques, the excessive oxygen species can cause tau phosphorylation and neurofibrillary tangle formation. Moreover, the NMDA overactivity with the accumulation of amyloid plaques acts to develop neuronal death and toxicity [10]. In this regard, in silico experiments play an invaluable role; especially, with the current advancement of high-end computer power devices and the development of chemoinformatic methods to screen and optimize a large number of drug molecules in a time- and cost-effective manner [11–14].

Saponins Against Age-Related Neurological Disorders

Why are natural products or functional foods preferred to treat neurodegenerative diseases? Rather than synthetic drugs, medicine foods are much favored; particularly, when a protective effect is required over a long period of time. Almost all humans need protective measures against neurodegenerative ailments. Plants have always been the repository of powerful secondary metabolites including alkaloids, phenolics, steroids, and saponins [15,16]. Saponins are bioactive natural products with a characteristic triterpene or steroidal aglycone core, and their effectiveness was long recognized in traditional Chinese medicine (TCM) [17]. Sun et al. referred to 33 saponins possessing a neuroprotective activity, either acting as antioxidants, inhibiting tau phosphorylation or ACHe inhibitors, or enhancing neuronal reconstruction and modulation of the GABA or NMDA receptors [17]. Ginseng was locally used in China to treat symptoms of AD [18], and its saponins effect on neuronal activity was recognized in several studies. A mixture of *Panax notoginseng* saponins composed of 14.5% notoginsenoside R1, 27.7% ginsenoside Rb1, and 28.0% ginsenoside Rg1 improved mitophagy and reduced cerebral oxidative stress in mice [19]. Ginsenoside Rg1 thwarted neuronal senescence via reducing NOX2-mediated ROS generation in H_2O_2-treated hippocampal neurons [20]. While short-term use of ginsenoside Rg1 reduced the level of amyloid β peptide with a notable memory improvement in OVX rats [21], their long-term use ameliorated redox stress and enhanced the plasticity proteins and activated progenitor cells in the hippocampus [22,23]. Moreover, ginsenoside Rg3 decreased amyloid β proteins 40 and 42 via a 2.9-fold increase in neprilysin gene expression [18]. Despite their violation of ADMET criteria, ginsenoside Rg3 proved to be effective in liposomal drug delivery systems instead of cholesterol to suppress the brain tumors in C6 glioma cells [24,25]. Furthermore, the drug delivery through an intranasal route together with β-cyclodextrin inclusion was efficient more than 150-fold to raise the brain concentration of gensinoside F1 [25]. Saponins of *P. grandiflorum* mainly comprised oleanolic acid with several sugar units attached; while platycodin (PD) possessed five sugars and PD3 contained six sugars, which contributed to the difference in their NO production or TNF-α activity. According to Wang et al., the less sugar units provided more potent pharmacological actions [26].

In this study, molecular docking and quantum chemistry simulations play an important role in predicting the binding capacity and interactions of *P. grandiflorum* saponins in the binding sites of key proteins involved in the neurodegenerative pathways. ADMET analysis was employed to indicate the oral bioavailability of these saponins and their nutraceutical use; dimethyl 2-O-methyl-3-O-a-D-glucopyranosyl, platycogenate A, polygalacin D2, platycodin C, platycodin D, prosapogenin D, deapioplatycodin D, platycodin D2, platycodin D3, deapioplatycodin D3, deapioplatycoside E, platyconic acid B lactone, platycodin A, 3″-O-acetylplatyconic acid A, polygalacin D, polygalacin D3, platycodin V, platycoside E, and polygalacin XI comprised the investigated list in this article (Figure 1).

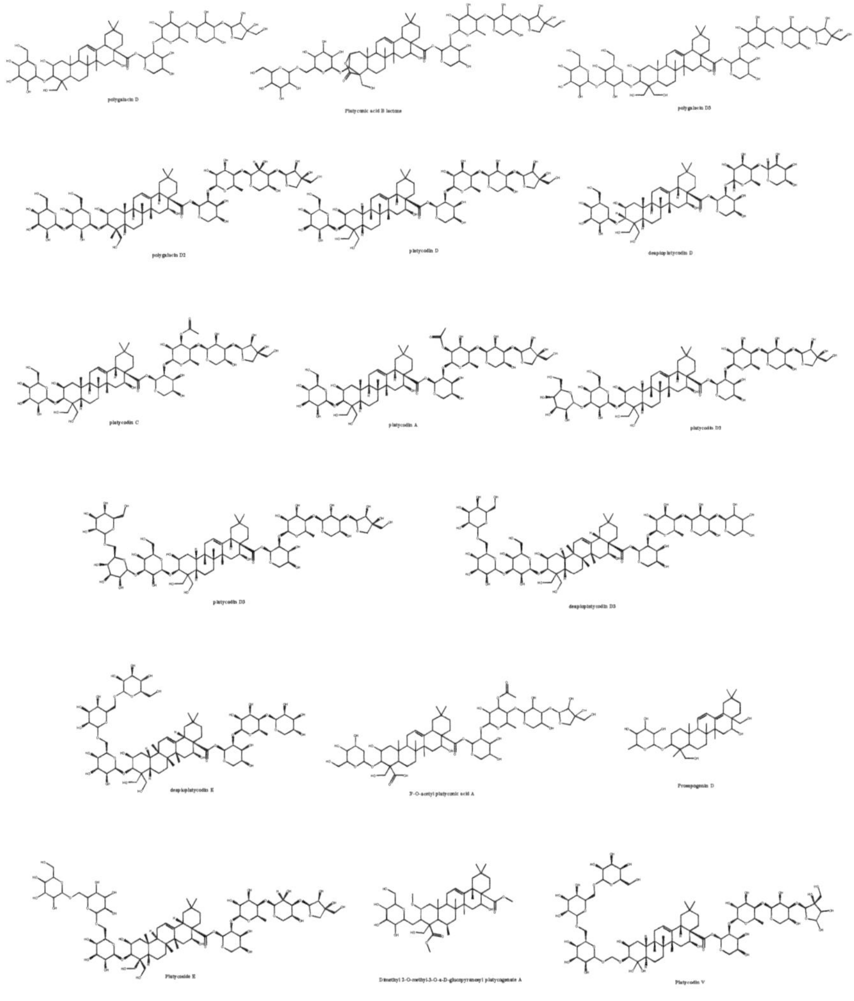

Figure 1. Chemical library of platycodin saponins subjected to computational analysis.

2. Results

2.1. Selection of the Best Inhibitors for Proteins

The structures of the compounds were acquired from our previous work [24] based on their physicochemical properties. The top 15 compounds were selected for each protein target, synapsin I, synapsin II, synapsin III, BACE1, GSK-3β, and NMDA receptor. The selected compounds underwent a pharmacokinetic analysis. Table 1 presents the inhibitors chosen for the targeted proteins.

Table 1. Selection of platycodon saponins in protein inhibitors using pharmacokinetic analysis.

No.	PubChem CID	Compounds	MW (g/mol)	HBD	HBA	logP	Drug-like
1	101596922	Prosapogenin D	603.817	6	8	3.62969	TRUE
2	162859	Platycodin D	1225.335	17	28	−5.3782	TRUE
3	385678065	Polygalacin D2	1371.477	19	32	−6.5264	FALSE
4	385678073	Polygalacin D3	1371.477	19	32	−6.5264	FALSE
5	46173919	Platycodin C	1267.372	16	29	−4.8074	TRUE
6	50900852	Platyconic acid B lactone	1383.444	18	33	−6.9818	FALSE
7	53317652	Platycodin D2	1387.476	20	33	−7.554	FALSE
8	70698202	Platycoside E	1549.617	23	38	−9.7298	FALSE
9	70698266	Deapioplatycodin D	1093.22	15	24	−3.8431	TRUE
10	70698289	Deapioplatycoside E	1417.502	21	34	−8.1947	FALSE
11	75251137	Platycodin D3	1387.476	20	33	−7.554	FALSE
12	96023791	Polygalacin D	1209.336	16	27	−4.3506	TRUE
13	70698190	Deapi-platycodin D3	1255.361	18	29	−6.0189	TRUE
14	46173910	Platycodin A	1267.4 g	16	29	−3.1	TRUE

2.2. ADMET Results

All compounds were subjected to pharmacokinetics and ADMET criteria determination [27] where most of them showed poor BBB penetration and oral bioavailability, yet improvements in drug delivery systems might present a promising solution to enhance their therapeutic potential as was reported for ginsenosides.

2.3. Molecular Docking

Maestro 2018 conducted protein–ligand docking, which was further evaluated through MD simulation, RMSD, RMSF, radius of gyration, and the number of hydrogen bonds in protein–ligand interactions. A model system was formulated for polygalacin D2 as an inhibitor for synapsin I and BACE1 and platycodin D as the NMDA inhibitor. Polygalacin D was employed as an inhibitor of synapsin II, III, and GSK-3β. The type of interaction between the residues of the ligand–protein complex is shown in Table 2.

Table 2. Docking interactions and binding free energies of platycodin saponins in the investigated enzymes.

Synapsin I	Binding Free Energy (Kcal/mol)	H-Bonds	Hydrophobic and VDW
polygalacin D2	−6.742	leu394, Arg186, Asp120, Glu121, Asn 334, Trp335, Asp283, Lys281 **(10)**	ile395, Pro393, Trp126, Met277, Gly276, Met297, Ser275. His123, Thr124, Asn338
3′-O-acetyl platyconic acid	−6.361	Asp309, Lys281, Gly333, Asn334, Asn 338, Thr339 **(8)**	Phe307, Pro265, Lys378, Trp335, Lys336, Gly340, Met277, Gly276, Ser275
LY281137	−6.916	Trp335	Arg328, Lys336, Thr337, Asn 338, Gly276, Ile385, Glu386, Lys279, Lys269, Leu375, Val267, Phe307, Ile308, Ala310, Asp313

Table 2. *Cont.*

Synapsin I	Binding Free Energy (Kcal/mol)	H-Bonds	Hydrophobic and VDW
PF-0480	−6.703	Lys269, Glu373, Ile308	Ala310, Leu375, Trp335, Asp309, Asn338, Phe307, Gly276
donepezil	−5.951	Lys279, Gly276, Trp335	Val388, Ile385, Leu375, Ala310, Ile308, Phe307, Pro306, Ile247, Val267
ifenoprodil	−4.672	Lys279	Trp335, Leu375, Ala310, Met392, Pro393, Val388, Ile385, Val267
synapsin II			
polygalacin D	−6.101	Lys337, trp336, asn335, glu122, asp121, arg187 **(7)**	ile396, leu395, pro304, ser392, met278, val281, lys282, gly279, met278, gly277, asn339, ser276
Polygalacin D2	−6.011	asp310(1.85, 2.50), lys379(2.05), lys282(1.89, 1.85, 2.08), asn339, ser276(1.88), glu197(1.65, 1.72) **(11)**	phe308, ala311, gly314, trp336, val281, met278, gly341, gly277, ala194, lys337, lys280, gly277, gly341
PF-0480	−6.839	ile309, phe386	val 376, trp 336, pro307, phe308, ile248, ala311, val268, lys379, lys374, lys337, lys280, ser276, asn335, glu306, asp310, asp314, ser331, arg329, lys270
LY-2811376	−5.727		phe386, val268, ile248, pro307, phe308, ile309, ala311, val376, trp336, lys280, lys270, lys337, lys374, arg329, gly277, ser276, asn339, thr338, glu306, asp314
ifenoprodil	−5.187	lys374, glu306, phe386, asp314	ile248, val376, pro307, phe308, ile309, val268, phe386, trp336, met389, pro394, lys270, lys337, arg329, arg316, thr338, asn339, ser276, gly277
donepezil	−4.879	lys374, glu306, phe386, asp34	ile248, val376, pro307, phe308, ile309, val268, phe386, trp336, met389
synapsin III			
polygalacin D	−6.145	ala254, asn313, gly312, lys260, phe286, asp288, asp358 **(8)**	met256, trp314, pro244, ala285, ile287, lys357, ser356
polygalacin D2	−5.882	ala254, asn313, lys260, asp288, asp358 **(8)**	met256, trp314, gly312, pro244,lys260, phe286, ile287, ser356, lys357, arg360, phe243
pf-0480	−6.361	lys248, ala254, lys352, trp314, ile287	met367, ile364, ala252, pro372, ala310, val354, phe286, ala285, val226, val246
ifenoprodil	−5.345	trp314	ala252, met367, ala254, ile364, val354, pro372, ala316, val246, ala285, phe286, ile287, val354
donepezil	−3.513	lys352, ser289	ala254, phe286, ile287, tyr291, ala316, trp314, val354, tyr291, val246, ile364
GSK-3B			
platycodin D	−7.147	asp200, asp181, lys183, ser66, asn186 **(6)**	val70, phe67, cys199, ile62, leu188, ala83, leu132, val110, tyr134, val135, tyr140
polygalacin D	−6.873	arg148, lys183, ile62, asp200, gly202, asn95 **(9)**	Tyr140, tyr222, ile217, val87, lys85, arg96, ser203, phe67, ser66, gly 65, gly63
platycodin D3	−6.514	ser66, asn64, gln185, arg141, thr138, pro136, tyr134 **(7)**	ile217, cys218, phe67, leu188, tyr140, pro136, val135, tyr134, ser219, asp181, lys183, phe67, gly65, gly62, ile62, lys60, gln72
platyconic acid B-lactone	−6.418	arg148, arg144, arg141, tyr140, glu137, tyr221, ser203, asp200, asp181, lys183 **(13)**	arg223, tyr222, tyr221, arg220, ser219, cys218, ser66, phe67, gly202, lys85, glu185, asn186
	−6.297	asp200, lys183, thr138, arg141, ser66, lys60, gln72 **(9)**	asp181, gln185, asn186, tyr140, phe67, gly65, asn64, gly63, ile62, val61, arg 148
PF-04820	−8.373	lys85, asp200, val135, arg141	phe67, val70, ala83, cys199, leu132, leu188, val110, tyr134, pro126, ile62
ifenprodil	−5.956	asn186	cys199, val110, ala83, leu132, ile62, tyr134, val135, pro134, phe67, val70, leu188

Table 2. *Cont.*

Synapsin I	Binding Free Energy (Kcal/mol)	H-Bonds	Hydrophobic and VDW
donepezil	−5.918	lys85	phe67, val70, ile62, phe201, cys199, met101, ala83, leu130, val110, leu132, tyr134, leu188, val135, pro136
BACE1			
polygalacin D2	−7.298	phe109, lys107, asn111, arg307, lys321, gly264, glu265, gln73, pro70, ile126, ser36, tyr198 **(15)**	phe47, phe108, ile110, leu263, tyr71, val69, ala127, gly34
polygalacin D	−5.607	Arg307, lys321, arg235, asp223, lys224 **(9 bonds)**	val309, tyr71, pro70, tyr198, Gly11, Gln12, Ser10, Lys107, Glu265, Thr72, Gln73, Ser327, Ser328, Thr329
3′-O-acetyl platyconic acid	−7.716	tyr198, tyr71, lys75, asn233, thr232, thr 231, gly230 **(7)**	trp197, lys224, arg128, pro70, thr72, gln73, gly74, thr329, ser328, gln326, er325, arg236, leu263, gly264, leu30, ile110, arg307, gln12
platyconic acid -B lactone	−6.93	lys75, tyr71, thr72, gln326, glu265, gly264, lys321, arg307, asn111 **(9)**	gly74, pro70, leu263, val309, phe47, ile110, phe109, val309, thr329, ser328, ser327, ser325, arg235, asn233, ser10, gly11, gln12, lys107
platycodin D3	−6.891	asp106, lys107, lys75, gly264, gln73, gly11, thr232, gly230 **(10)**	phe108, ile110, leu263, gly74, arg307, lys321, ser325, glu265, arg235, asn233, thr231, gln12, gly13, leu30, trp115
platycodin D	−6.758	gly11, thr232, the231, thr72, asp106 **(5)**	pro46, phe47, lys107, phe108, phe109, ile110, asn111, ile118, val332, ile226, tyr198, tyr71, gln73,gln12, arg235, lys321, arg307, glu265
ifenoprodil	−5.726	pro70, trp115, asp228	tyr71, val69, ile118, phe108, ile110, tyr198, ile226, leu30, val332
PF-04820	−5.147	tyr198, tyr71, asp228,	ile126, val332, phe108, ile110, trp115, ile118, pro70, val69, ile226, leu30
donepezil	−5.052	thr72	ile126, tyr198, ile226, val332, leu30, pro70, tyr71, ile118, phe108, ile110, trp115, arg128, gly34, ser35, asp228, asp32
LY-2811376	−5.018	gly11, thr72	trp115, ile118, leu30, tyr198, tyr71, phe108, ile110, asp228, ser35, asp32
NMDA			
polygalacin D2	−7.755	Gln48, tyr351, asp353, lys296, gly354, gln291, arg380 **(10)**	leu280, leu377, met375, ile275, leu292, ile293, val355, ala307
platycodin D	−7.296	leu382, gln371, gly354, asp353, tyr351, ser303 **(7)**	val383, ile293, Met375, val355, ala307, gln384, gln291, ser373, asn297, thr356, arg52, asp304
3′-O-acetyl platyconic acid	−7.214	pro95, thr122, lys37,glu298, lys296, asn294 **(10)**	pro96, pro95, ile41, ile293, ser93, ser34, thr35, ile293, ser299, asn297, gln291
platycodin D3	−7.183	glu246, tyr144, lys296, leu292, ser373, gln384 **(10)**	ser245, arg273, ser276, gly277, met125, arg124, thr123, thr122, asn297, gly295, asn294, ile293, gln291
polygalacin D	−6.736	pro95, glu298, lys296, leu292, asn294, val355 **(8)**	pro96, ser93, thr 122, ser34, asn297, gly295, ile293, gln 291, glu272, gln384, ser373
ifenoprodil	−5.29	ser393, gln291 **(2)**	leu292, ile293
prosapogenin D	−4.415	lys296, gln291**(4)**	ser299, glu298, asn297, asn294, ile293, ile387, gln384, ser373, met375
donepezil	−4.354	lys37, ser299 **(3)**	pro95, pro96, ile41, gly295
ly-2811375	−4.269	lys37, ser303 **(2)**	val355, ala45, ile41
pf-04820	−4.193	lys296 **(2)**	ile41, pro95, pro96

Green colour represents a heatmap that indicates the strength of bindings. Underline indicates matching between the native ligands and the docked compounds.

2.4. Molecular Quantum Calculations

Elucidating the quantum chemical characteristics of drug molecules is essential more than ever before to predefine interactions and reactivities especially before chemical synthesis [28]. The four compounds of platycodin D, polygalacin D, polygalacin D2, and prosapogenin D were selected to conduct the DFT calculations via the 6–31G (dp) basis set to reveal their key structural and electronic features (Table 3).

Table 3. Density functional theory calculations for selected compounds.

Compound	Polygalacin D	Prosapogenin D	Platycodin D	Polygalacin D2
E_{LUMO} (eV)	1.1064155	−0.02387	−0.026395057	0.122995523
E_{HOMO} (eV)	−6.7048887	−0.21197	−6.11385713	−6.03249504
ΔE	5.5984732	0.18826	6.087462073	5.909499517
χ	3.9056521	0.23538	3.070126094	2.954749759
η	3.9056521	−0.09405	3.043731037	3.077745282
σ	1.716714271	−2.25188569	2.008671941	0.324913178
ω	7.840523073	−124.3888	8.516100697	10.58547613
dipole moment (DEBYE)	7.8259	4.8371	7.2001	8.0721
chemical potential (μ)	2.7992366	0.11792	3.070126094	2.954749759
electronic energy	−17,794.64455	−7977.932898	−22,130.38172	−25,782.40283

Electronegativity (χ) was determined as (I + A)/2 and the chemical potential (μ) was obtained as −(I + A)/2. Moreover, the global hardness (η) was attained as (I − A)/2, and the global softness was I/η. While ionization potential (I) was computed as $-E_{HOMO}$, electron affinity (A) was the $-E_{LUMO}$ value based on Koopman's theorem.

$$\mu = \left(\frac{\partial E}{\partial N}\right)v(r) = -\left[\frac{IP + EA}{2}\right] = -\left[\frac{EHOMO + ELUMO}{2}\right]$$

ω or the electrophilicity index was realized from this equation ω = χ2/(2η), which describes the tendency of a molecule to accept electrons. Compounds chemical stability and reactivity were deduced from their energy band gap (ΔE) from the HOMO to the LUMO [29].

Polygalacin D showed a higher ΔE of 5.5984732 eV compared to prosapogenin D whose ΔE scored 0.18826 eV; thus, the former was of higher stability and lower chemical reactivity than the latter. Polygalacin D demonstrated a higher kinetic stability and lower reactivity with adequate hardness. Thus, it manifested a greater tendency to donate electrons with an EHOMO = −6.7048887 eV. On the other hand, the ELUMO was of lower energy 1.1064155 eV than −0.02371 eV, indicating its lower tendency to accept electrons. The ability of the selected compounds to acquire electron charges was noted from their electrophilicity index (ω), which was lower for polygalacin D than prosapogenin D. The electronegativity (χ) value indicating the affinity to attract electrons was 3.9056521 and 0.23538 for polygalacin D and prosapogenin D, respectively. The close values of global softness (σ) and hardness (η) in both compounds indicated their balanced electron density. Moreover, polygalacin D showed better charge separation signified by its high dipole moment (μ) of 7.8259 debye compared to prosapogenin D 4.8371 debye. Polygalacin D revealed its enhanced energetic stability based on its electronic energy value of (−17,794.6445474876) compared to prosapogenin D (−7977.93289844684). Polygalacin D2 and platycodin D were closely related to polygalacin D in most of their quantum chemical values; especially, their ΔE and electronic energy values of −25,782.40 and −22,130.38 Kcal/mol, respectively, denoting their stable energetic fitting, which matched their binding features in molecular docking and MD simulation results. Prosapogenin descriptors were not favored for good binding interactions inside the enzyme pockets.

2.5. Frontier Molecular Orbitals (FMOs)

By analyzing the highest occupied molecular orbital (HOMO) and the least unoccupied molecular orbital (LUMO) of one of the promising compounds polygalacin D and prosapogenin D as a negative control, it was possible to infer the difference in their chemical stability and reactivity. This orbitals study promoted our understanding of a compound's reactivity, electronic structure, and chemical interactions with other molecules. As far as HOMO is concerned, it controls the tendency to donate electrons in areas of lone pairs or bonding areas while the LUMO being vacant of electrons indicates the ability of the compound to accept electrons. The ΔE between the HOMO and LUMO is called the "energy band gap" and is given by the equation $\Delta E = |E_{LUMO} - E_{HOMO}|$. ΔE is a key indicator of the molecular reactivity during the quantum mechanical studies. Polygalacin D and D2 demonstrated its HOMO electron cloud throughout the C12 double bond. On the other hand, the LUMO featured a distribution over the C19-ester bond in half of the molecule. This arrangement manifested the possibility of electron donation from the HOMO in the central part of the molecule and the LUMO electron gain (Figure 2).

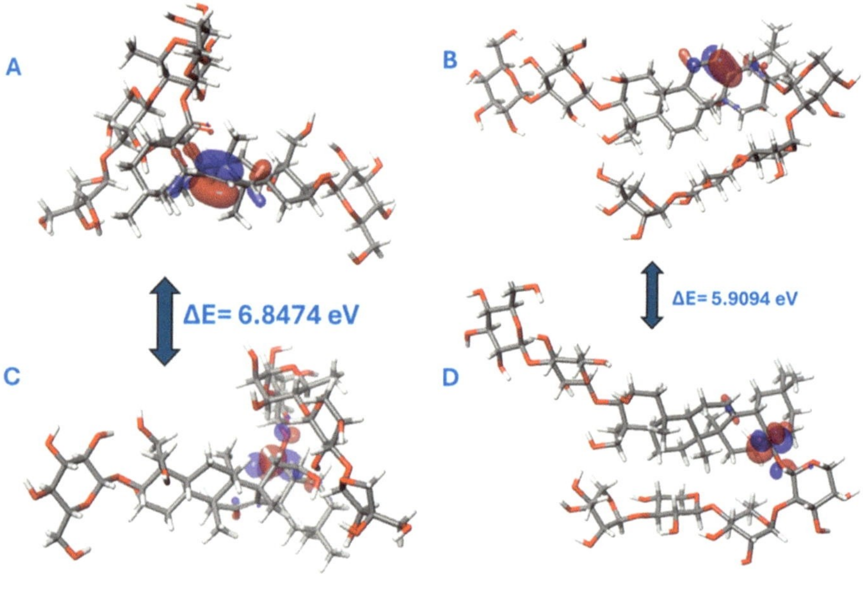

Figure 2. (**A**) Polygalacin D HOMO-LUMO, (**B**) prosapogenin D HOMO-LUMO, (**C**) platycodin D HOMO-LUMO, (**D**) polygalacin D2 HOMO-LUMO.

The three selected compounds varied in their 1- or 2-position substitutions in the pentacyclic aglycone core. After conducting docking analysis, this moiety was shown to fit in the same groove as the controls ifenoprodil and PF-04802, while the tetrasaccharide moiety might have rendered more stability via its prominent polar interactions outside the groove. The diglucoside in position 2 was favored as in polygalacin D2 in enzymes synapsin I, NMDA, and BACE1 followed by polygalacin D in synapsin II and III, which possessed one 2-glucose moiety with a single hydroxy ethyl group in position 1. These subtle disparities rendered different values of quantum chemical energies. Remarkably, polygalacin D2 showed the best interaction with all AD key enzymes, excluding GSK-3β, which indicated the importance of the second glucose unit in position 2 of the pentacyclic system. Platycodin D HOMO orbital was allocated to C11-C12 unsaturation, while the LUMO was detected in the C19-ester bond, and it possessed the highest ΔE energy gap of value 6.087462073 eV, denoting its stability.

In contrast, prosapogenin D HOMO orbitals were distributed over the double bonds C11-C12, C13-C18, and the C17 hydroxy methanol group single bond. The LUMO electron cloud was allocated over the single bond C12-C13 as well as atoms C11 and C17 in the attached methyl hydroxide. This electronic environment suggested that the HOMO orbital can donate electrons from the region of the conjugated double bonds C11-C12, C13-C18, and the LUMO can gain electrons in the region of the C12-C13 single bond and the C17 attached hydroxy methyl, which shed light over a high expected reactivity of this molecule (Figure 2).

The HOMO-LUMO gap was seen to be large in the three target compounds indicating their chemical stability and selectivity for the receptor properties.

From Table 3, Polygalacin D was the least prone to accept electrons due to its high LUMO orbital energy being the least nucleophilic. The HOMO-LUMO energy gap manifested that prosapogenin D is the most reactive and the softest compound with ΔE (0.1882 eV). Both polygalacin D2 and platycodin D were close in their ΔE values denoting favorable stability. The highest χ value was reported for polygalacin D, indicating its powerful ability to attract electrons compared to the non-docked compound prosapogenin D. The hardness index η was essential to indicate the compounds resistance to electronic distribution changes where polygalacin D was the hardest compared to the soft prosapogenin D ($\eta = -0.0945$). The Electrophilicity Index (ω) also matched with other descriptors pointing out polygalacin D2 as the most electrophilic candidate and the more prone to conduct intermolecular interactions with its high dipole moment of 8.0721 D and most negative electronic energy. On the other hand, platycodin D showed moderate stability with balanced nucleophilicity and electrophilicity.

2.6. Molecular Electrostatic Potential (MEP)

The MEP data of the three promising molecules were depicted compared to prosapogenin D as a negative control, being only docked by the NMDA protein, by employing DFT optimization with the B3LYP/6-311G (2d, p) functional and basis set (see Figure 3).

Platycodin D revealed multiple nucleophilic sites around the hydrogen atoms of the 2-glucoside moiety, likewise, around the hydroxy ethyl groups attached to position 1 of the aglycone. The electrophilic sites were noted around the heterocyclic oxygen of sugar. This part of the molecule is the same, which fitted in most of the bioactive sites in the enzymes investigated within the same docking location of the standards, which might refer to the potential significance of this 2-position attachment particularly with the planar conformation of the 2-glycoside rendering its insertion within the groove of interest feasible. The molecular electrostatic surface was displayed with color codes for polygalacin D, polygalacin D2, and prosapogenin D to unveil the active sites inside molecules. The green code manifested a neutral area. Whereas, the positive blue color indicated low electron density and regions favoring nucleophilic attacks, and the red color showed negative potential values and sites subjected to electrophilic attacks (Figure 3).

In both compounds, the nucleophilic regions were centered around oxygen atoms in the sugar rings or the oxygen substituents in C17 attached side chain, and the electrophilic sites were around the hydrogens attached to the hydroxyl groups. The rest of the molecule's parts including the pentacyclic aglycone were mostly neutral. The MEP allows the prediction of hydrogen bonding and hence biological efficacy. According to the Mulliken-type charge distribution, the blue color denoted strong attraction around the alkyl groups and the red region indicated strong repulsion around the electronegative oxygen atoms, respectively. Consequently, polygalacin D and D2, and platycodin D, were suitable and compatible for AD enzyme binding.

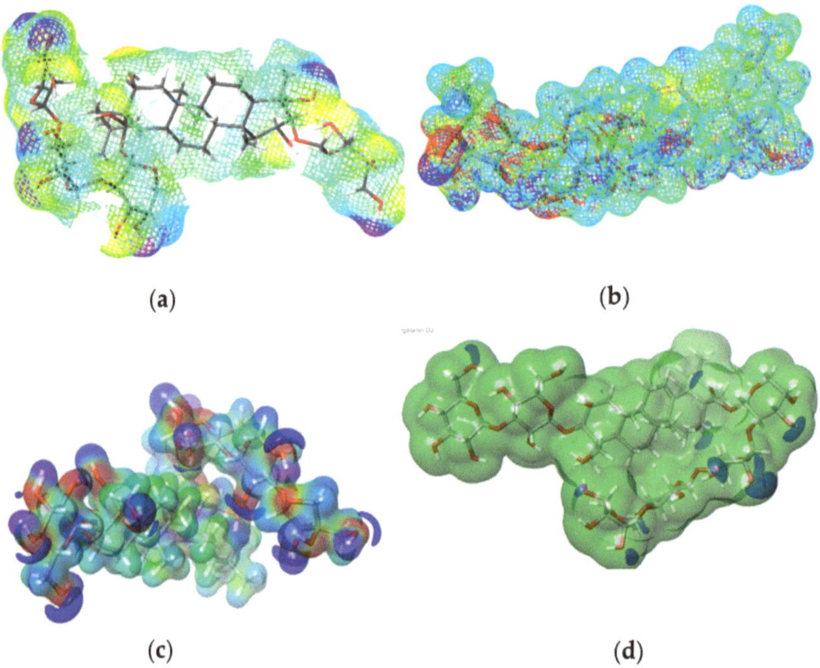

Figure 3. (**a**) Polygalacin D electrostatic potential, (**b**) prosapogenin D electrostatic potential, (**c**) platycodin D electrostatic potential, (**d**) polygalacin D2.

2.7. Docking Interactions in Selected Proteins

2.7.1. Synapsin I Receptor

The saponins library was docked using AutoDock Vina software (v1.2.x) and ranked according to their binding free energies (ΔG) with the respective proteins. A standard molecule with reported anti-Alzheimer activity was used to compare the results as shown in Table 3 where the more negative values indicated a stronger and more stable binding interaction. All the investigated platycodin saponins exhibited a ΔG value below −4.67 Kcal/mol in synapsin I bind pocket with polygalacin D2 and 3″-O-acetyl platyconic acid scoring −6.74 and −6.36 Kcal/mol, respectively (Figure 4). These two compounds were elected to validate their binding interactions via molecular dynamics simulations. LY2811376, donepezil, and ifenoprodil were utilized as reference positive drugs. Polygalacin D2 superseded 3″-O-acetylplatyconic acid regarding the number of polar and nonpolar interactions (Table 2) [30].

The RMSD of Cα atoms of Synapsin I bound to polygalacin D2 was measured and compared with the RMSD of the apo protein structure. The RMSD of the apo protein and complex deviated to 0.4 nm at 20 ns and then showed deviations in this range until the end, indicating the similar behavior in protein upon binding of the ligand (Figure 5). The RMSF plots revealed that the RMSF values of both the apo protein and complex were similar to most residues that exhibited values less than 0.2 nm (see Figure 5). Figure 5 shows the behavior of Rg during simulation. The plot revealed that the Rg values of the apo protein and complex were similar until 30 ns and then the Rg values of the complex showed deviations of 0.10 nm until the end of simulation. Figure 5 depicts the hydrogen bonds exhibited by the complex during the simulation run. At the beginning of the simulation, the protein–ligand complex formed six hydrogen bonds, which decreased to zero at some frames. The average number of hydrogen bonds was three during the simulation. The total binding free energy of the complex was −67.41 ± 1.64 Kcal/mol. The contribution of other energy parameters is shown in Figure 5.

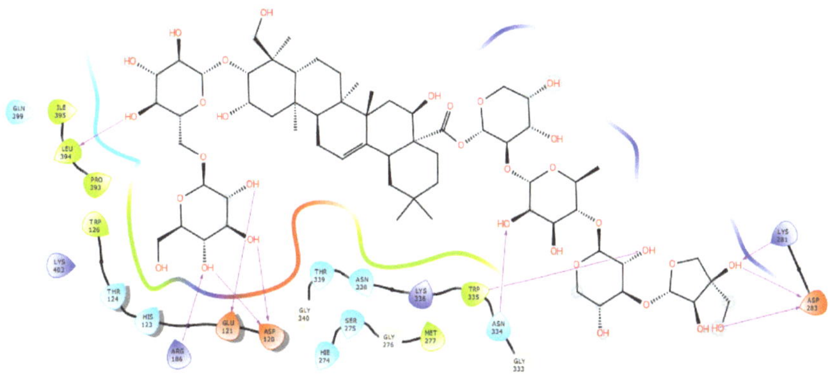

Figure 4. Polygalacin D2 in the binding site of synapsin I.

Polygalactia D2 revealed around 10 hydrogen bonds to several amino acid residues as Asn 334, Trp335, and Lys281 in a manner that resembles the standard drugs LY281137 and donepezil: additionally, nonpolar interactions with the hydrophobic bed formed by ile395, Pro393, Trp126, Met277, Gly276, Met297, and Ser275. His123, Thr124, and Asn338 explained its favorable binding free energy −6.742 Kcal/mol that surpassed the controls (see Figure 4). Different descriptors such as RMSD, RMSF, and Rg were measured and all indicated the stability of ligands binding. The Rg values indicated the compactness of the protein after binding to polygalacin D2 and 3″-O-acetyl platyconic acid (Figure 5).

2.7.2. Synapsin II Receptor

The results of docking, MD simulations, and MMGBSA of synapsin II receptor with polygalacin D, polygalacin D2, and donepezil (control) are shown below. As shown in Table 4, the binding free energies for the docked molecules ranged between −3.712 and −6.101 Kcal/mol. Polygalacin D and polygalacin D2 were the candidates selected for further molecular dynamics measurements since they scored the lowest energy values −6.011 and −6.101 Kcal/mol, respectively. The docked complex of synapsin II receptor–polygalacin D2 is shown in Figure 6.

Polygalacin D2 formed polar interactions with residues asp310, lys379, lys282, asn339, ser276, and glu197 via 11 hydrogen bonds that ranged in their length between 1.65 and 2.50 A. Despite the variation in molecular size, polygalacin D2 effectively fitted in the 1i7L binding site as standards PF-0480, LY-2811376, and ifenoprodil with amino acid contacts comprising phe308, trp336, lys280, ala311, gly277, and lys337, which formed both van der Waals and hydrophobic interactions.

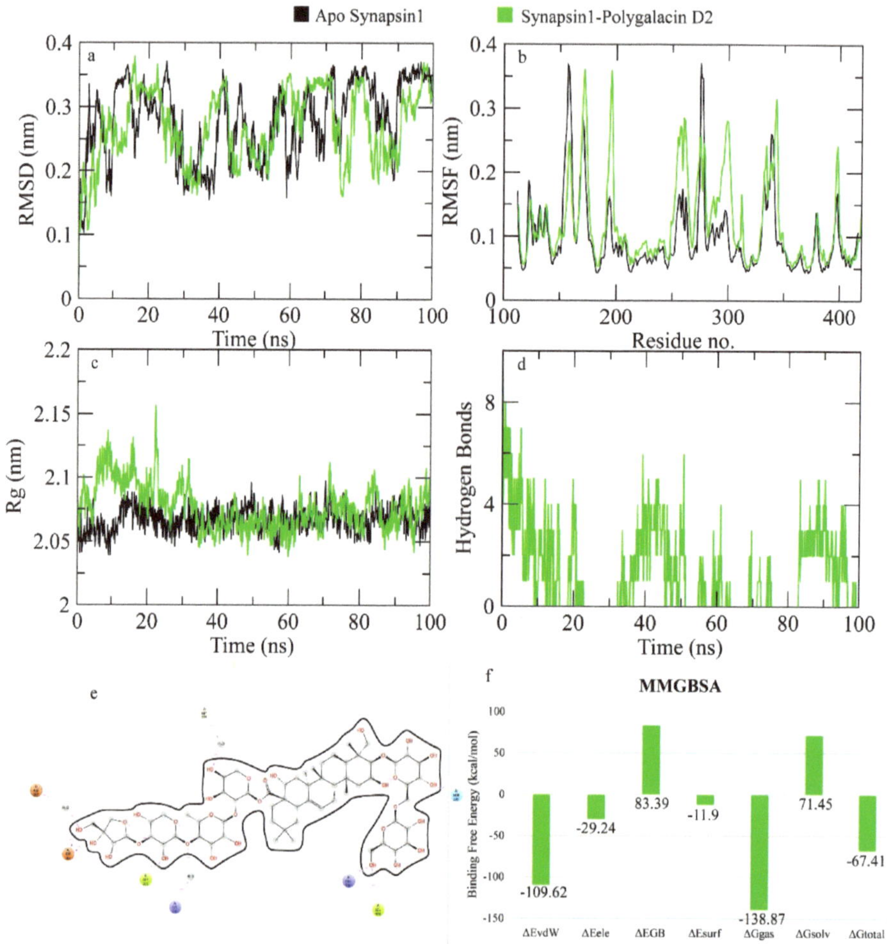

Figure 5. The simulation analysis. (**a**) The RMSD of Cα atoms of the apo protein and complex. (**b**) The flexibility of protein residues upon binding of the ligand. (**c**) Radius of gyration calculation of protein upon binding the ligand. (**d**) The hydrogen bonds calculation between the protein and ligand. (**e**) The molecular simulation result between polygalacin D2 and synapsin I. (**f**) The total binding free energy of energy components in the synapsin I-polygalacin D2 complex.

The RMSD of Cα atoms of synapsin II bound to polygalacin D was measured and compared with the RMSD of the apo protein structure. The RMSD of the apo protein and complex deviated to 0.4 nm at 20 ns and then showed deviations in this range until the end, indicating the similar behavior in protein upon binding of the ligand (Figure 7). The RMSF plots revealed that the RMSF values of both the apo protein and complex were similar as most residues exhibited values less than 0.2 nm. Figure 7 shows the behavior of Rg during simulation. The plot revealed that the Rg values of the apo protein and complex were similar until 20 ns and then the Rg values of the apo protein showed deviations of 0.15 nm until the end of the simulation. Figure 7 depicts the hydrogen bonds exhibited by the complex during the simulation run. At the beginning of the simulation, the protein–ligand complex formed six hydrogen bonds, which decreased to three at some frames. The average number of hydrogen bonds was four during the simulation. The total binding free energy of the complex was −45.61 ± 1.69 Kcal/mol. The contribution of other energy parameters is shown in Figure 7.

Table 4. The glide scores of the saponin compounds against protein targets.

Compounds	GSK-3β	NMDA Receptor	BACE1	Synapsin I	Synapsin II	Synapsin III
Platycodin A	0	0	0	0	0	0
Platycodin C	0	0	0	0	0	0
Platycodin D	−7.147	−7.296	−6.758	−4.898	−3.712	−3.202
Deapioplatycodin D	0	0	0	0	0	0
Platycodin D2	0	0	0	0	0	0
Platycodin D3	−6.514	−7.183	−6.891	−5.995	−5.513	0
Deapioplatycodin D3	0	0	0	0	0	0
Deapioplatycoside E	0	0	0	0	0	0
Platyconic acid B lactone	−6.418	−6.623	−6.93	−6.058	−5.973	−4.801
3″-O-acetylplatyconic acid A	−6.297	−7.214	−7.716	−6.361	−6.078	−6.018
Prosapogenin D	−5.456	−4.415	−4.407	0	0	0
Polygalacin D	−6.873	−6.736	−5.607	−4.841	−6.101	−6.145
Polygalacin D3	0	0	0	0	0	0
Platycoside E	0	0	0	0	0	0
Polygalacin D2	−5.532	−7.755	−7.298	−6.742	−6.011	−5.882
Dimethyl 2-O-methyl-3-O-a-D-glucopyranosyl platycogenate A	−4.79	−3.918	−4.282	0	0	0
Platycodin V	0	0	0	0	0	0
PF-04802367	−8.373	/	/	/	/	/
Ifenprodil	/	−5.29	/	/	/	/
LY2811376	/	/	−5.018	/	/	/
Staurosporine	/	/	/	−3.52	/	/
Donzepil	/	/	/	/	−4.879	/
Methylphenidate (MPH)	/	/	/	/	/	−3.942

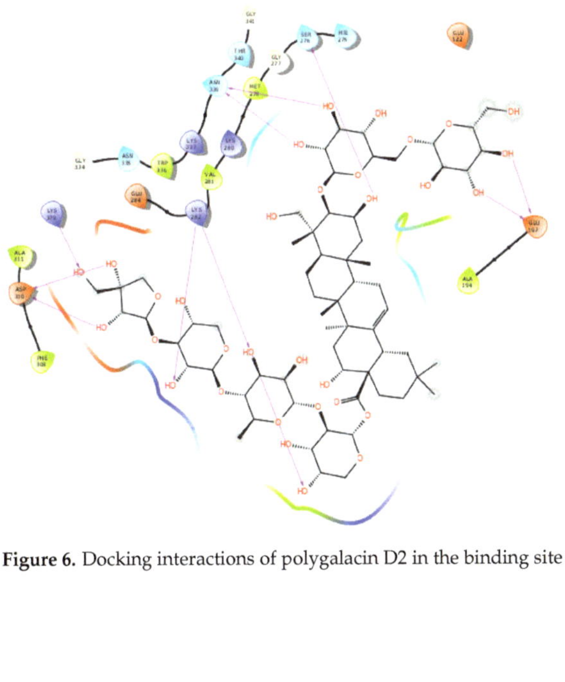

Figure 6. Docking interactions of polygalacin D2 in the binding site of synapsin II.

2.7.3. Synapsin III Receptor

The docking results, MD simulations, and MMGBSA of synapsin III receptors with polygalacin D and polygalacin D2 compared to ifenoprodil, donepezil, and PF-0480 (control) are shown below.

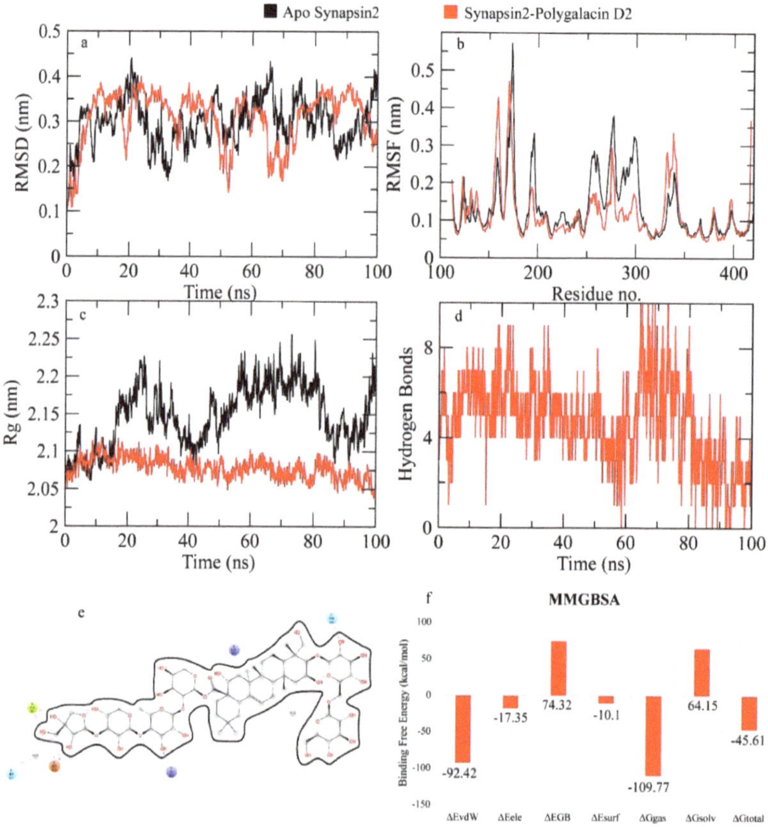

Figure 7. The simulation analysis. (**a**) The RMSD of Cα atoms of the apo protein and complex. (**b**) The flexibility of protein residues upon binding of the ligand. (**c**) Radius of gyration calculation of protein upon binding the ligand. (**d**) The hydrogen bonds calculation between the protein and ligand. (**e**) The molecular simulation result between polygalacin D and synapsin II. (**f**) The total binding free energy of energy components in the synapsin II–polygalacin D complex.

All the selected saponins were docked into the binding site of synapsin III and scored between −3.202 and −6.145 Kcal/mol. Based on the calculation of the binding free energies, the candidates polygalacin D and polygalacin D2 were employed for further molecular dynamics studies (see Table 4). The positive controls were known as CNS stimulants and dopamine reuptake inhibitors. Upon investigating the binding modes, we could infer that both molecules formed several hydrogen bond interactions and were accessible to the hydrophobic bed of the enzymatic pocket but in a different way from PF-04802367, which reflects the different orientation inside the receptor groove (see Table 2, Figure 8). Common residues between the control and the two binding ligands were phe286, trp314, ala254, and ile287 (Figure 8). Both docked compounds fitted horizontally without entering the narrow hole characteristic for the controls, yet they revealed better stability and enhanced binding features, which might be justified by the larger surface area and greater number of hydrogen bonds as well as nonpolar interactions (see Figure 8).

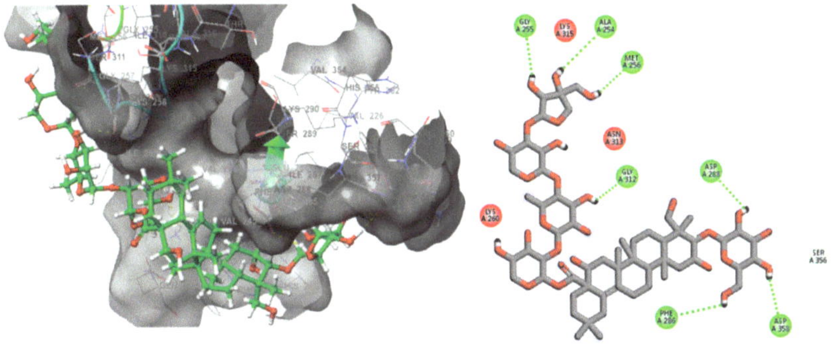

(**a**) (**b**)

Figure 8. (**a**) The docked complex of synapsin III receptor–polygalacin D. (**b**) Docking interactions of polygalacin D in the binding site of synapsin III.

The RMSD of Cα atoms of Synapsin III bound to polygalacin D was measured and compared with the RMSD of the apo protein structure. The RMSD of the apo protein and complex deviated to 0.3 nm at 20 ns and then showed deviations in the range of 0.3–0.4 nm until the end, indicating the similar behavior in protein upon binding of the ligand (Figure 9). The RMSF plots revealed that the RMSF values of both apo proteins were less than the complex, where the RMSF of N terminal reached 2.5 nm while the RMSF values of N terminal in the apo protein were around 1.5 nm. Figure 9 shows the behavior of Rg during simulation. The plot revealed that the Rg values of the apo protein and complex were similar until 40 ns and then the Rg values of the apo protein and complex showed deviations of 0.06 nm in the second half of the simulation. Figure 9 depicts the hydrogen bonds exhibited by the complex during the simulation run. At the beginning of the simulation, the protein–ligand complex formed three hydrogen bonds, which decreased to one at some frames. The average number of hydrogen bonds was two during the simulation. The total binding free energy of the complex was -41.23 ± 0.05 Kcal/mol. The contribution of other energy parameters is shown in Figure 9.

2.7.4. N-Methyl-D-Aspartate (NMDA) Receptor

The results of docking, MD simulations, and MMGBSA of the NMDA receptor with polygalacin D and ifenprodil (control) are shown below.

The NMDA receptor is comprised of two subunits: the GluN1 and the GluN2 assembling as a heterodimer. While GluN1 binds glycine or D-serine as a co-agonist, GluN2 binds glutamate as an agonist [31]. The domains of each subunit can be categorized into four types: the ligand-binding domain (LBD), the amino-terminal domain (ATD), the cytoplasmic C-terminal domain, and the transmembrane domain (TMD) [32].

The docked complexes and residue-specific interactions of the NMDA receptors with polygalacin D2, polygalacin D, and donepezil (control) are illustrated in Table 2, respectively. Polygalacin D2 and platycodin D were docked in a similar lateral orientation featuring a new location distinct from the standards, yet displayed better glide docking scores of −7.755 and −7.296 Kcal/mol, respectively. Polygalacin D2 manifested hydrogen bonds to the key residue lys296 with distance 2.51 A and gln291 with distances 2.31 and 2.34 A (Figure 10) and extended hydrophobic forces to leu280, leu377, met375, ile275, and leu292 as well as ile293, val355, and ala307, which are characteristic to the binding features of ifenoprodil and ly-2811375.

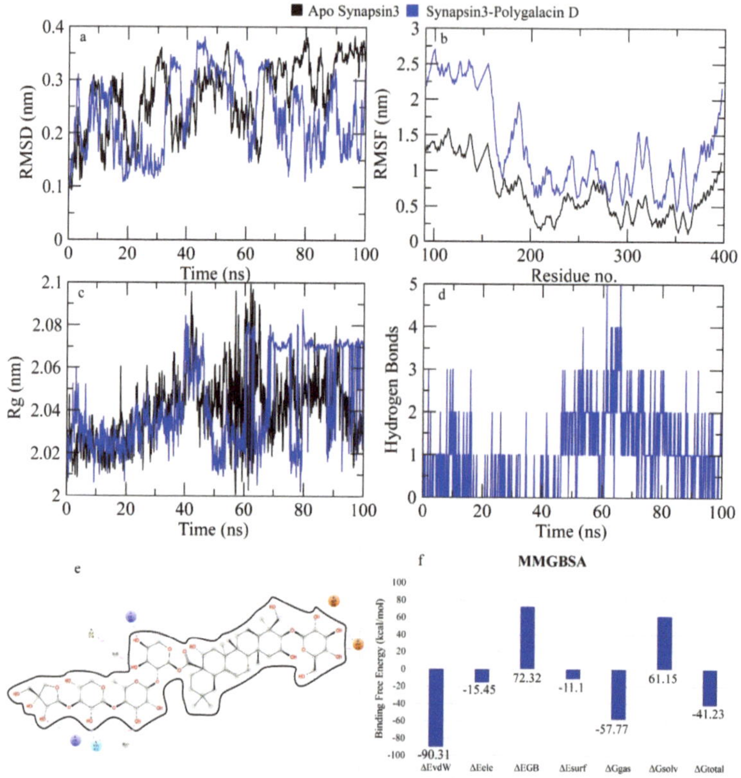

Figure 9. The simulation analysis. (**a**) The RMSD of Cα atoms of the apo protein and complex. (**b**) The flexibility of protein residues upon binding of the ligand. (**c**) Radius of gyration calculation of protein upon binding the ligand. (**d**) The hydrogen bonds calculation between the protein and ligand. (**e**) The molecular simulation result between polygalacin D and synapsin III. (**f**) The total binding free energy of energy components in the synapsin III–polygalacin D complex.

The RMSD of Cα atoms of NMDA bound to polygalacin D was measured and compared with the RMSD of the apo protein structure. The RMSD of the apo protein deviated to 0.45 nm at 10 ns and then showed a decline in values of 0.2 nm during the 60 to 80 ns interval. After 80 ns, it showed deviations in the range of 0.4 nm, while the RMSD of the complex was comparable to the apo protein but it was more than the apo protein in the last part of the simulation (Figure 11). The RMSF plots revealed that the apo protein showed more fluctuations than the complex as the complex values remained less than 0.2 nm while apo protein values were around 0.3 nm. Figure 11 shows the behavior of Rg during simulation. The plot revealed that the Rg values of both the apo protein and complex demonstrated stability throughout the simulation beginning from ~2.25 nm and consistently ranged from ~2.25 to 2.3 nm throughout the simulation run, indicating the stability of the protein structure upon binding of the ligand. Figure 11 depicts the hydrogen bonds exhibited by the complex during the simulation run. At the beginning of the simulation, the protein–ligand complex formed seven hydrogen bonds, which decreased to two at some frames. The average number of hydrogen bonds was three during the simulation. The total binding free energy of the complex was −44.38 ± 1.23 Kcal/mol. The contribution of other energy parameters is shown in Figure 11.

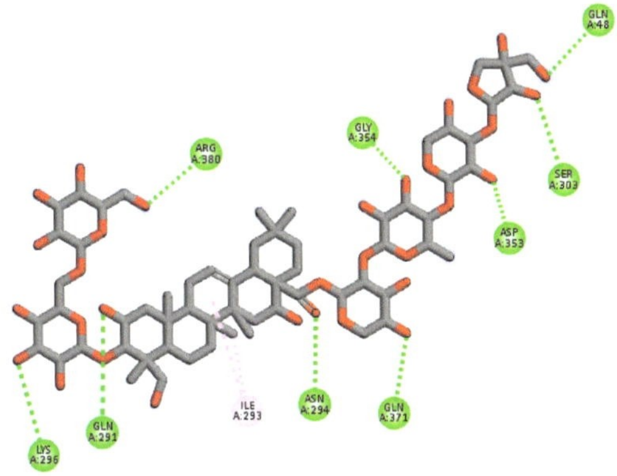

Figure 10. Polygalacin D2 in the binding site of the NMDA receptor.

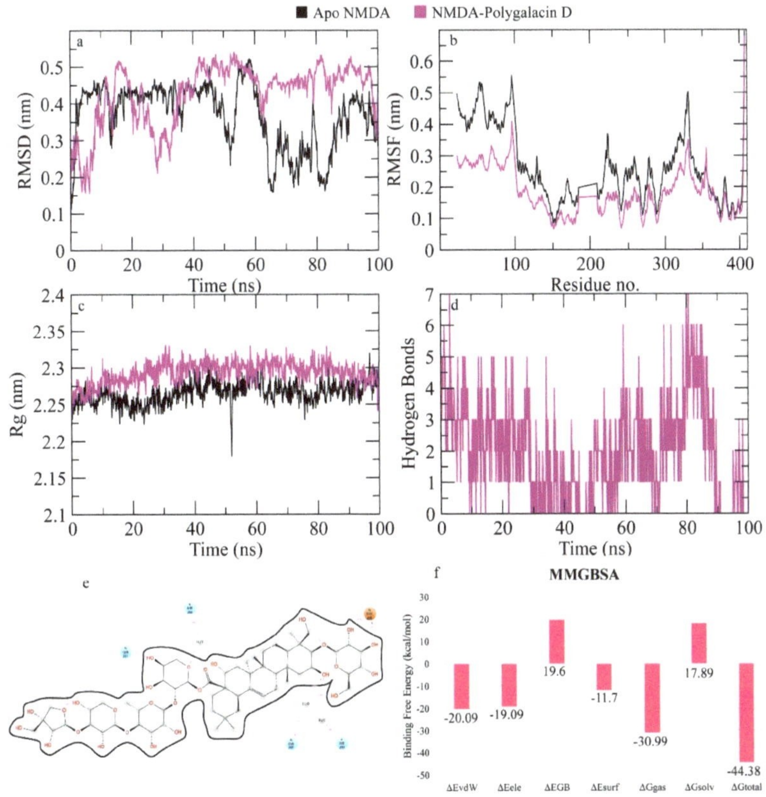

Figure 11. The simulation analysis. (**a**) The RMSD of Cα atoms of the apo protein and complex. (**b**) The flexibility of protein residues upon binding of the ligand. (**c**) Radius of gyration calculation of protein upon binding the ligand. (**d**) The hydrogen bonds calculation between the protein and ligand. (**e**) The molecular simulation result between polygalacin D and NMDA. (**f**) The total binding free energy of energy components in the NMDA–polygalacin D complex.

2.7.5. GSK-3β

Inhibition of GSK-3β might improve aging brain symptoms and reduce memory deficits and detrimental AD events; especially, the neuronal death induced by amyloid β and tau hyperphosphorylation. GSK-3β with its matching oligonucleotide structure as tau protein kinase I plays a focal role in forming the tau helical filaments that develop into neurofibrillary tangles disrupting brain function [33,34].

The best binding scores were reserved for platycodin D, which successfully inserted its pentacyclic ring into the nonpolar enzymatic groove like the controls used (Figure 12). The attached sugar residues extended to form many of the polar contacts outside this hole with a binding score of −7.147 Kcal/mol compared to PF-04820. It seemed that this hydrophobic groove represented a crucial bioactivity requirement since all the controls satisfied this condition. Moreover, combining both internal and external interactions exhibited the best-fitting scores and was only effectively achieved by platycodin D. It is worth noting that prosapogenin D bound inside the hydrophobic hole but without a polar external side chain and polygalacin D as well as platycodin D3 passed around the hydrophobic groove favoring the more polar hydrogen bond interactions outside.

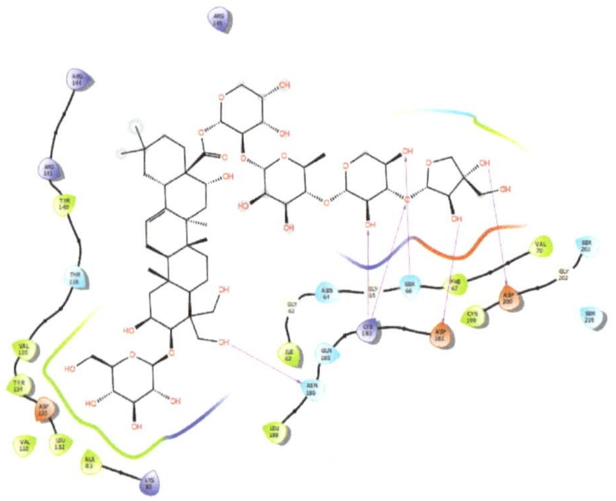

Figure 12. Docking interactions of platycodin D in the binding pocket of GSK-3β.

The platycodin D sugar part formed stable polar bonds with asp200, asp181, lys183, ser66, and asn186; whereas, its aglycone interacted with val70, phe67, cys199, ile62, leu188, ala83, leu132, val110, tyr134, val135, and tyr140, where most of them matched the used standard drugs.

The RMSD of Cα atoms of GSK-3β bound to polygalacin D was measured and compared with the RMSD of the apo protein structure to find the complex stability during simulation. The RMSD of the apo protein deviated to 0.4 nm at 10 ns and then showed a decline in value of 0.2 nm during the 10 to 20 ns intervals. After 20 ns, it showed deviations in the range of 0.2–0.4 nm throughout the time, while the RMSD of complex was comparable to the apo protein but it was lower than the apo protein in the last part of the simulation (Figure 13). The RMSF plots revealed that the apo protein showed more fluctuations than the complex as the complex values remained less than 0.5 nm while the apo protein values were around 1 nm (Figure 13). Figure 3c shows the behavior of Rg during simulation. The plot revealed that the Rg values of the apo protein showed deviations throughout simulation while the Rg values of the complex demonstrated stability throughout the simulation beginning from ~2.0 nm and consistently ranged from ~2.1 to 2.2 nm throughout the simulation run, indicating the stability of the protein structure upon binding of the ligand. Figure 13 depicts the hydrogen bonds exhibited by the complex during the simulation run. At the beginning of the simulation, the protein–ligand complex formed six hydrogen bonds, which decreased to one at some frames. The average number of hydrogen bonds was three during the simulation. Moreover, the binding free energy of the complex was calculated by employing the MMGBSA methods. It is calculated by extracting the sum of the receptor and ligand free energy from the total energy of the complex. Several energy components such as electrostatic contribution, van der Waals contribution, and solvation energy in a generalized Born environment were measured. The total binding free energy of the complex was -50.88 ± 1.07 Kcal/mol. The contribution of other energy parameters is shown in Figure 13.

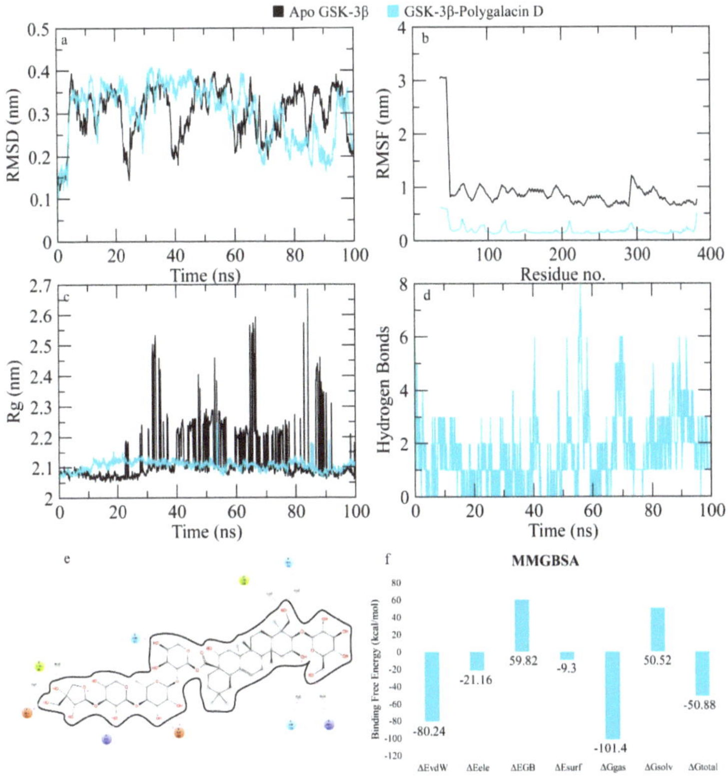

Figure 13. MD simulation measurements of the GSK-3β–polygalacin D complex. (**a**) The RMSD of Cα atoms of the apo protein and complex. (**b**) The flexibility of protein residues upon binding of the ligand. (**c**) Radius of gyration calculation of protein upon binding the ligand. (**d**) The hydrogen bonds calculation between the protein and ligand. (**e**) The molecular simulation result between polygalacin D and GSK-3β. (**f**) The total binding free energy of energy components in the GSK-3β–polygalacin D complex.

2.7.6. BACE1

The native co-crystallized ligand of the BACE1 receptor was redocked and illustrated in Figure 14 with an RMSD value of 0.23.

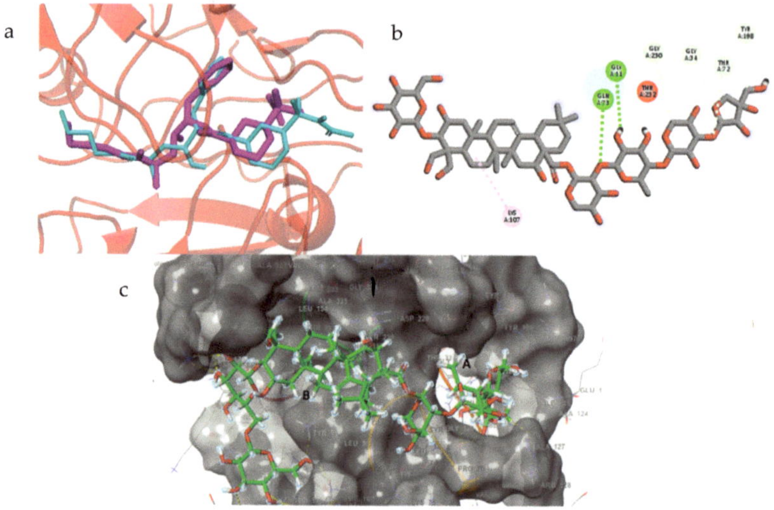

Figure 14. Redocking of the co-crystallized native ligand in the BACE1 enzyme. (**a**) Superimposition of the docked and co-crystallized native ligand (**b**) Docking interactions of platycodin D in the binding pocket of BACE1. (**c**) Polygalacin D2 in the binding pocket of BACE1.

Polygalacin D2 scored the lowest binding free energy of −7.298 Kcal/mol on account of its larger molecule that spanned a wider surface area of the enzyme and interacted with tyr198, arg128, pro70, and ser36 in the same way as the controls as well as with additional hydrogen bonds to asn111, phe109, arg307, asp228, and lys107 (see Table 2) in a neighboring binding groove (Figure 14). Briefly, a larger number of polar interactions might explain the polygalacin D2 better fitting ability and leave it to be investigated via in vitro studies for its possible biological effects (Figure 14). Polygalacin D2 was the only molecule that displayed sufficient polar contacts matching those of the standards. Further analysis of the rest of the docked structures revealed their location far from groove A or even both grooves A and B; thus, lacking an adequate distance to interact appropriately with the enzyme.

The RMSD of Cα atoms of BACE1 bound to polygalacin D was measured and compared with the RMSD of the apo protein structure. The RMSD of the apo protein deviated to 0.5 nm at 20 ns and then showed deviations in this range until the end, while the RMSD of the complex was lower than the apo protein (Figure 15). The RMSF plots revealed that the RMSF values of both the apo protein and complex were similar as most residues exhibited values less than 0.2 nm. Figure 15 shows the behavior of Rg during simulation. The plot revealed that the Rg values of the apo protein were more than the complex, but both demonstrated stability throughout the simulation. Figure 15 depicts the hydrogen bonds exhibited by the complex during the simulation run. At the beginning of the simulation, the protein–ligand complex formed five hydrogen bonds, which decreased to two at some frames. The average number of hydrogen bonds was three during the simulation. The total binding free energy of the complex was −67.86 ± 1.67 Kcal/mol. The contribution of other energy parameters is shown in Figure 15.

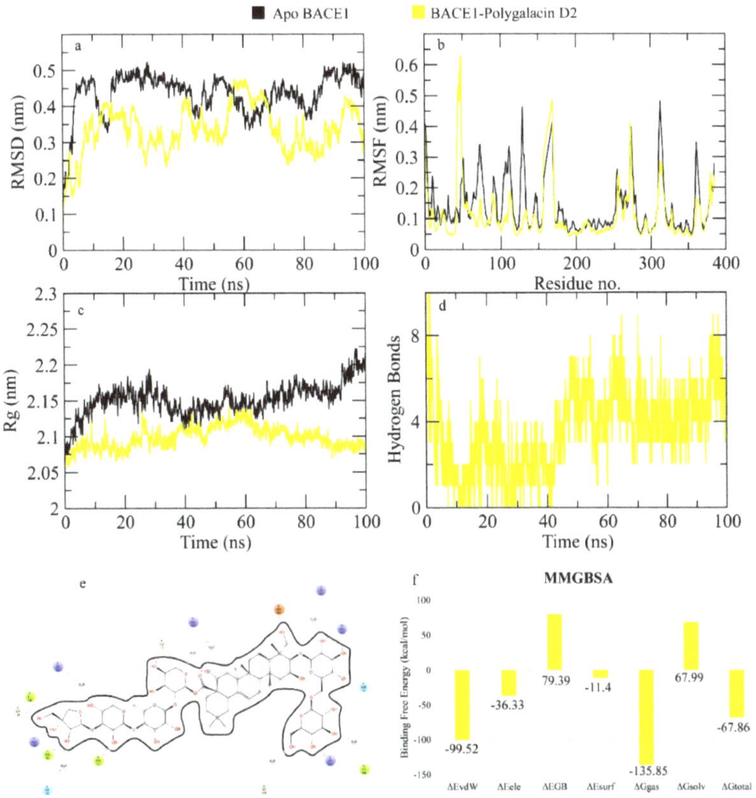

Figure 15. MD simulation data of the BACE1– polygalacin D complex. (**a**) The RMSD of Cα atoms of the apo protein and complex. (**b**) The flexibility of protein residues upon binding of the ligand. (**c**) Radius of gyration calculation of protein upon binding the ligand. (**d**) The hydrogen bonds calculation between the protein and ligand. (**e**) The molecular simulation result between polygalacin D and BACE1. (**f**) The total binding free energy of energy components in the BACE1–polygalacin D complex.

3. Discussion

P. grandiflorum is one of the famous medicine food homology plants in the *Campanulaceae* family, growing mainly in Korea, Japan, East Siberia, and China, whose potential in treating AD has not been investigated before despite its wide availability and popular use [35]. This highly nutritional popular appetizer comprises many beneficial compounds such as polyphenols, flavonoids, and saponins, which display activities such as hepatoprotective, antioxidant, hypolipidemic, anti-inflammatory, and anticancer effects [35]. Moreover, *P. grandiflorum* was previously described in Japan medicinal history and used as food and medicine during the long Chinese and East Asian traditional medicine records [36]. Whereas, the polar extracts of *P. grandiflorum* ameliorated amnesia and reversed memory impairment, few studies dealt with purified saponins and their activity against neuronal inflammation or amnesia [37], among which only platycodin D showed an antioxidant effect and neuroprotective role by elevating AMPK activation both in vivo and in vitro [38]. Moreover, 2″-*O*-acetyl-polygalacin D2 and platycodin D protected against ischemic reperfusion injury in the hippocampus [39]. In this study, we aimed to virtually analyze a list of

18 saponins isolated from *P. grandiflorum* by means of molecular docking, density functional theory, MD simulations, and MMGBSA energy calculations in the binding sites of several critical enzymes involved in AD pathology—the NMDA; BACE1; synapsin I, II, and III; and GSK-3B—consequently, paving the way to the development of multitarget molecules that can treat, ameliorate, or stop the progression of AD. This saponins list bears structural similarity to the ginsenosides of *Panax ginseng,* in that neuroprotective and antioxidant effects were proven both in vitro and in vivo [22].

With about 2% saponin content, the genus *Platycodon* is rich in oleanane triterpenoidal saponins comprised of C30 structures arranged in 4 or 5 ring configurations. The C3 and C28 formed ether and ester links, respectively. While various sugar types formed the hydrophilic side of the molecule, aglycones shaped the hydrophobic side. The attached sugars are mostly D-rhamnose, D-glucose, D-xylose, D-arabinose, and D-apiose [40]. This chemical skeleton largely matched ginseng saponins; particularly, ginsenoside Ro with its oleanolic acid core structure [26], which motivated the authors to embark on this study. De-glycosylated platycosides were more active than their glycosides due to their better bioavailability and BBB accessibility [41,42]. Yet, advancements in the field of drug delivery and inclusion opened the way for investigating the whole saponin molecules as well. C4 substituents such as methyl, carboxylic, or hydroxy methyl groups provided more diversity of the platycoside backbone forming up to 30 saponins including deapiplatycodin D3, deapiplatycoside E, platycoside E, platycodin A, platycodin D3, dPD, PD, polygalacin D, 2″-O-acetyl polygalacin D, and 3″-O-acetyl polygalacin D [43].

Synapsin I, II, and III interactions with the functional α-helical αSyn can aggravate AD symptoms and motor disabilities. Bioactive molecules as MPH act as synapsin silencing agents to destabilize fibril formation [44]. Many protein–ligand complexes in a variety of docked conformations were generated using AutoDock Vina V1.12 and AutoDock 4 as tools for computer-aided drug design to serve for the prediction of the receptor binding interactions. Selection of the final poses was conducted based on the free energy scores of ΔG values. Ligand molecules comprised of the 15 saponins list and positive controls were docked in the selected targets of BACE1, synapsin I, II, III, GSK-β, and NAMD to accurately select the poses with highest binding affinities, which can modulate the target proteins.

Polygalacin D was reported as an apoptotic and antitumor agent against hepatocellular carcinoma cells and the HSC-T6 cell line [45,46]. The mechanism was inferred to proceed via the PI3K/Akt pathway in non-small cell lung cancer [47]. Its anti-inflammatory effect was revealed by suppressing NF-kB activation [48]. In this study, polygalacin D manifested promising computational results that support its prospected neurodegenerative and anti-AD activity. Polygalacin D showed stable interactions with the six enzymatic targets. In the synapsin II binding pocket, PGD assumed a molecular orientation that largely matched donepezil with hydrophobic contact to Ser276 and hydrogen bonds to Asn339 (2.97 Å) and Asn339 (2.86 Å) residues. The ΔG value was among the best scores achieved in the whole saponin screened library −12.3 Kcal/mol. Molecular dynamics simulations showed a larger extent of persistent interactions such as Glu122, Glu148, Gln188, Asn339, and Arg187 with 2-hydroxyl glucosyl and 2-hydroxy apiosyl groups featuring the core interactions of donepezil in the same subunit.

The FMOs calculations also displayed a low ΔE value between the HOMO and LUMO, suggesting the stability of polygalacin D to interact with the enzymatic pocket residues. The MEP revealed a complementary electrostatic potential between the candidate molecule polygalacin D and many of the standard molecules.

Recently, the treatment direction of AD was shifted towards combination therapies of both amyloid- and tau protein–directed drugs, which is regarded as a breakthrough particularly in late stages when both proteins are abundant [21]. Although the accurate role of any of these proteins is still undiscovered, their implication in AD pathology is well confirmed. Therefore, double-targeting and applying preventive measures as well might represent a successful approach.

An intricate relation exists between synapsin I, BACE1, and NMDA, where synapsin I upregulates BACE1/β-secretase activity enhancing Aβ production. This commences via the α-secretase cleavage of APP releasing sAPPα, which is further modified by γ-secretase to generate the intracellular domain of APP [49]. BACE1 as an aspartate protease acts on the aforementioned domain to produce sAPPβ, the precursor of Aβ. The abnormal levels of Aβ bind with the voltage-gated Ca^{2+} channel NMDA receptor to block Ca^{2+} flow and express the etiology of AD [50]. BACE1 enzyme is an important target to ameliorate AD with its higher levels supporting higher Aβ production in the diseased brain compared to the normal one [51,52].

Upon activation, BACE1 displayed its four N-glycosylation sites and cysteine residues that manifested three disulfide bonds necessary for enzymatic activity. Moreover, the BACE1 conformation suitable for substrate binding was the flap-closed conformation, which was stabilized via substrate binding to a cleft in BACE1 with amino acids Thr72, Arg235, Ser328, and Thr329. These contacts were clearly noticed in polygalacin D justifying its binding energy of −5.607 Kcal/mol; yet, in polygalacin D2, further polar contacts were formed up to 15 hydrogen bonds (see Table 4) giving rise to a Gibbs free energy of −7.298 Kcal/mol. Fortunately, polygalacin D2 interacted with the same amino acids as the controls ifenprodil, PF-04820, and donepezil in the same groove and with a matching stable orientation. For instance, the 1-ethyl hydroxy group with glu265; the 2 and 3-OH of the first glucose unit in position 2 with lys321, gly264, and arg307; as well as the 3,4,6-OH of the second glucose unit in position 2 with asn111, phe109, lys107, and gln73. Further interactions as hydrogen bonds between the xylosyl unit with tyr198, and the apiosyl unit with arg128, ile126, pro70, and ser36 were only seen in polygalacin D2. Similar residue contacts were seen in rosmarinic acid, donezepil, and ursolic acid binding with BACE1; particularly, with gly235, Thr329, Arg235, and Phe113, respectively [53]. Mirza et al. pointed out good binding interactions of rosmarinic, ursolic, and carnosic with synapsins I, II, and III in a manner highly matching polygalacin D2 and close binding energy values [53].

Polygalacin D2 showed an anti-neuroinflammatory effect by reducing Th1 and Th2 cytokines in microglial cells and protected against reperfusion injury in the gerbil hippocampus [31,54]. Moreover, its activity was prominent in cancer research as a kinase inhibitor and apoptotic agent in Chinese hamster and lincomycin resistant cancer tissues [48]. Polygalacin D2 demonstrated promising binding to key proteins involved in the etiology of AD such as NMDA, BACE1, and synapsin III, II, and I, showing the lowest Gibbs free energy value of −7.755, −7.298, −5.882, −6.011, and −6.742 Kcal/mol, respectively.

Further confirmation was established by MD simulation analysis based on the RMSD and RMSF values. Polygalacin D2 possesses major functional groups as hydroxyl groups in positions 1, 3, and 16, as well as those in the glucose disaccharide interacting with lys107, asn111, phe109, arg307, and lys321, and the tetra saccharide units comprised of arabinose, mannose, xylose, and apiose interacting with tyr198, arg128, ile126, and ser36. Additionally, the 12,13-unsaturation in ring C was the contact point with several hydrophobic residues. These outcomes refer strongly towards the BACE1 modulating effect of polygalacin D2, which might contribute to the AD disease-altering activity of this molecule.

From all platycodon saponins, polygalacin D and D2 demonstrated the most favorable binding affinity towards synapsin proteins I, II, and III. Synapsins are involved in proper synaptic maturation and plasticity, and their loss of function or expression impairment is a necessary concern after AD development [49]. Herein, we introduced polygalacin D and D2 as two promising molecules with a combined promising effect in BACE1 and synapsin proteins in AD; yet, further studies in in vitro and in vivo models are demanded for a holistic approach. Despite its high computational costs, MD simulation results provided a more accurate drug design model through its flexible binding modules with the receptor and adding up the effect of surrounding water molecules.

NMDA represented one enzyme that is affected by high levels of tau pathology and is one of the alternative targets in the case of inefficient symptomatic relief on targeting Aβ. Many medications were introduced to target the inhibition of phosphorylation, glycosylation, or nitration of tau proteins to enhance their clearance, but the efficiency is still very low and many fail in phase I clinical trials [55]. Tau-targeted drugs are promising in the view of their expected AD disease-modifying effect; yet, are not achievable so far [3,55]. Being an essential glutamatergic cation channel gated receptor, NMDA mediated synaptic plasticity and transmission and acted in response to the Aβ binding to trigger neurotoxicity and neuronal death. Pallas-Bazarra et al. demonstrated that this action of NMDA was conducted via Tau protein, which transports tyrosine kinase Fyn to phosphorylate the GluN2B subunit of NMDA in synapses [55]. Drugs that combine both an action against tau proteins and Aβ are highly demanded, particularly, in the stage of preclinical AD when the neuronal changes are still reversible [56]. In the binding site of NMDA, six platycodin saponins revealed a better binding affinity than the controls, polygalacin D2, platycodin D, 3′-O-acetyl platyconic acid, platycodin D3, polygalacin D, and prosapogenin D with polygalacin D2 scoring the best Gibbs free energy of −7.755 Kcal/mol. It is worth mentioning that the same six molecules also superseded controls in the binding pockets of GSK-3β and BACE1.

A thorough conformational and docking analysis manifested the interaction of polygalacin D2 with the hydrophobic bed formed of leu280, leu377, met375, ile275, leu292, ile293, val355, and ala307, from which the last four amino acids exactly bonded to ifenprodil and ly-2811375 (Table 4). Polar interactions were displayed through 10 hydrogen bonds to residues Gln48, tyr351, asp353, lys296, gly354, gln291, and arg380, in which lys296 and gln291 also bonded to the controls. The hydroxyl groups in the tetrasaccharide side revealed contacts with gln48, tyr351, asp353, gly354, and asn294, while the aglycone core was located close to the nonpolar residues leu377 and met375. The diglucoside attached to position 3 formed two hydrogen bonds with lys296 and arg380, while gln 291 was unique in conducting another hydrogen bond to the 3-OH group and the anomeric oxygen of its attached glucose moiety. In most of the successfully NMDA docked molecules with a better than ifenprodil docking score, we noted the following interactions: hydrogen bonds with lys296, tyr351, gln291, and gly354. This primarily agreed with the results of Vashisth et al. [57].

Another tau pathology targeting protein is the serine threonine kinase GSK-3β that is involved in the hyperphosphorylation of tau protein, the neurofibrillary tangles' main component and hallmark of AD [58]. Additionally, it contributes to the control of amyloid-β (Aβ) peptide cell death and abnormal synaptic function [59]. Iwaloye et al. reported significant amino acids working for ATP-ligand recognition in the active site of GSK-3B as polar residues of LYS85, ASP200, and GLU51 as well as ASN186, ARG141, GLN185, LYS183, and ILE62. Asp200 was associated with the ATP phosphate interaction and was visible in most of the best-fitting platycodin saponins docked in the GSK-3B binding pocket [60]. Five docked saponins scored better binding free energy than the used controls of donepezil and ifenprodil; namely, platycodin D, polygalacin D, platycodin D3, platyconic acid B-lactone, and 3′-*O*-acetyl platyconic acid. Platycodin D attained a binding score of −7.147 Kcal/mol, which surpassed that of ifenoprodil and donepezil but not PF-04802, which revealed a ΔG value of −8.738 Kcal/mol. This was explained by the PF-04802 unique polar and nonpolar interactions to core residues in the GSK-3B pocket such as asp200, lys85, and val135 hydrogen bonds with distances of 2.64, 2.76, 2.29, and 2.41, respectively (Table 4). Halogen bonds were detected between chlorine and lys85 (3.75 A°) and asp200 (2.64 A°). Moreover, the hydrophobic bed in its vicinity comprised val70, phe67, cys199, ile62, leu188, ala83, leu132, val110, tyr134, val135, and tyr140 amino acids, among which ile62, leu188, cys199, and phe67 were clearly identified as powerful contacts in donepezil and ifenoprodil binding sites. Pi–cation interaction (4.39 A°) was noted between arg141 and the triazole ring. Furthermore, the orientation of platycodin D demonstrated that the aglycone part with the attached 3-glucose moiety was docked in the same side of the pocket as PF-04802; whereas, the trisaccharide attached unit was kept outside. The orientation of the two molecules despite varying sizes matched, where the aglycone part was closest to Arg141 in platycodin D instead of the triazole ring in PF-04802 with its Pi–cation interface. Furthermore, the chlorinated ring in PF-04802 occupied the same slot as the trisaccharide moiety in platycodin D but differed in forming two halogen carbon bonds and one hydrogen bond. In contrast, platycodin D trisaccharide extended five hydrogen bonds to core amino acids in the same slot, which justifies their binding energy values (Table 3). Polygalacin D assumed the same general orientation and interactions of platycodin D with the trisaccharide moiety extending outside the narrow groove and the 2-glucosyl aglycone impeded inside (Figure 16). However, arg148 showed interaction with the 6-hydroxy group of the glucose moiety, indicating a reversed aglycone positioning. This could contribute to the slight decrease in the ΔG value of polygalacin D and highlights the consequence of hydrophobic contacts in this enzymatic pocket.

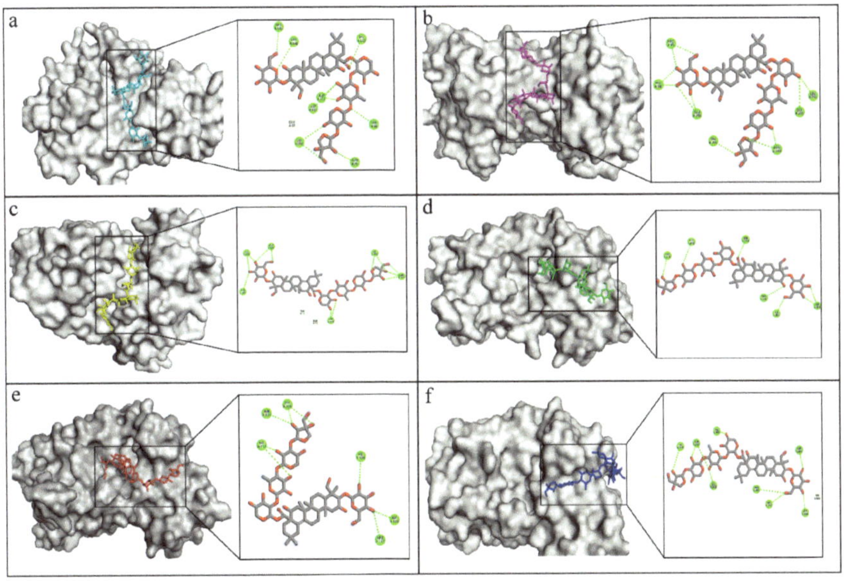

Figure 16. The molecular interactions of polygalacin D with protein targets. (**a**) GSK-3β, (**b**) NMDA receptor, (**c**) BACE1, (**d**) Synapsin I, (**e**) Synapsin II, (**f**) Synapsin III.

4. Materials and Methods

4.1. Protein and Ligand Structures Retrieval

The Protein Data Bank was used for the retrieval of protein structures of synapsin I, synapsin II, synapsin III, BACE1, NMDA Receptor, and GSK-3β with PDB IDs 1PK8, 1I7L, 2P0A, 3K5F, 5EWJ, and 5K5N respectively. The missing residues and charges were detected and repaired by Modeller v9.22 [61]. PyMol 3.1 was used to visualize 3D structures of the proteins, ligands, and protein–ligand complexes. PubChem database [62] provided structural and functional aspects of compounds platycodin D, platycodin D3, 3″-O-acetylplatyconic acid A, prosapogenin D, and polygalacin D, polygalacin D2 with PubChem IDs 162859, 75251137, 101495503, 101596922, 96023791, and 70698218, respectively. The controls used with receptors GSK-3β, NMDA receptor, BACE1, synapsin I, synapsin II, and synapsin III were PF-04802367, ifenprodil, LY2811376, staurosporine, donzepil, and methylphenidate, respectively [63].

4.2. Ligand Preparation

The saponins compounds were retrieved from PubChem and processed using the LigPrep program in Schrödinger's Maestro 2018 [64]. The OPLS_2005 force field was used to optimize the geometry of the ligands, ensuring they achieved energetically favorable conformations [21]. Energy minimization was applied to remove any unfavorable interactions or strained geometries.

4.3. Molecular Docking

The prepared saponins compounds were docked to BACE1 (PDB ID: 3K5F), GSK3β (PDB ID: 5K5N), NMDA receptor (PDB ID: 5EWJ), Synapsin I (PDB ID: 1PK8), Synapsin II (PDB ID: 1I7L), and Synapsin III (PDB ID: 2P0A). The crystal structures were prepared for docking using Protein Preparation Wizard of Maestro-2018 [65]. During receptor preparation, several stages were performed including the generation of disulfide bonds, the assignment of zero-order metal bonds, and the addition of hydrogens. The additional ligands and crystal water were also removed. In the optimization step, the pKa values of the ionizable group were optimized at pH 7.0 utilizing the PROPKA program in Schrödinger's Maestro 2018 [62]. Finally, the OPLS_2005 forcefield was used for energy minimization. After the protein preparation, a three-dimensional grid was constructed at each receptor for site specific docking. The compounds were docked to the protein using the SP mode of glide [66].

4.4. MD Simulation

To analyze the protein ligand stability, an MD simulation of 100 ns was executed on the complexes of polygalacin D, polygalacin D2, and respective proteins. Solvation of the complexes was performed in a periodic box with a 10 Å size containing the TIP3P water molecules [67]. Counter ions of Na^+ and Cl^- were introduced into the system to neutralize it. The minimization of the system was performed using the steepest decent method of 5000 steps following neutralization to remove steric conflicts. After minimization, the systems were prepared for the production run by equilibrating for 50,000 and 100,000 steps, respectively, at 310 K temperature at the NVT and NPT ensembles. The simulation was conducted with the Berendson thermostat and Parrinello-Rahman algorithms to maintain a constant temperature (310 K) and pressure (1 atm). By adjusting the time at τ P = 2.0 ps and τ T = 0.1 ps, the system was relaxed, and by applying the LINCS algorithm, the hydrogen atoms' bond lengths were kept at their ideal lengths [68], whereas Verlet computed the non-bonded interactions [69]. To compute the electrostatic interactions beyond the short-range limit, the particle mesh Ewald approach was used [70]. In the x, y, and z dimensions, the periodic boundary conditions were imposed, and a production run was conducted on the system. Every 10 ps, the production run's trajectory was saved and examined using the R BIO3D package 2.4-1.9000 and gromacs commands [71]. The CHARMM36 forcefield and the Gromacs simulation program 2024.2 were used to execute the simulation [72].

4.5. Quantum Chemical Investigation

We based our study primarily on the docking results, which revealed the interactions of the whole list of platycodin saponins to six key enzymes linked to AD pathology. In depth after-docking quantum chemical analysis was focused on chemical candidates with the highest binding free energy by conducting density functional theory (DFT) within Gaussian software 6.0 [73]. Geometry optimization was employed for the best two compounds, polygalacin D and polygalacin D2, by utilizing theory level B3LYP and the basis set of 6-31G, which allowed electronic energy calculations afterwards. Frontier molecular orbitals (FMOs) and molecular electrostatic potential (MEP) analysis were encompassed to showcase a thorough understanding of compounds reactivity in the biological system [74].

5. Conclusions

Currently, the scientific community is still searching for a cure for Alzheimer's disease. With the current available computational advancements offering fast and cheaper approaches compared to the experimental one, platycodin saponins were investigated as promising drug candidates acting against neurodegenerative diseases, particularly, AD based on their previous bioactive profile and safety indexes. The studied compounds of polygalacin D and D2, and platycodin D, successfully inhibited AD enzymes like synapsins, NMDA, GSK-3B, and BACE1. Among the best results was polygalacin D2, which demonstrated a glide docking score of −7.755 Kcal/mol in the NMDA binding pocket verified through docking interactions and molecular dynamics simulation. Furthermore, the DFT quantum and electronic calculations were carried out for the three compounds to characterize their structures, which will assist their downstream development and formulation. In platycodin D, the moderate reactivity and electrophilicity was shown through its hydroxy ethyl group at carbon 1. The polygalacin D2 carbon2 diglucoside group was deemed responsible for its favored intermolecular interactions compared to polygalacin D. However, more in vitro studies, in vivo studies, and formulation studies are still needed to further study its effect, prove its therapeutic benefits, and improve its pharmacokinetics properties.

Author Contributions: Conceptualization, B.X.; methodology, A.Y.M.; software, A.Y.M., A.R.A. and J.L.; validation, A.Y.M., A.R.A., J.L. and W.S.C.; formal analysis, A.Y.M., A.R.A., J.L. and J.W.; investigation, A.Y.M., A.R.A., J.L. and J.W.; resources, B.X.; data curation, A.Y.M., A.R.A., J.L. and J.W.; writing—original draft preparation, A.Y.M., A.R.A., J.L. and J.W.; writing—review and editing, A.Y.M., W.S.C. and B.X.; visualization, A.Y.M., A.R.A., J.L. and J.W.; supervision, A.Y.M. and B.X.; project administration, B.X.; funding acquisition, B.X. All authors have read and agreed to the published version of the manuscript.

Funding: This work was jointly supported by Guangdong Higher Education Upgrading Plan (2021–2025) with No. of UICR0400015-24 and UICR0400016-24 at Beijing Normal-Hong Kong Baptist University, Zhuhai, China, and a grant (RSPD2025R885) from King Saud University, Riyadh, Saudi Arabia.

Institutional Review Board Statement: Not applicable.

Informed Consent Statement: Not applicable.

Data Availability Statement: Data will be available upon request.

Conflicts of Interest: The authors declare no conflicts of interest.

References

1. Elhawary, E.A.; Moussa, A.Y.; Singab, A.N.B. Genus Curcuma: Chemical and ethnopharmacological role in aging process. *BMC Complement. Med. Ther.* **2024**, *24*, 31. [CrossRef] [PubMed]
2. Better, M.A. 2023 Alzheimer's disease facts and figures. *Alzheimers Dement.* **2023**, *19*, 1598–1695.
3. Limantoro, J.; de Liyis, B.G.; Sutedja, J.C. Akt signaling pathway: A potential therapy for Alzheimer's disease through glycogen synthase kinase 3 beta inhibition. *Egypt. J. Neurol. Psychiatry Neurosurg.* **2023**, *59*, 147. [CrossRef]
4. Abeysinghe, A.A.D.T.; Deshapriya, R.D.U.S.; Udawatte, C. Alzheimer's disease; a review of the pathophysiological basis and therapeutic interventions. *Life Sci.* **2020**, *256*, 117996. [CrossRef]
5. Hampel, H.; Mesulam, M.M.; Cuello, A.C.; Farlow, M.R.; Giacobini, E.; Grossberg, G.T.; Khachaturian, A.S.; Vergallo, A.; Cavedo, E.; Snyder, P.J.; et al. The cholinergic system in the pathophysiology and treatment of Alzheimer's disease. *Brain* **2018**, *141*, 1917–1933. [CrossRef]
6. Liu, J.; Chang, L.; Song, Y.; Li, H.; Wu, Y. The role of NMDA receptors in Alzheimer's disease. *Front. Neurosci.* **2019**, *13*, 43. [CrossRef]
7. Dou, K.X.; Tan, M.S.; Tan, C.C.; Cao, X.-P.; Hou, X.-H.; Guo, Q.-H.; Tan, L.; Mok, V.; Yu, J.-T. Comparative safety and effectiveness of cholinesterase inhibitors and memantine for Alzheimer's disease: A network meta-analysis of 41 randomized controlled trials. *Alzheimers Res. Ther.* **2018**, *10*, 126. [CrossRef]
8. Iqbal, D.; Alsaweed, M.; Jamal, Q.M.S.; Asad, M.R.; Rizvi, S.M.D.; Rizvi, M.R.; Albadrani, H.M.; Hamed, M.; Jahan, S.; Alyenbaawi, H. Pharmacophore-based screening, molecular docking, and dynamic simulation of fungal metabolites as inhibitors of multi-targets in neurodegenerative disorders. *Biomolecules* **2023**, *13*, 1613. [CrossRef]
9. Deng, M.; Yan, W.; Gu, Z.; Li, Y.; Chen, L.; He, B. Anti-neuroinflammatory potential of natural products in the treatment of Alzheimer's disease. *Molecules* **2023**, *28*, 1486. [CrossRef]
10. Back, M.K.; Ruggieri, S.; Jacobi, E.; von Engelhardt, J. Amyloid beta-mediated changes in synaptic function and spine number of neocortical neurons depend on NMDA receptors. *Int. J. Mol. Sci.* **2021**, *22*, 6298. [CrossRef]
11. Ugale, V.G.; Bari, S.B. Identification of potential Gly/NMDA receptor antagonists by cheminformatics approach: A combination of pharmacophore modelling, virtual screening and molecular docking studies. *SAR QSAR Environ. Res.* **2016**, *27*, 125–145. [CrossRef] [PubMed]
12. Clyde, A. Ultrahigh throughput protein-ligand docking with deep learning. *Methods Mol. Biol.* **2022**, *2390*, 301–319. [PubMed]
13. Moussa, A.Y.; Alanzi, A.R.; Riaz, M.; Fayez, S. Could mushrooms' secondary metabolites ameliorate Alzheimer disease? A computational flexible docking investigation. *J. Med. Food* **2024**, *27*, 775–796. [CrossRef]
14. Moussa, A.Y.; Alanzi, A.; Luo, J.; Chung, S.K.; Xu, B. Potential anti-obesity effect of saponin metabolites from adzuki beans: A computational approach. *Food Sci. Nutr.* **2024**, *12*, 3612–3627. [CrossRef] [PubMed]
15. Moussa, A.Y.; Labib, R.M.; Ayoub, N.A. Isolation of chemical constituents and protective effect of *Pistacia khinjuk* against CCl_4–induced damage on HepG2 cells. *Phytopharmacology* **2013**, *2*, 1–9.
16. Torky, Z.A.; Moussa, A.Y.; Abdelghffar, E.A.; Abdel-Hameed, U.K.; Eldahshan, O.A. Chemical profiling, antiviral and antiproliferative activities of the essential oil of *Phlomis aurea Decne* grown in Egypt. *Food Funct.* **2021**, *12*, 4630–4643. [CrossRef]
17. Sun, A.; Xu, X.; Lin, J.; Cui, X.; Xu, R. Neuroprotection by saponins. *Phytother. Res.* **2015**, *29*, 187–200. [CrossRef]
18. Yang, L.; Hao, J.; Zhang, J.; Xia, W.; Dong, X.; Hu, X.; Kong, F.; Cui, X. Ginsenoside Rg3 promotes beta-amyloid peptide degradation by enhancing gene expression of neprilysin. *J. Pharm. Pharmacol.* **2009**, *61*, 375–380. [CrossRef]
19. Yang, Y.; Chen, W.; Lin, Z.; Wu, Y.; Li, Y.; Xia, X. Panax notoginseng saponins prevent dementia and oxidative stress in brains of SAMP8 mice by enhancing mitophagy. *BMC Complement. Med. Ther.* **2024**, *24*, 144. [CrossRef]
20. Xu, T.Z.; Shen, X.Y.; Sun, L.L.; Chen, Y.; Zhang, B.; Huang, D.; Li, W. Ginsenoside Rg1 protects against H_2O_2-induced neuronal damage due to inhibition of the NLRP1 inflammasome signalling pathway in hippocampal neurons in vitro. *Int. J. Mol. Med.* **2019**, *43*, 717–726. [CrossRef]
21. Shivakumar, D.; Harder, E.; Damm, W.; Friesner, R.A.; Sherman, W. Improving the prediction of absolute solvation free energies using the next generation OPLS force field. *J. Chem. Theory Comput.* **2012**, *8*, 2553–2558. [CrossRef] [PubMed]
22. Shen, L.H.; Zhang, J.T. Ginsenoside Rg1 promotes proliferation of hippocampal progenitor cells. *Neurol. Res.* **2004**, *26*, 422–428. [CrossRef] [PubMed]
23. Zhao, H.F.; Li, Q.; Li, Y. Long-term ginsenoside administration prevents memory loss in aged female C57BL/6J mice by modulating the redox status and up-regulating the plasticity-related proteins in hippocampus. *Neuroscience* **2011**, *183*, 189–202. [CrossRef] [PubMed]

24. Zhu, Y.; Liang, J.; Gao, C.; Wang, A.; Xia, J.; Hong, C.; Zhong, Z.; Zuo, Z.; Kim, J.; Ren, H.; et al. Multifunctional ginsenoside Rg3-based liposomes for glioma targeting therapy. *J. Control. Release* **2021**, *330*, 641–657. [CrossRef]
25. Mao, Y.; Yuan, W.; Gai, J.; Zhang, Y.; Wu, S.; Xu, E.-Y.; Wang, L.; Zhang, X.; Guan, J.; Mao, S. Enhanced brain distribution of Ginsenoside F1 via intranasal administration in combination with absorption enhancers. *Int. J. Pharm.* **2024**, *654*, 123930. [CrossRef]
26. Wang, C.; Schuller Levis, G.B.; Lee, E.B.; Levis, W.R.; Lee, D.W.; Kim, B.S.; Park, S.Y.; Park, E. Platycodin D and D3 isolated from the root of *Platycodon grandiflorum* modulate the production of nitric oxide and secretion of TNF-alpha in activated RAW 264.7 cells. *Int. Immunopharmacol.* **2004**, *4*, 1039–1049. [CrossRef]
27. Kiranmayee, M.; Rajesh, N.; Vidya Vani, M.; Khadri, H.; Mohammed, A.; Chinni, S.V.; Ramachawolran, G.; Riazunnisa, K.; Moussa, A.Y. Green synthesis of Piper nigrum copper-based nanoparticles: In silico study and ADMET analysis to assess their antioxidant, antibacterial, and cytotoxic effects. *Front. Chem.* **2023**, *11*, 1218588. [CrossRef]
28. Kaya, S.; Putz, M.V. Atoms-in-molecules' faces of chemical hardness by conceptual density functional theory. *Molecules* **2022**, *27*, 8825. [CrossRef]
29. Branches, A.D.S.; da Silva, J.N.; de Oliveira, M.D.L.; Bezerra, D.P.; Soares, M.B.P.; Costa, E.V.; Oliveira, K.M.T. DFT calculations, molecular docking, binding free energy analysis and cytotoxicity assay of 7,7-dimethylaporphine alkaloids with methylenedioxy ring in positions 1 and 2. *Comput. Theor. Chem.* **2024**, *1233*, 114483. [CrossRef]
30. Yamari, I.; Abchir, O.; Nour, H.; Khedraoui, M.; Rossafi, B.; Errougui, A.; Talbi, M.; Samadi, A.; El Kouali, M.; Chtita, S. Unveiling Moroccan nature's Arsenal: A computational molecular docking, density functional theory, and molecular dynamics study of natural compounds against drug-resistant fungal infections. *Pharmaceuticals* **2024**, *17*, 886. [CrossRef]
31. Yovanno, R.A.; Chou, T.H.; Brantley, S.J.; Furukawa, H.; Lau, A.Y. Excitatory and inhibitory D-serine binding to the NMDA receptor. *eLife* **2022**, *11*, e77645. [CrossRef] [PubMed]
32. Mayer, M.L. The challenge of interpreting glutamate-receptor ion-channel structures. *Biophys. J.* **2017**, *113*, 2143–2151. [CrossRef] [PubMed]
33. Takashima, A. GSK-3 is essential in the pathogenesis of Alzheimer's disease. *J. Alzheimers Dis.* **2006**, *9* (Suppl. S3), 309–317. [CrossRef]
34. Wang, C.; Cui, Y.; Xu, T.; Zhou, Y.; Yang, R.; Wang, T. New insights into glycogen synthase kinase-3: A common target for neurodegenerative diseases. *Biochem. Pharmacol.* **2023**, *218*, 115923. [CrossRef]
35. Casiraghi, A.; Longhena, F.; Faustini, G.; Ribaudo, G.; Suigo, L.; Camacho-Hernandez, G.A.; Bono, F.; Brembati, V.; Newman, A.H.; Gianoncelli, A.; et al. Methylphenidate analogues as a new class of potential disease-modifying agents for Parkinson's disease: Evidence from cell models and alpha-synuclein transgenic mice. *Pharmaceutics* **2022**, *14*, 1595. [CrossRef]
36. Zhan, Q.; Zhang, F.; Sun, L.; Wu, Z.; Chen, W. Two new oleanane-type triterpenoids from *Platycodi Radix* and anti-proliferative activity in HSC-T6 cells. *Molecules* **2012**, *17*, 14899–14907. [CrossRef]
37. Nan, F.; Nan, W.; Yu, Z.; Wang, H.; Cui, X.; Jiang, S.; Zhang, X.; Li, J.; Wang, Z.; Zhang, S.; et al. Polygalacin D inhibits the growth of hepatocellular carcinoma cells through BNIP3L-mediated mitophagy and endogenous apoptosis pathways. *Chin. J. Nat. Med.* **2023**, *21*, 346–358. [CrossRef]
38. Seo, Y.S.; Kang, O.H.; Kong, R.; Zhou, T.; Kim, S.-A.; Ryu, S.; Kim, H.-R.; Kwon, D.-Y. Polygalacin D induces apoptosis and cell cycle arrest via the PI3K/Akt pathway in non-small cell lung cancer. *Oncol. Rep.* **2018**, *39*, 1702–1710. [CrossRef]
39. Kim, M.; Hwang, I.G.; Kim, S.B.; Choi, A.J. Chemical characterization of balloon flower (*Platycodon grandiflorum*) sprout extracts and their regulation of inflammatory activity in lipopolysaccharide-stimulated RAW 264.7 murine macrophage cells. *Food Sci. Nutr.* **2019**, *8*, 246–256. [CrossRef]
40. Maesako, M.; Zoltowska, K.M.; Berezovska, O. Synapsin 1 promotes Aβ generation via BACE1 modulation. *PLoS ONE* **2019**, *14*, e0226368. [CrossRef]
41. Yan, R.; Fan, Q.; Zhou, J.; Vassar, R. Inhibiting BACE1 to reverse synaptic dysfunctions in Alzheimer's disease. *Neurosci. Biobehav. Rev.* **2016**, *65*, 326–340. [CrossRef] [PubMed]
42. Egbertson, M.; McGaughey, G.B.; Pitzenberger, S.M.; Stauffer, S.R.; Coburn, C.A.; Stachel, S.J.; Yang, W.; Barrow, J.C.; Neilson, L.A.; McWherter, M.; et al. Methyl-substitution of an iminohydantoin spiropiperidine β-secretase (BACE-1) inhibitor has a profound effect on its potency. *Bioorg Med. Chem. Lett.* **2015**, *25*, 4812–4819. [CrossRef] [PubMed]
43. Joseph, O.A.; Babatomiwa, K.; Niyi, A.; Olaposi, O.; Olumide, I. molecular docking and 3D osar studies of C000000956 as a potent inhibitor of Bace-1. *Drug Res.* **2019**, *69*, 451–457.
44. Han, B.; Luo, J.; Xu, B. Revealing molecular mechanisms of the bioactive saponins from edible root of *Platycodon grandiflorum* in combating obesity. *Plants* **2024**, *13*, 1123. [CrossRef]
45. Ji, M.Y.; Bo, A.; Yang, M.; Xu, J.-F.; Jiang, L.-L.; Zhou, B.-C.; Li, M.-H. The pharmacological effects and health benefits of *Platycodon grandifloras*—A medicine food homology species. *Foods* **2020**, *9*, 142. [CrossRef]

46. Moon, M.K.; Ahn, J.Y.; Kim, S.; Ryu, S.Y.; Kim, Y.S.; Ha, T.Y. Ethanol extract and saponin of *Platycodon grandiflorum* ameliorate scopolamine-induced amnesia in mice. *J. Med. Food* **2010**, *13*, 584–588. [CrossRef]
47. Zhang, J.T.; Xie, L.Y.; Shen, Q.; Liu, W.; Li, M.-H.; Hu, R.-Y.; Hu, J.-N.; Wang, Z.; Chen, C. Platycodin D stimulates AMPK activity to inhibit the neurodegeneration caused by reactive oxygen species-induced inflammation and apoptosis. *J. Ethnopharmacol.* **2023**, *308*, 116294. [CrossRef]
48. Choi, J.H.; Yoo, K.Y.; Park, O.K.; Lee, C.H.; Won, M.-H.; Hwang, I.K.; Ryu, S.Y.; Kim, Y.S.; Yi, J.-S.; Bae, Y.-S.; et al. Platycodin D and 2″-O-acetyl-polygalacin D2 isolated from *Platycodon grandiflorum* protect ischemia/reperfusion injury in the gerbil hippocampus. *Brain Res.* **2009**, *1279*, 197–208. [CrossRef]
49. Choi, Y.; Kang, S.; Cha, S.H.; Kim, H.-S.; Song, K.; Lee, Y.J.; Kim, K.; Kim, Y.S.; Cho, S.; Park, Y. Platycodon saponins from Platycodi Radix (*Platycodon grandiflorum*) for the green synthesis of gold and silver nanoparticles. *Nanoscale Res. Lett.* **2018**, *13*, 23. [CrossRef]
50. Ha, Y.W.; Kim, Y.S. Preparative isolation of six platycosides from *Platycodi Radix* by high-speed counter-current chromatography with evaporative light scattering detection. *Planta Med.* **2008**, *74*, PC89. [CrossRef]
51. Shin, K.C.; Oh, D.K. Biotransformation of platycosides, saponins from balloon flower root, into bioactive deglycosylated platycosides. *Antioxidants* **2023**, *12*, 327. [CrossRef] [PubMed]
52. Nyakudya, E.; Jeong, J.H.; Lee, N.K.; Jeong, Y.S. Platycosides from the roots of *Platycodon grandiflorum* and their health benefits. *Prev. Nutr. Food Sci.* **2014**, *19*, 59–68. [CrossRef] [PubMed]
53. Mirza, F.J.; Zahid, S.; Amber, S.; Sumera, J.H.; Asim, N.; Ali Shah, S.A. Multitargeted molecular docking and dynamic simulation studies of bioactive compounds from *Rosmarinus officinalis* against Alzheimer's disease. *Molecules* **2022**, *27*, 7241. [CrossRef] [PubMed]
54. Ji, Y.J.; Kang, M.H.; Kim, G.S.; Kim, H.D.; Jang, G.Y. *Platycodon grandiflorum* exhibits anti-neuroinflammatory potential against beta-amyloid-induced toxicity in microglia cells. *Front. Nutr.* **2024**, *11*, 1427121. [CrossRef] [PubMed]
55. Pallas-Bazarra, N.; Draffin, J.; Cuadros, R.; Antonio Esteban, J.; Avila, J. Tau is required for the function of extrasynaptic NMDA receptors. *Sci. Rep.* **2019**, *9*, 9116. [CrossRef]
56. Pluta, R.; Ułamek-Kozioł, M. Tau protein-targeted therapies in Alzheimer's disease: Current state and future perspectives. In *Alzheimer's Disease: Drug Discovery*; Huang, X., Ed.; Exon Publications: Brisbane, Australia, 2020.
57. Vashisth, M.K.; Hu, J.; Liu, M.; Basha, S.H.; Yu, C.; Huang, W. In-Silico discovery of 17alpha-hydroxywithanolide-D as potential neuroprotective allosteric modulator of NMDA receptor targeting Alzheimer's disease. *Sci. Rep.* **2024**, *14*, 27908. [CrossRef]
58. Dajani, R.; Fraser, E.; Roe, S.M.; Young, N.; Good, V.; Dale, T.C.; Pearl, L.H. Crystal structure of glycogen synthase kinase 3 beta: Structural basis for phosphate-primed substrate specificity and autoinhibition. *Cell* **2001**, *105*, 721–732. [CrossRef]
59. Sayas, C.L.; Ávila, J. GSK-3 and tau: A key duet in Alzheimer's disease. *Cells* **2021**, *10*, 721. [CrossRef]
60. Iwaloye, O.; Elekofehinti, O.O.; Oluwarotimi, E.A.; Kikiowo, B.I.; Fadipe, T.M. Insight into glycogen synthase kinase-3β inhibitory activity of phyto-constituents from *Melissa officinalis*: In silico studies. *In Silico Pharmacol.* **2020**, *8*, 2. [CrossRef]
61. Webb, B.; Sali, A. Protein structure modeling with MODELLER. *Methods Mol. Biol.* **2021**, *2199*, 239–255.
62. Kim, M.O.; Nichols, S.E.; Wang, Y.; McCammon, J.A. Effects of histidine protonation and rotameric states on virtual screening of M. tuberculosis RmlC. *J. Comput. Aided Mol. Des.* **2013**, *27*, 235–246. [CrossRef] [PubMed]
63. Alanzi, A.; Moussa, A.Y.; Mothana, R.A.; Abbas, M.; Ali, I. In silico exploration of PD-L1 binding compounds: Structure-based virtual screening, molecular docking, and MD simulation. *PLoS ONE* **2024**, *19*, e0306804. [CrossRef]
64. Yang, Y.; Yao, K.; Repasky, M.P.; Leswing, K.; Abel, R.; Shoichet, B.K.; Jerome, S.V. Efficient exploration of chemical space with docking and deep learning. *J. Chem. Theory Comput.* **2021**, *17*, 7106–7119. [CrossRef] [PubMed]
65. Lu, C.; Wu, C.; Ghoreishi, D.; Chen, W.; Wang, L.; Damm, W.; Ross, G.A.; Dahlgren, M.K.; Russell, E.; Von Bargen, C.D.; et al. OPLS4: Improving force field accuracy on challenging regimes of chemical space. *J. Chem. Theory Comput.* **2021**, *17*, 4291–4300. [CrossRef] [PubMed]
66. Friesner, R.A.; Banks, J.L.; Murphy, R.B.; Halgren, T.A.; Klicic, J.J.; Mainz, D.T.; Repasky, M.P.; Knoll, E.H.; Shelley, M.; Perry, J.K.; et al. Glide: A new approach for rapid, accurate docking and scoring. 1. Method and assessment of docking accuracy. *J. Med. Chem.* **2004**, *47*, 1739–1749. [CrossRef]
67. Liu, Y.; Grimm, M.; Dai, W.T.; Hou, M.C.; Xiao, Z.X.; Cao, Y. CB-Dock: A web server for cavity detection-guided protein-ligand blind docking. *Acta Pharmacol. Sin.* **2020**, *41*, 138–144. [CrossRef]
68. Qureshi, K.A.; Al Nasr, I.; Koko, W.S.; Khan, T.A.; Fatmi, M.Q.; Imtiaz, M.; Khan, R.A.; Mohammed, H.A.; Jaremko, M.; Emwas, A.-H.; et al. In vitro and in silico approaches for the antileishmanial activity evaluations of actinomycins isolated from novel *Streptomyces smyrnaeus* strain UKAQ_23. *Antibiotics* **2021**, *10*, 887. [CrossRef]
69. Hess, B.; Bekker, H.; Berendsen, H.J.C.; Fraaije, J.G.E.M. LINCS: A linear constraint solver for molecular simulations. *J. Comput. Chem.* **1997**, *18*, 1463–1472. [CrossRef]

70. Grubmüller, H.; Heller, H.; Windemuth, A.; Schulten, K. Generalized verlet algorithm for efficient molecular dynamics simulations with long-range interactions. *Mol. Simul.* **1991**, *6*, 121–142. [CrossRef]
71. Kholmurodov, K.; Smith, W.; Yasuoka, K.; Darden, T.; Ebisuzaki, T. A smooth-particle mesh Ewald method for DL_POLY molecular dynamics simulation package on the Fujitsu VPP700. *J. Comput. Chem.* **2000**, *21*, 1187–1191. [CrossRef]
72. Grant, B.J.; Skjaerven, L.; Yao, X.Q. The Bio3D packages for structural bioinformatics. *Protein Sci.* **2021**, *30*, 20–30. [CrossRef]
73. Huang, J.; MacKerell, A.D. CHARMM36 all-atom additive protein force field: Validation based on comparison to NMR data. *J. Comput. Chem.* **2013**, *34*, 2135–2145. [CrossRef]
74. Frisch, R.; Gary, M.J.; Trucks, G.; Trucks, S.H.B. *Gaussian 6.0*; Gaussian, Inc.: Wallingford, CT, USA, 2016.

molecules

MDPI

Review

Flaxseed in Diet: A Comprehensive Look at Pros and Cons

Sara Duarte [1], Muhammad Ajmal Shah [2] and Ana Sanches Silva [1,3,4,*]

[1] University of Coimbra, Faculty of Pharmacy, Polo III, Azinhaga de Santa Comba, 3000-548 Coimbra, Portugal; sara.alves.duarte@gmail.com
[2] Department of Pharmacy, Hazara University, Mansehra 21300, Pakistan; ajmalshah@hu.edu.pk
[3] Centre for Animal Science Studies (CECA), Instituto de Ciências e Tecnologias Agrárias e Agro-Alimentares (ICETA), University of Porto, 4501-401 Porto, Portugal
[4] Associate Laboratory for Animal and Veterinary Sciences (Al4AnimalS), 1300-477 Lisbon, Portugal
* Correspondence: asanchessilva@ff.uc.pt

Abstract: Flaxseeds, which have been consumed for thousands of years, have recently gained increasing popularity due to their rich composition, including omega-3 fatty acids, lignans, proteins, and fibers. These components are strongly associated with various health benefits, such as improving cardiovascular health, preventing certain types of cancer, controlling diabetes, promoting gastro-intestinal well-being, and aiding in weight management. This monograph explores the role of flaxseeds in nutrition, as well as their potential risks. Despite their numerous health benefits, flaxseeds also represent concerns due to excessive consumption and possible contamination, particularly from cyanogenic glycosides. Therefore, the levels of these compounds must be controlled, and this monograph also analyzes the available methods to detect and reduce these contaminants, ensuring the safety of flaxseed and flaxseed products consumers. Flaxseed is considered a valuable addition when incorporated into the diet, but it is necessary to continue research and promote technological improvements to maximize their benefits and minimize their risks.

Keywords: flaxseed; diet; benefits; cyanogenic glycosides; determination methods; detoxification methods

1. Introduction

Flaxseed (or linseed, as it is known in British English) is an ancient crop with a rich history, dating back nearly 12,000 years [1]. Originally, flaxseed was likely used for its laxative properties before its widespread consumption as a food ingredient. In recent years, there has been a marked increase in health-oriented products containing flaxseed, driven by growing research that highlights its potential health benefits. Therefore, dietary flaxseed has been the object of a growing research literature supporting its use, as well as trials to help and provide reliable information to the general public, especially to those in a disease-compromised condition [2], but also to the rest of the population who would like to introduce flaxseed in their diet daily. It is increasingly evident that pharmacologic therapy must be complemented by other interventions, such as dietary strategies, to effectively treat diseases or even prevent or delay their onset. Growing awareness of flaxseed's health benefits is driving increased demand, which is likely to impact farmers, food processors, and retailers in the coming years [2]. However, this rising popularity may also lead to overconsumption, with potential adverse consequences for consumers. In addition, environmental conditions such as heat damage, frosts, and drought can affect the composition and quality

 https://doi.org/10.3390/molecules30061335

of flaxseed. Flaxseed contamination is also possible and can lead to detrimental effects on the consumers' health, so it is possible contaminants must be known and controlled by safety measures.

Consequently, flaxseed is widely recognized for its health benefits, particularly due to its components, such as alpha-linolenic acid (ALA), proteins, lignans, and fiber. Many reviews focus on its positive effects in disease prevention, treatment, and management, offering insights into its role in promoting human health, which can lead to an oversimplified view of flaxseed's overall impact and may inadvertently encourage overconsumption, without taking into account the potential risks associated with flaxseed, such as the presence of toxic components.

This paper aims to provide a novel perspective on flaxseed by presenting a comprehensive examination that goes beyond the existing literature. While previous studies have primarily focused on the well-documented health benefits of flaxseed, this work also critically assesses the potential risks associated with its consumption, particularly those arising from improper use, excessive intake, and contamination with cyanogenic glycosides. What sets this study apart is its emphasis on practical, innovative solutions to mitigate these risks, including advanced detection and reduction techniques for toxic compounds, thus, ensuring safer consumption. By integrating these new insights, this article aspires to furnish a more balanced understanding of flaxseed, encompassing both its advantages and drawbacks, ultimately contributing to a more informed perspective on its role in health and nutrition.

2. Flaxseeds (*Linum usitatissum* L.)

Flaxseeds is the common name of *Linum usitatissum* L., also known as common flax when grown for the fiber extracted from its stem, and as linseed or oilseed flax when cultivated for the oil extracted from its seeds. It belongs to the genus *Linum* L. and the family *Linaceae*. Despite the numerous species in this family, flax is the only one cultivated. Flax is an erect, herbaceous annual plant that branches above the main stem [3]. Cultivated flax reproduces by its seed. Because of its perfect flower structure and the composition of its pollen, it is rarely transferred by insects, which makes flax a highly self-pollinated species. Its leaves are simple, sessile, linear-lanceolate with entire margins, borne on stems and branches. Flowers are borne on long erect pedicels, are hermaphroditic, hypogynous, and are composed of five sepals, five petals (blue, pink, white), five stamens, and a compound pistil or five carpels, each separated by a false septum. The fruit is a capsule with five carpels and may contain up to ten seeds. The flax seeds are formed by the seed and the shell, which are oval and flat, lenticular, and 4–6 mm long with pointed tips. The seed contains lignans, digestible proteins, oil rich in omega-3 fatty acids, and phenolics [3]. The shell comprises high-quality fiber with low density and mechanical qualities and holds a percentage of mucilage. It is smooth and glossy and shows a specific color from brown to dark gold [4] (Figure 1).

Figure 1. (a) Whole golden flaxseed; (b) milled golden flaxseed.

Flax has been an integral part of human culture for thousands of years, with its origins likely tracing back to the regions east of the Mediterranean, extending toward India. From there, it gradually spread across Asia and Europe before being introduced to the Americas. The domestication of flax is believed to have first occurred in the Fertile Crescent, a region known for its early advancements in agriculture [5]. The use of the crop spread in Europe 5000 years ago and domesticated flax was also cultivated in China and India [3]. The flax industry spread with European powers' colonization of the New World and other regions [6]. According to the Centre for Agriculture and Biosciences International (2018), flax is grown commercially mainly in Western (France, Belgium, the Netherlands) and Eastern Europe (Russia, Ukraine, Belarus, Poland), the latter has the largest land areas dedicated to flax, but with the best agronomic yields and processing technology in the former. The latest data of FAOSTAT, corresponding to the year 2022, confirms that Europe is the continent with the major production share of flax, i.e., 95.1% of total production. The top three producers worldwide are European countries, which are France, Belgium, and Belarus [7].

Flax thrives in moderate to cool climates, primarily in northern latitudes that receive 150 to 200 mm of rainfall during the main growing season (April–June). It grows best in fertile, well-drained soils with a medium to heavy texture and a pH range of 5.5 to 7.0 [6]. In addition to cultivated fields, flax can also be found in disturbed areas, along roadsides, and in abandoned homesteads at elevations ranging from 0 to 2400 m [3]. Its life cycle consists of a 45–60-day vegetative phase, followed by a 15–25-day flowering period, a reproductive phase, and a maturation stage lasting 30–40 days [3].

Currently, flax is used for food, feed, and industrial purposes. Focusing on the seeds, many food products are manufactured from them, including bread, cereals, crackers, energy bars, omega-3 eggs, and pasta [3]. Flaxseed can change quality factors like moisture, color, and hardness, which can significantly influence consumer acceptance of a product [8]. For people with celiac disease or gluten sensibility, flaxseed powder is a substitute for baking ingredients; in addition, results show that adding flaxseed powder enhances the number of calories, protein, ash, acidity, and antioxidant qualities [9]. Besides its use as seed for planting, the primary industrial application of flaxseed is processing it to produce flaxseed oil and meal. Due to its high protein content, a byproduct of flaxseed can be used as a supplement in livestock, including poultry feed. Linseed oil is highly valued in paints and varnishes due to its exceptional drying properties, which stem from its unique fatty acid composition. This characteristic makes it an essential ingredient in protective coatings and artistic applications. Additionally, the plant's stem fibers are processed to produce high-quality linen textiles and fine paper, further showcasing the versatility and economic importance of flax [3].

Although this section addresses the presence of flax and its seeds in various contexts, this paper will specifically focus on the composition of flaxseeds and their potential benefits and drawbacks for human health.

3. Composition of Flaxseed

Flaxseeds are primarily categorized into two types: yellow (or golden) flaxseeds and brown flaxseeds. The main difference between them lies in their color and potential nutrient profiles, although both types are rich in omega-3 fatty acids, lignans, and dietary fiber. Yellow flaxseeds tend to have a slightly milder flavor and are often favored by those looking for a more neutral taste in recipes. Brown flaxseeds, on the other hand, are more commonly found in grocery stores and have a more robust, nutty flavor. Some studies suggest that there may be slight variations in their nutritional content, with brown flaxseeds potentially boasting a higher lignan concentration [10]. However, both varieties offer similar health benefits, making them interchangeable in most culinary uses. Additionally, there are specialty flaxseed varieties, such as high-lignan flaxseeds, bred for increased health benefits, which can further diversify options for consumers.

3.1. Nutrients

According to the Portuguese Food Composition Database (INSA-National Health Institute Dr. Ricardo Jorge), 100 g of raw flaxseed contains 487 kcal. The energetic distribution is 56.7% lipids, 21.0% protein, 15.2% carbohydrates, and 7.1% fiber. The corresponding composition includes 100 g of raw flaxseed, 31 g of lipids, 25 g of protein, 18.1 g of carbohydrates, 18 g of fiber, 4.9 g of water, and 3.0 g of others (inorganic compounds) [10].

The lipids fraction (per 100 g) comprehends 2.7 g of saturated fatty acids, 5.5 g of monounsaturated fatty acids, and 21 g of polyunsaturated fatty acids, 6.3 g of which is linoleic acid. Sugars, sucrose, and starch constitute the carbohydrates in the following amounts: 5.2 g, 2.7 g, and 10.7 g, per 100 g, respectively [11].

Other studies have reported different values and variations in composition, which may result in slight discrepancies among studies. The natural variability in components arises from several factors, including differences in geographic origin, soil quality, climate conditions, agricultural practices, and seasonal variations [3,4]. These factors can influence the nutrient composition, bioactive compounds, and overall quality of food, leading to variability in the results of food analysis. Research has shown that factors such as heat damage can lead to alterations in oil yield and fatty acid profiles, while frost events can negatively affect seed viability and nutrient content [12,13]. Additionally, prolonged drought conditions have been linked to reduced seed weight and overall crop health. For example, studies by Melelli et al. [12] and Zare et al. [13] highlight the detrimental effects of extreme temperatures and water stress on flaxseed quality. By integrating these findings, it becomes evident that understanding and mitigating the impacts of environmental stresses is vital for optimizing flaxseed production and ensuring product quality.

The sections below provide a more detailed description of the key constituents of flaxseed and explain why they contribute significantly to its value.

3.1.1. Alpha-Linolenic Acid

Alpha-linolenic acid (ALA) is a component of flaxseed oil, which contains a mixture of mono- and polyunsaturated, as well as saturated, fatty acids. The unsaturated fraction is the largest (87.8 to 89.8%), with ALA representing the majority of it [14], according to a study carried out by Qiu et al., while 47.44–53.67% of flaxseed oil is formed by linolenic acid, 19.56–24.33% by oleic acid, 12.8–15.01% by linoleic acid, 5.01–9.57% by stearic acid, and 5.14–6.43% by palmitic acid [15].

ALA is an essential omega-3 fatty acid that serves as a precursor to eicosapentaenoic acid (EPA) and docosahexaenoic acid (DHA), both of which are linked to various health benefits. Since the human body cannot synthesize ALA, it must be obtained through diet. While seafood, particularly fatty fish, is the primary source of EPA and DHA, these beneficial omega-3s are also present in oilseeds like flaxseed. Epidemiological studies and randomized controlled trials consistently highlight the positive impact of omega-3 polyunsaturated fatty acids (PUFAs), particularly EPA and DHA, on long-term health. These benefits include reduced cardiovascular disease morbidity and mortality, improved visual and neurological development, and better outcomes in inflammatory conditions such as arthritis and asthma. However, the conversion of ALA to EPA and DHA in the body is limited in humans, with the conversion of ALA to EPA estimated at 8–12% and to DHA less than 1%. The efficiency of this conversion may vary by gender, with women typically exhibiting higher conversion rates than men. Research has reported that in men, the conversion of ALA to EPA ranges from 0.3% to 8%, whereas in women, it can be as high as 21%. Similarly, the conversion of ALA to DHA is less than 1% in men, but in women, this efficiency can reach up to 9%. This highlights the need for continued research into factors that influence this conversion and the potential health benefits of ALA supplementation and, consequently, diets including flaxseed, given its clear promise for delivering significant health advantages [16].

3.1.2. Proteins

Flaxseed is a plenteous protein source, representing 23% of the overall seed mass, increasing to 35 to 40% after oil extraction. The protein provided by flaxseed is complete, meaning that all the essential amino acids are present in adequate amounts and its consumption can overcome protein deficiency. One of the latest studies revealed which amino acids are present in high content in flaxseed plants; this diversity includes arginine, alanine, glycine, isoleucine, histidine, lysine, and leucine [17]. The sulfur-based amino acids, cysteine, and methionine are also present in significant amounts, and when compared to other plenteous proteins, such as soy, flaxseed can be a preferable choice as an animal protein substitute, for drug formulations and infant formula, for example. Furthermore, proteins play a crucial role in cardiovascular health by helping to reduce cholesterol levels, thereby exerting a positive impact on cardiovascular disease management [4].

3.1.3. Fiber

Flaxseed has been found to have up to 40% dietary fiber in its composition, 25% of which is insoluble fiber majorly represented by cellulose, lignin, and hemicellulose [4]. The remaining 75% is the portion of soluble fiber that includes gums, pectin, and β-glucan. The dietary fiber in flaxseed can lower blood sugar levels, absorb cholesterol and triglycerides, enhance serum lipid profiles, and raise basal metabolism [4]. These effects of its composition in fiber can have an important role in the prevention of diabetes and cardiovascular disease. Moreover, flaxseed, as a source of soluble fiber, constitutes a dietary source for bacteria in the gut, consequently enhancing gut flora and improving gastrointestinal health [4].

3.1.4. Vitamins

Flaxseed contains both water (ascorbic acid, thiamine, and riboflavin and vitamin C, B1, and B2, respectively) and fat-soluble vitamins. The last ones, although in small amounts (mg/100 g), are present in greater quantities than the first ones. According to the Portuguese Table of Food Composition, raw flaxseed (per 100 g) has in its composition different vitamins, including α-tocopherol (8.2 mg), thiamine (0.79 mg), riboflavin (0.25 mg), niacin (7.9 mg), vitamin B6 (0.79 mg), and folates (97 μg) [11].

The fat-soluble vitamins in flaxseed are represented by the abundant vitamin E (569 mg/100 g), fundamentally, as γ-tocopherol (552 mg/100 g), but also α-tocopherol (7 mg/100 g) and δ-tocopherol (10 mg/100 g), a known antioxidant that protects cell proteins and fats from oxidation, regulates sodium elimination within the urine, which might assist lowering its blood load, which is important in heart disease or its prevention. Vitamin K, a basic component in the blood-clotting mechanism, is also present in flaxseed but only when it is milled (0.3 mg/tablespoon) [4].

3.1.5. Minerals

Minerals are essential in human nutrition since they regulate various metabolic processes and their importance in immunity, growth, and development. Flaxseed is not only a rich source of macronutrients (carbohydrates, proteins, and fats) but also micronutrients, which are vitamins and minerals. Magnesium, potassium, phosphorus, and sodium are the macro-minerals present in flaxseed, while zinc, copper, and iron are the most relevant micro-nutrients. A recent study by Noreen et al. gave values for some of the minerals present in flaxseed: potassium (763.7 mg/100 g), phosphorus (581.5 mg/100 g), magnesium (406.6 mg/100 g), iron (5.13 mg/100 g), and zinc (3.30 mg/100 g) [17].

According to the Portuguese Table of Food Composition, the inorganic compounds of flaxseed (per 100 g) are sodium (11 mg), potassium (470 mg), calcium (200 mg), phosphorus (550 mg), magnesium (350 mg), iron (15 mg), zinc (7.8 mg), and selenium (28 μg) [11].

3.2. *Phenolic Compounds*

Phenolic compounds can be categorized into several main classes: phenolic acids, flavonoids, and lignans.

3.2.1. Lignans

Flaxseed is the richest plant-based source of lignans, providing up to 0.7–1.5% of its dry weight in these compounds, which are highly beneficial to human health [18]. Lignans present in flaxseed are essentially matairesinol, pinoresinol, diphyllin, (in minor quantities), and secoisolarisericinol [16]. The primary lignan in flaxseed is secoisolariciresinol diglycoside (SDG), a phytoestrogenic lignan. It has antioxidant properties that can protect the body from free radicals' production and act as an anticarcinogenic agent [4]. This lignan is converted by the gut microbiota into enterolignans, which show the characteristic properties of mammalian lignans enterodiol (END) and enterolactone (ENL). Flaxseed lignans possess both estrogenic and antiestrogenic properties, along with antioxidant effects [19,20], which contribute to their potential to inhibit the growth of cancerous tumors. Their impact is particularly notable in hormone-sensitive cancers, such as those affecting the breast, endometrium, and prostate [20]. Additionally, flaxseed lignans, particularly SDG, have shown promise as an adjuvant in the prevention and management of diabetes. Studies conducted in vitro and on animal models have demonstrated their potential benefits. Wang et al. investigated the effects of SDG on glucose homeostasis in obese mice and found that it significantly reduced fasting blood glucose, insulin, and free fatty acid levels while enhancing oral glucose tolerance and insulin responsiveness [21].

It also can be responsible for gene expression modulation, impacting enzyme activity. Not only can SDG influence gene expression and the activity of genes involved in metabolic processes related to diabetes, but it also exerts effects on both normal and cancerous tissues. This offers an additional mechanism beyond its antioxidant properties for combating cancer. SDG also showed potential anti-inflammatory properties in vitro on lipoxygenase (LOX) and human cyclooxygenase (COX-2) enzyme inhibition, known as pro-inflammatory enzymes [19].

SDG, a recognized phytoestrogen found in flaxseed, has been reported to exhibit mild antihypertensive effects. It functions as a long-acting hypotensive agent by activating guanylate cyclase pathways, which are essential for regulating blood pressure [22].

3.2.2. Phenolic Acids and Flavonoids

Phenolic compounds are commonly found in plants and are frequently associated with numerous health benefits, primarily due to their strong antioxidant effects. Phenolic acids and flavonoids demonstrate a wide range of therapeutic effects, including antimicrobial, anti-inflammatory, antithrombotic, antiallergenic, anti-atherogenic, antioxidant, cardioprotective, and vasodilatory properties [23]. These antioxidant components also prevent the photo-oxidation in whole and ground flaxseed, preserving these seeds for longer times [4].

The values of phenolic acids depend on the cultivar and environment. Flaxseed phenolic acids primarily consist of p-hydroxybenzoic acid (1719 mg/100 g), with substantial contributions from chlorogenic acid (720 mg/100 g), ferulic acid (161 mg/100 g), and coumaric acid (87 mg/100 g) [23].

Flavonoids include various subgroups such as flavonols, flavones, flavanones, isoflavones, anthocyanins, and flavanols. Their content in flaxseed also depends on the cultivar and growing conditions [4]. The highest reported content of flavonoids in flaxseed was around 71 mg/100 g and the lowest was around 35 mg/100 g [24]. The bioactivity of flavonoids depends on gut bacteria, such as *Lactobacilli* and *Bifidobacteria*, which metabolize them into bioactive compounds like herbacitin, quercetin, quercetagetin, kaempferol, naringenin, and eridictyol, which are antioxidants and may have a positive impact on health [4].

Even in minor amounts, the phenolic compounds and flavonoids found in flaxseed play a crucial role in promoting antioxidant activity, regulating blood pressure, and protecting against chronic diseases. This makes flaxseed a valuable addition to any balanced diet aimed at supporting overall well-being.

4. Benefits of Its Use in Diet

Flaxseed offers numerous health benefits when included in a diet (Figure 2). Rich in omega-3 fatty acids, particularly ALA, it supports heart health by reducing inflammation and improving cholesterol levels. Its high fiber content aids digestion and promotes a healthy gut, while the lignans present in flaxseed provide antioxidant and hormone-balancing effects. Some specific components of flaxseed, such as ALA, lignans, and fiber, present a significant effect on the prevention and management of chronic conditions like cardiovascular disease and cancer. Research has shown that ALA, an omega-3 fatty acid, can help reduce blood pressure, lower cholesterol levels, and improve overall heart health, while lignans possess antioxidant and anti-inflammatory properties that may play a role in cancer prevention, particularly breast and prostate cancers [25–27]. Additionally, the high fiber content in flaxseed supports digestive health and helps regulate blood sugar levels, further contributing to cardiovascular well-being. Data from clinical studies highlight that regular flaxseed consumption may reduce the risk of these diseases, making it a valuable dietary addition for promoting long-term health. Additionally, flaxseed's antioxidant properties help protect against chronic diseases, making it a valuable addition to a balanced diet. In the following subsections, these benefits are explored.

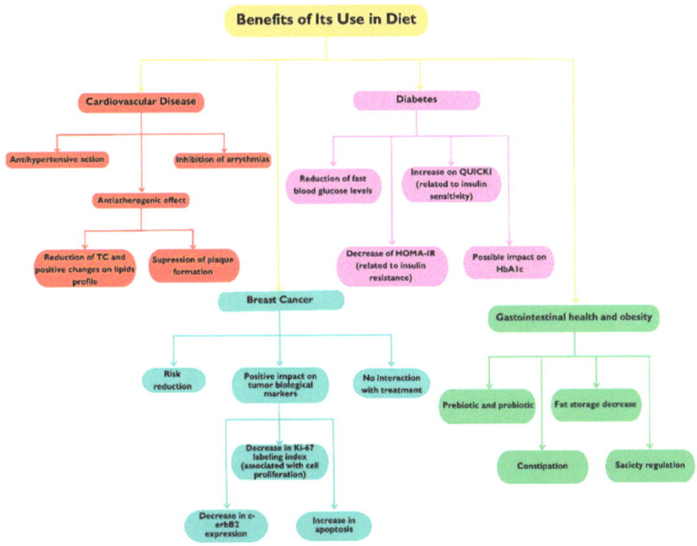

Figure 2. Diverse health benefits associated with incorporating flaxseed into the diet. c-erB2: human epidermal growth factor receptor 2; HbA1c: glycated hemoglobin; HOMA-IR: homeostasis model assessment-estimated insulin resistance; Ki-67: antigen kiel 67; QUICKI: quantitative insulin sensitivity check index; TC: total cholesterol.

4.1. Cardiovascular Disease

Cardiovascular diseases continue to be the leading cause of death and illness globally. Functional foods are gaining increasing recognition as key components of lifestyle changes in the fight against various health conditions. These foods offer proven health benefits beyond their basic nutritional value, as they contain nutraceuticals–nutrients and bioactive compounds that contribute to these positive effects [25].

Multiple preclinical and clinical studies have demonstrated significant benefits of dietary flaxseed supplementation in this context [26]. Its consumption acts in the different aspects of cardiovascular disease, which intensifies its action against it, as flaxseed has an antihypertensive action, antiatherogenic effects, an anti-inflammatory action, and an inhibition of arrhythmias [26].

ALA is the bioactive responsible for the antihypertensive effects of flaxseed. The mechanism behind this action involves the effect of ALA on the concentration of oxylipins [27]. Oxylipins are oxygenated metabolites of PUFAs and have been implicated in the pathogenesis of hypertension since they take part in controlling inflammation and vascular tone when they are derived from omega-6 fatty acids. When oxylipins are derived from omega-3 PUFAs, they tend to have the opposite effects reducing vasoconstriction and inflammation. PUFAs can be metabolized by cyclooxygenase, lipoxygenase, and CYP450. This last one catalyzes the formation of an endothelium-derived hyperpolarizing factor associated with vasodilation through the activation of nitric oxide (NO) synthase. By enhancing vasodilation, this factor promotes a drop in blood pressure. The ALA content of these seeds can also compete with the same enzymes and receptors as omega-6 fatty acids, inhibiting its hypertensive effects [27].

While not the primary component responsible for flaxseed's antihypertensive effects, SDG plays a significant role due to its antioxidant properties. It can mitigate the pathogenic impact of reactive oxygen species, which are involved in the development of hypertension.

SDG also has other proposed mechanisms to be able to reduce blood pressure, such as the activation of guanylate cyclase and angiotensin I inhibition [26].

Where flaxseed peptides are concerned, they may be capable of inhibiting the angiotensin-converting enzyme and renin. They are also rich in arginine, an amino acid converted in vascular endothelium to nitric oxide and citrulline, causing vasodilation and contributing to flaxseed hypotensive action [26].

The FLAX-PAD (FLAX Effects in Peripheral Arterial Disease) trial was a randomized, double-blind, placebo-controlled study conducted over one year to assess the impact of daily flaxseed supplementation (30 g) on 110 patients with Peripheral Arterial Disease (PAD). Participants, all over 40 years old with an ankle–brachial index of less than 0.9, were observed for more than six months. The trial found a significant reduction in blood pressure among those receiving flaxseed, with a decrease of 10 mmHg in systolic blood pressure (SBP)—a 6.5% reduction—and 7 mmHg in diastolic blood pressure (DBP)—a 9.8% reduction. Notably, a subgroup of patients who started with elevated blood pressure ($\geq$140 mmHg) experienced even greater reductions, with a 15 mmHg decrease in SBP and a 7 mmHg decrease in DBP, both after six months [28].

Consuming flaxseed can positively influence the serum lipid profile in humans, offering significant benefits, particularly for individuals at risk of cardiovascular disease or those already affected by it. With whole flaxseed, the majority of human studies reported reductions in TC (total cholesterol) by 6–11% and low-density lipoprotein cholesterol (LDL-c) by 9–18% in normolipidemic subjects, and TC by 5–17% and LDL-c by 4–10% in hypercholesterolemic subjects. The levels of atherogenic lipoproteins and apoproteins A-1 and B after 12 weeks of dietary intervention, 38–40 g/day in food, can be reduced. However, the FLAX-PAD trial also evaluated the effect of daily 30 g of flaxseed in foods in plasma lipids (measured at 0, 1, 6, and 12 months). A reduction of 15% and 11% in LDL-c and TC, respectively, was observed at 1 and 6 months (differences within the treatment group). At 12 months, the differences were much less significant. In a subgroup of patients taking cholesterol-lowering medication, LDL-c decreased by 8.5 $\pm$ 3.0% at 12 months. This trial also concluded that the LDL-lowering effects of flaxseed were additive to medication and a healthy lifestyle [29].

In addition, both animal studies and a meta-analysis of 28 clinical trials concluded that, unfortunately, flaxseed does not produce any significant changes in HDL-c and triglyceride levels [30]. Animal studies were also conducted to determine if flaxseed has any effects on atherosclerosis. Dupasquier et al. reported that flaxseed supplementation led to a reduction in atherosclerosis and suppression of atherosclerotic plaque formation, with these effects being dose-dependent in the diet [31]. It is also crucial to determine whether established atherosclerotic plaques can regress during a 14-week period of dietary flaxseed supplementation following plaque stabilization, resulting in, approximately, a 40% reduction in the affected area [32]. Furthermore, in another study by Dupasquier et al., a clinical setting characterized by increased vascular reactivity, the flaxseed-enriched diet significantly improved both vascular contraction and endothelium-dependent vessel relaxation [33].

Of all the main components of flaxseed, flax oil rich in ALA indicates no consistent beneficial effect on lipid-lowering. While flax fiber has been suggested to potentially increase fat excretion in feces and thereby help lower lipid levels, the effects have not been consistently significant. In contrast, SDG and other flax lignans, which are bioactive compounds, have demonstrated notable effects on lipid levels, with significant results observed in both animal and human studies. The first one demonstrated the reduction in hepatic cholesterol in a dose-dependent manner in hypercholesterolemia and the slowing of aortic atherosclerosis and atherosclerotic plaque progression [26]. Moreover, in an 8-

week, randomized, double-blind, placebo-controlled study conducted in 55 patients, all hypercholesterolemic, previously divided into three groups of daily supplementation of SDG from flaxseed of 0 mg (placebo), 300 mg, and 600 mg, this compound demonstrated a possible beneficial effect on plasma lipids. At weeks 6 and 8, the 600 mg SDG group showed a reduction of approximately 22% in total cholesterol (TC) and 24.38% in LDL cholesterol (LDL-c) compared to the placebo group. In the 300 mg SDG group, significant reductions in TC and LDL-c were only observed when compared to baseline levels. It was concluded that cholesterol levels relate to dietary flaxseed SDG in a dose-dependent manner [34].

The lignan content also had a beneficial effect on the lipid profile of participants. A 6-week, randomized, double-blind, placebo-controlled study examined the impact of lignans on cardiovascular risk factors. A total of 37 participants (13 men and 24 women, average age of 54 ± 7 years, BMI of 29.7 ± 1 kg/m^2) consumed nutrition bars with similar macronutrient profiles (including 3.0 g of ALA) but different lignan contents, 0.15 g versus 0.41 g. The high-lignan flaxseed bars resulted in significant reductions of 12% in total cholesterol (TC), 15% in LDL cholesterol (LDL-c), and 25% in oxidized LDL (Ox-LDL). In contrast, the regular flaxseed bars showed a tendency to increase Ox-LDL levels. The difference between the high-lignan and regular-lignan bars on Ox-LDL was statistically significant, which is particularly important because Ox-LDL is an independent cardiovascular risk factor, considered more atherogenic than LDL-c itself [35].

The phytosterols activate another lipid-lowering mechanism of flaxseed, present in the amount of 338 mg for every 100 g of flaxseed; these bioactive compounds are structurally similar to cholesterol, which makes them compete with it to be absorbed in the intestinal tract, and this leads to an increase in the fecal excretion of cholesterol and a reduction in its levels [25].

Flaxseed fiber has a lipid-lowering effect because its consumption creates satiety; therefore, the caloric intake is reduced and so is the lipid one. Fiber also reduces transit time, increases bile acid excretion, and decreases bile acid reabsorption via increased fecal cholesterol excretion [26].

Antiplatelet effects would also be a beneficial outcome of the consumption of flaxseed; however, there is a lack of studies on this matter both animals and humans. Flaxseed oil rich in ALA has demonstrated a significant reduction in platelet aggregation, including inhibition of thrombin- and collagen-induced platelet aggregation, which is the case of ALA in animal studies [36]. In contrast, human studies have shown no effects on platelets or isolated cases of significant inhibition of aggregation or even significant effect, but in small sample sizes; therefore, it is important to study the impact of flaxseed in this area and its mechanisms [26].

In the reported animal studies, a whole flaxseed-supplemented diet resulted in a lower incidence of arrhythmias and a shorter QT interval after the heart was subjected to ischemia and reperfusion [37]. Flax oil rich in ALA revealed a cardioprotective effect in ischemia and reperfusion. ALA decreased infarct size, apoptosis, and TNF-α and IL-6 concentrations, and increased the anti-inflammatory cytokine IL-10 and cardiac antioxidant enzymes, and, consequently, inflammation and apoptosis [38]. ALA may also be the bioactive responsible for the antiarrhythmic effects of flaxseed because of omega-3 PUFA's action on the phospholipid bilayer of the myocardial cells and its role in depolarization [39]. SDG was shown to significantly enhance the expression of vascular endothelial growth factor (VEGF), angiopoietin-1, and phosphorylated endothelial nitric oxide synthase in the myocardium. VEGF plays a crucial role in stimulating blood vessel formation and may help improve endothelial cell survival. Endothelial nitric oxide synthase is thought to mediate VEGF's effects. Additionally, angiopoietin-1 promotes endothelial cell survival by reducing apoptosis. These factors collectively contribute to the restoration of vascular responsiveness in ischemic tissue, leading to improved myocardial perfusion and cardiac function [26].

4.2. Diabetes

Although there are two types of diabetes mellitus, the effects of flaxseed were mainly evaluated in diabetes mellitus type 2 (DMT2). This endocrine metabolic disorder affects the metabolism of carbohydrates, primarily glucose, but also lipids and proteins. Glucose metabolism is affected by insulin secretion and action, which in DMT2 are compromised. The main characteristic of all types of diabetes is hyperglycemia or fast blood glucose with values above 126 mg/dL. Long-term hyperglycemia leads to macrovascular and microvascular complications such as diabetic retinopathy, nephropathy, neuropathy, or atherosclerosis, which are great causes of morbidity and mortality worldwide. So, to complement pharmacological therapy, there is a growing interest in what role functional foods, like flaxseed, can play in these types of diseases [40].

Many randomized clinical trials have demonstrated the benefits of flaxseed on glycemic control and insulin sensitivity, even though more data are needed, and the research has been increasing because of the potential that flaxseed has revealed in improving glycemic levels.

The aforementioned 8-week, randomized, double-blind, placebo-controlled study involved 55 hypercholesterolemic patients who were divided into three groups receiving daily supplementation of SDG from flaxseed at doses of 0 mg (placebo), 300 mg, or 600 mg. The study also evaluated the impact of SDG on fasting glucose levels. Significant reductions of 25.56% and 24.96% were observed at weeks 6 and 8, respectively, compared to both placebo and baseline. These decreases were greater in individuals with baseline glucose levels of 105 mg/dL. In the group with the 300 mg daily supplementation, the lowering effect of fasting glucose levels was noticeable, but not statistically significant. As the plasma cholesterol, the decrease in glucose concentration is related to dietary flaxseed SDG in a dose-dependent manner [34]. An updated systematic review and meta-analysis of 53 randomized clinical trials revealed significant effects on reducing fasting blood glucose, HOMA-IR (homeostasis model assessment of insulin resistance), and serum insulin levels, while increasing QUICKI (quantitative insulin sensitivity check index). These effects were notably more pronounced when compared to control groups or placebo, particularly within subgroups [40]. However, a significant reduction in HbA1c was not found, which could be due to the short duration of some studies, since HbA1c is the index for evaluating glycemic response in the last three months. Nevertheless, in a subgroup analysis where studies had a duration superior to 12 weeks, whole flaxseed significantly reduced these levels compared to placebo or control. Despite the good results, in all the subgroups' analyses, heterogeneity was high, which does not permit an accurate interpretation, as wished, of these results. It is important to note that most of these studies were conducted in patients whose pathogenesis of the disease was related to insulin resistance, so the results have shown progress in reducing insulin resistance and improving insulin sensitivity.

Various mechanisms were mentioned to explain how flaxseed produces its effect on glycemic indices, these being the α-glucosidase and α-amylase inhibition (studies in vivo and in vitro), and flaxseed polyphenols helped in the recovery of pancreas, liver, pancreatic β-cells, and kidney functions (in vivo study in diabetic rats and histopathological investigations). Whole flaxseed has a great percentage of fiber, which delays glucose absorption and improves hormone response. Fibers can also produce short-chain fatty acids and regulate gut microbiota, positively affecting glycemic levels [40]. Flaxseed oil, rich in omega-3 fatty acids, has been shown to modulate anti-inflammatory responses and influence gut microbiota in vivo [41]. Additionally, polyunsaturated fatty acids (PUFAs) act as natural agonists of proliferator-activated receptor-γ (PPARγ), which plays a crucial role in glucose metabolism and insulin sensitivity. PUFAs also activate G-protein-coupled receptors (GPRs), which stimulate the increased secretion of glucagon-like peptide-1 (GLP-1) [40].

SDG may have an important role, too, as its hypoglycemic effect is played via the inhibition of PEPCK (phosphoenolpyruvate carboxykinase) gene expression; this way, PEPCK, an enzyme involved in the glucose production by the liver (glucogenesis), is not coded and there is less production of glucose, which decreases glycemic levels [19].

If some systematic reviews and meta-analyses show no effects of flaxseed in HbA1c, others affirm the opposite. Of 13 studies included, the results indicate a decrease in HbA1c in participants with a baseline of more or equal to 7.0% after flaxseed supplementation, especially with poorly controlled DMT2, which may mean that this supplementation is an ally in the control of the disease, even in worsen periods [42].

Effective diet management is widely recognized as the cornerstone of treating type 2 diabetes mellitus (DMT2) and its associated complications. Although further randomized clinical trials and consistent results are necessary to fully understand the potential of flaxseed supplementation on DMT2 and related cardiometabolic parameters, existing data already provide valuable insights. These findings offer promising recommendations that could be beneficial for complementing the diets of diabetic patients [42].

4.3. Breast Cancer

Cancer poses one of the most significant public health challenges due to its high and rising prevalence. As a leading cause of morbidity and mortality globally, it not only significantly impacts life expectancy but also greatly diminishes quality of life. Nutrition is fundamental during cancer treatment and can prevent complications and hopefully reduce the severity of side effects, improving the patients' well-being.

In this context, flaxseed has been one of the most studied foods regarding its role in breast cancer, one of the most common cancers, and the second with the biggest mortality rate worldwide. This happens because the seeds are characterized by their lignans content, 95% of which is SDG, and may have a protective effect against breast cancer due to their low estrogenic activity and antioxidant properties [43]. The former property is related to enterolactone and enterodiol, mammalian lignans that result from the conversion of SDG in the colon by bacteria [19]. There are studies that demonstrate a significant association between serum enterolactone levels and the risk reduction in breast cancer [43]. These metabolites share structural similarities to the main form of estrogen, which allows their binding to the cell receptors, displacing estrogen from cells. Therefore, these lignans work as antiestrogens, preventing estrogen from binding to the receptors and inhibiting the growth of cancer cells. Estrone, the biologically active form of estrogen, is converted into two different metabolites: 2-hydroxyestrone (2OHE1) (with small biological activity) and 16α-hydroxyestrone (16OHE1), which enhances estrogen activity and promotes cell proliferation, particularly in estrogen receptor-positive (ER+) breast tumors. The displacement of estrogen by lignans, such as those from flaxseed, can help reduce the production of 16OHE1, potentially decreasing the risk or severity of estrogen-dependent cancers [43].

A randomized crossover trial involving 28 postmenopausal women, aged 53 to 82 years, was conducted to assess the effects of ground flaxseed supplementation. The study comprised three 7-week feeding periods: a control period where participants followed their usual diet, and two experimental periods where the usual diet was supplemented with either 5 g or 10 g of ground flaxseed. Each feeding period was separated by a washout period of at least 7 weeks. Despite the small biological activity of 2OHE1, it was suggested that this metabolite is protective against breast cancer, suppressing the growth and proliferation of breast cancer cells. So, the ratio of 2/16-OHE1 was used as a biomarker for breast cancer risk, with an increase in its value considered protective. And 2OHE1 excretion showed a linear, dose–response increase between the control, the 5 g, and 10 g feeding periods, and significantly higher for the 10 g feeding period. The same was observed for the 2/16-OHE1 ratio. The 16OHE1 excretion was not significantly affected. This study concluded that flaxseed supplementation may be protective against breast cancer [44].

A randomized, double-blind, placebo-controlled clinical trial was conducted to examine the effects of dietary flaxseed on tumor biological markers and urinary lignan excretion in postmenopausal women newly diagnosed with breast cancer. The treatment group (n = 19) consumed a daily muffin containing 25 g of flaxseed, while the control group (n = 13) consumed a placebo muffin. Tumor tissue was analyzed at the time of diagnosis and again at the time of definitive surgery for key markers, including the tumor cell proliferation rate (Ki-67 labeling index, the primary endpoint), apoptosis, c-erbB2 expression, and estrogen and progesterone receptor levels. Additionally, 24 h urine samples were collected and analyzed for lignan levels. The average treatment duration was 39 days for the placebo group and 32 days for the flaxseed group. Results showed a significant reduction in the Ki-67 labeling index and c-erbB2 expression, along with an increase in apoptosis in the flaxseed group. Furthermore, urinary lignan excretion was notably higher in this group. The study concluded that dietary flaxseed may have the potential to reduce tumor growth [45].

Moreover, ingestion of ALA, of which flaxseed is one of the best sources, has been associated in studies with the reduction in the risk of breast cancer, and animal studies demonstrated suppression of growth, size, cell proliferation, and an increase in its death [43].

It is also crucial to explore whether flaxseed interacts with drugs used in breast cancer treatment, such as tamoxifen. While there are no clinical trials available, experimental studies have not demonstrated any adverse interactions, and flaxseed may offer protective effects. In animal studies, flaxseed kept or increased tamoxifen action in decreasing tumor growth, cell proliferation, and increased apoptosis [43]. A pilot study including postmenopausal women with ER+ breast cancer was also conducted to evaluate the effects of flaxseed in the treatment with anastrozole, and, although further studies are still needed, no interaction was noticed [46].

At first sight, based on existing clinical trials that examine the relationship between flaxseed intake and its effects on breast cancer, the results are encouraging to include flaxseed in the diet due to its apparent beneficial role, owing to its lignan composition. However, these phytoestrogens, apart from their anti-estrogenic activity, can also have estrogenic properties, which raises concerns among healthcare professionals and scientists since further studies are still needed and the current evidence is mixed. That being said, the safety of phytoestrogens in people with a history of hormone-positive cancers is not well established and it is not possible to predict how each individual organism with variable characteristics is going to react, so it is important to consult a healthcare provider before incorporating flaxseed into the diet to balance potential benefits and risks.

4.4. Gastrointestinal Health and Obesity

Gastrointestinal health plays a crucial role in maintaining human health, immunity, cognitive function, and mental well-being, as its microbiota is involved in the production of various substances, including neurotransmitters, specific metabolites, brain-derived neurotrophic factors, short-chain fatty acids, tryptophan, and gamma-aminobutyric acid (GABA) [47]. These substances are essential for nervous system coordination, as the brain, nervous, digestive, and endocrine systems are interconnected through sympathetic and parasympathetic nerves, forming the brain–gut axis. This axis directly influences the physiological functioning of bodily organs and the immune system. Therefore, maintaining proper regulation of the gut can help prevent stress-related disorders, endocrine imbalances, and inflammatory diseases, and improve the prognosis of various health conditions [47].

To guarantee gastrointestinal health, guaranteeing the preservation of the gut microbiome or its enhancement is crucial. One of the most significant genera and a major component of the gut microbiota is *Bifidobacteria*. These beneficial bacteria play a crucial role in the digestion and assimilation of carbohydrates and lipids, contributing to overall metabolic health [47]. Other generas are *Bacillus, Clostridium, Klebsiella, Eubacterium, Peptostreptococccus, Ruminococcus, Nocardia,* and *Streptomyces*.

The consumption of flaxseed can positively impact the gut microbiome since this functional food can work as a prebiotic. Prebiotics are nutrients that feed the intestinal microbiota and give it the substrate to support its activity. The maintenance of the gut microbiome activity and thus regulating and stabilizing these microorganisms can bring benefits to human health. When consumed, flaxseed polysaccharides can gradually alter the composition of the intestinal microbiota, promoting the growth of bacteria that metabolize these compounds, such as Akkermansia, Bifidobacterium, Clostridium, Enterococcus, Lactobacillus, Megamonas, Phascolarctobacterium, and Prevotella species. This shift has been shown to play a significant role in reducing inflammation and colitis, repairing the gut lining, enhancing insulin sensitivity, protecting against intestinal tumors, and slowing the progression of disease [47].

SDG, as mentioned above, is also transformed into enterodiol and enterolactone by the action of various bacteria (*Prevotella, Akkermansia,* and *Bifidobacterium*), compounds that have a significant role in human defense mechanisms against disease [47].

It was also found that flaxseed promotes the increase in the percentage of omega-3 fatty acids in the diet, which leads to an abundance of *Bifidobacteria* in the colon. Gut bacteria synthesize fatty acids that will be used in different processes or become part of triacylglycerols (TAG), energy-storage molecules present in the cell cytoplasm [47]. However, ALA, an essential amino acid that composes TAG and is a building block of other long-chain polyunsaturated fatty acids, cannot be synthesized in the human body, and it has to be acquired from external sources, like flaxseed. The gut microbiota also participates in the metabolism of PUFAs and ALA.

Flaxseed also plays a role in obesity prevention. This functional food can positively influence the gut microbiota of individuals. A key factor is the balance between Firmicutes and Bacteroides in the intestines, with an increased ratio of Firmicutes to Bacteroides linked to greater energy extraction from food and higher triglyceride storage in tissues [48]. Flaxseed fiber and mucilage help reduce this ratio, promoting a healthier balance of bacteria. Additionally, flaxseed helps regulate blood sugar levels, decrease fat storage, and enhance satiety, all of which contribute to weight loss and obesity prevention [47].

However, further in vitro, in vivo, and nutritional intervention studies are necessary to better understand the effects of flaxseed on the gut microbiome and its potential to improve human health and prevent various diseases [47].

Constipation has become increasingly common, particularly among middle-aged and elderly populations and women. A clinical trial investigating the impact of flaxseeds on the gut microbiota in 60 elderly patients (54 men and 6 women, average age of 68.68 $\pm$ 8.73 years) with functional constipation showed that consuming 50 g of flaxseed daily for one month significantly improved defecation frequency, reduced abdominal distension, improved stool consistency, and decreased the need for laxatives compared to pre-treatment symptoms in patients with chronic constipation [49].

Fibers are well known for improving gut motility and reducing defecation time. However, flaxseed goes further by significantly enhancing the diversity of bacterial clusters in the gut. Specifically, it increases beneficial probiotics like *Akkermansia* and *Bifidobacterium* while decreasing harmful bacteria such as *Clostridium perfringens* [49]. This shift, along with the elevated content of short-chain fatty acids, can help reduce the risk of developing conditions such as colitis and colon cancer [49]. These findings indicate that flaxseed might regulate gut microbiota as a probiotic and showed no adverse reaction during the period of treatment, and its therapeutic efficacy was better than that of lactulose [49].

Diet intervention plays a crucial role in constipation management and gastrointestinal health, and flaxseed stands out as an optimal supplement because of its safety, effectiveness, and convenience.

5. Potential Toxicity

While the consumption of flaxseed offers significant health benefits, it is equally important to be aware of the potential risks associated with its chronic use and its interactions with medicines and other mechanisms in the human body.

The toxicity of flaxseed can come from overconsumption of its valuable compounds, interactions, toxic factors, or possible contaminants. Consulting a nutritionist, doctor, or pharmacist about incorporating flaxseed into one's regular diet can be beneficial for everyone, regardless of their medical history or current medication use. This personalized guidance ensures a thorough evaluation of potential health benefits.

The main counter-indications of flaxseed meal include lactating and pregnant women because of its activity as estrogens, responsible for inducing labor and low birth weight. Due to these effects, flaxseed can also interfere with birth control pills or hormone replacement therapy. In the case of ALA, which promotes blood flow, it can have a synergistic effect with blood-thinning medicines and increase the risk of bleeding. When consumed alongside antidiabetic medications or insulin, flaxseed may amplify the hypoglycemic effect more than anticipated, potentially leading to hypoglycemia and its associated risks, which can be both unwanted and hazardous [4].

In an experimental study, several participants decided to withdraw due to unpleasant gastrointestinal effects that occurred within 4 weeks caused by whole flaxseed and oil preparations. This highlights the potential for excessive flaxseed intake to cause gastrointestinal disturbances and underscores the varying effects of different forms of flaxseed. In fact, milled flaxseed was associated with fewer adverse effects compared to whole flaxseed and oil [50].

Apart from that, flaxseed also holds in its composition antinutritional compounds, mainly, cyanogenic glycosides transformed in hydrogen cyanide, phytic acid, and trypsin inhibitors.

5.1. Phytic Acid

Phytic acid is the main storage form of phosphorus not only in oil seeds, but also in cereals, legumes, and nuts, comprising 50–80% of the total phosphorus in these plants. Its main adverse effect is interfering with the absorption of essential minerals such as calcium, zinc, magnesium, copper, and iron, as well as proteins. This interference occurs because phytic acid chelates these minerals and, consequently, reduces their bioavailability to monogastric animals, including humans, who lack the enzyme phytase necessary to break down phytic acid [51].

Nissar et al. reported that phytic acid should be lower as much as possible and recommended consuming 25 mg or less of this antinutrient per 100 g of food. However, the recommended daily intake of phytic acid varies from country to country; for example, the United Kingdom and the United States present values that range between 631 and 746 mg/day, while, on the other side, Sweden recommends only 180 mg/day [52].

According to Khare et al., phytic acid was present in flaxseed in the range of 770–920 mg/100 g. In this study, four varieties of flaxseeds were analyzed [53].

5.2. Trypsin Inhibitors

Trypsin inhibitors are natural metabolites that protect flaxseed from biological stress. These compounds inhibit the activity of key pancreatic enzymes, trypsin, and chymotrypsin, resulting in a decrease in protein digestion and absorption by forming indigestible complexes [54]. The reduction in crucial proteins involved in growth in maintenance may lead to poor and diminished growth of humans and animals. In more extreme cases, to compensate for protein reduction, there is an overstimulation of the pancreas to produce more enzymes, which may cause pancreatic hypertrophy [55]. For decades, the presence of these antinutritional compounds has been studied, since their control should be a health concern.

The study by Khare et al. also determined the amount of trypsin inhibitors present in flaxseed, ranging from 22.78 to 28.85 mg/100 g [53].

To ensure the safety and health benefits of flaxseed, it is crucial to advance detection, quantification, and detoxification techniques. Despite the significant potential and numerous health advantages of flaxseed, its content of various antinutritional compounds requires careful consideration. This monograph will give special attention to the presence of cyanogenic glycosides; therefore, a special section will be dedicated to these contaminants.

6. Cyanogenic Glycosides

Cyanogenic glycosides are nitrogenous secondary plant metabolites that include linamarin, linustatin, lotaustralin, and neolinustatin [56]. These compounds consist of an α-hydroxynitrile aglycon and a sugar moiety (glucose or gentiobiose), which accumulate in vacuoles and, upon cell destruction by ingestion, the glycone portion is removed by hydrolysis performed by intestinal β-glucosidase [26]. The obtained cyanohydrin is degraded to hydrogen cyanide (HCN). This newly produced compound can be toxic to the respiratory, nervous, and endocrine systems, as HCN binds to iron, manganese, or copper ions in various enzymes, including those critical to the cytochrome respiratory chain [26,57]. Apart from being a respiratory inhibitor, HCN can also be converted into thiocyanates that interfere with iodine uptake by the thyroid gland, and long-term exposures may lead to iodine-deficiency disorders such as goiter or cretinism [58].

These compounds serve as a natural defense mechanism for flaxseed, and despite the calculations indicating minimal risk, it is essential to ensure that the remaining cyanogenic glycosides pose no threat to human health. This can be achieved through rigorous detection, quantification, and detoxification techniques.

6.1. Determination Techniques

Nine articles from 1996 to 2023 were analyzed and are detailed in Tables 1 and 2.

The extraction techniques used for cyanogenic glycosides (CNGs) in flaxseed primarily included solid–liquid extraction, solid-phase extraction (SPE), distillation, and derivatization to prepare trimethylsilyl (TMS) derivatives for further analysis. Except for distillation (which lacks available data), all these methods demonstrated good recovery values.

The determination and quantification methods for CNGs can be divided into two categories: direct and indirect methods. Direct methods allow for the differentiation and quantification of cyanogenic glycosides typically present in flaxseed (linustatin, neolinustatin, linamarin, and lotaustralin), as well as others (amygdalin, prunasin, epilotaustralin). These methods mainly involve liquid chromatography (LC), specifically high-performance liquid chromatography (HPLC) and ultra-high-performance liquid chromatography (UHPLC), often coupled with mass spectrometry (MS), often with electrospray ionization (ESI-MS). Quantitative nuclear magnetic resonance (qNMR) and gas chromatography (GC) have also been employed for these purposes. The advantages of liquid chromatography (LC) over gas chromatography (GC) for the determination of cyanogenic glycosides include LC's ability to analyze non-volatile, thermally labile compounds without the need for derivatization, offering higher sensitivity and specificity for these types of compounds.

Indirect techniques, such as ion chromatography and colorimetric methods, measure the hydrogen cyanide (HCN) content that results from the degradation of cyanogenic glycosides. These methods do not directly quantify the cyanogenic glycosides themselves but, rather, assess the HCN produced from their breakdown.

Evaluating the values of limit of detection (LOD), limit of quantification (LOQ), and relative standard deviation (RSD), alongside the accuracy (evaluated through recovery assays) of the corresponding extraction methods, liquid chromatography techniques stand out as the most sensitive, accurate, precise, and reliable. These techniques also exhibit good repeatability and have been validated and successfully applied not only to flaxseed but also to its products and other foods.

Between the methods that identified the major number of different cyanogenic compounds, one qNMR and a UHPLC-QqQ-MS/MS, the UHPLC-QqQ-MS/MS from Zhong et al. outperformed in terms of sensitivity, achieving better limits in terms of LODs (1–25 ng/g) and LOQs (5–100 ng/g), and demonstrating good recovery rates (89.88–125.69%) [59]. When applied to flaxseed samples, not all eight cyanogenic glycosides were detected. Only four compounds, linamarin, lotaustralin, linustatin, and neolinustatin, were found in all samples. Among these, linustatin (0.22–2.83 mg/g) and neolinustatin (0.29–3.65 mg/g) were identified as the main cyanogenic compounds [59]. The UHPLC-QqQ-MS/MS method provides the first validated method (as of the publication date) for direct and simultaneous quantification of these eight cyanogenic glycosides in cyanic agri-foods. It demonstrated good sensitivity, precision, and accuracy, and was successfully applied to various cyanic agri-foods, including flaxseed [60]. However, it should be noted that qNMR is an efficient and rapid (≈20 min) tool to quantify cyanogenic glycoside content in flaxseed, which makes it possible to incorporate it into the routine [61].

The UHPLC-MS/MS method from Cai et al. is more sensitive than the UHPLC-QqQ-MS/MS method, even though it identified only four cyanogenic compounds—linamarin, lotaustralin, linustatin, and neolinustatin—in flaxseed [62]. These four compounds were the only ones found in flaxseed in the previous method, too. Cai et al. compared their UHPLC-MS/MS method to UHPLC-QqQ-MS/MS and aimed to solve the matrix effects [62]. Instead of using a traditional Prime HLB (hydrophilic–lipophilic-balanced copolymer) column, they employed a cigarette filter for purification in the solid-phase extraction (SPE) before chromatography analysis. This approach resulted in limits of detection (LODs) between 0.13 and 0.44 pg/g and limits of quantification (LOQs) between 0.44 and 1.48 pg/g, indicating very high sensitivity. Additionally, the method achieved good recovery rates (113–133%) [62]. The UHPLC-MS/MS method was validated and found to be simple, easy to perform, accurate, reliable, highly sensitive, and stable, with broad applicability [62].

Table 1. Compilation of methods to determine cyanogenic compounds in flaxseed or flaxseed products.

Samples	Cyanogenic Compounds	Extraction	Analytical Technique	LOD	LOQ	Recovery	Precision (RSD)	Year	Ref.
Flaxseed from a previous study, ethanol-extracted ground flax, flax pressed cake, spent flaxseed, and CO_2-extracted flaxseed	Linamarin, linustatin, and neolinustatin	See Table 2.	HPLC-RI Analytical column: RP-18 (4 × 250 mm, 10 µm particle size). Mobile phase: 95/4.95/0.05 H_2O/methanol/H_3PO_4. Injection volume: 20 µL. Flow rate: 0.7 mL/min. Temperature: 21 °C. Isocratic mode.	N.A.	N.A.	N.A.	37.41%	1996	[63, 64]
Mature oil-type flaxseed from 7 different cultivars grown in 2 locations in 3 different years	Linustatin	Extraction solvent: methanol/H_2O 3:1 (v/v). Four successive extractions with a ratio of 6:1 of solvent volume to the mass of ground seed, with an ultrasonic bath for 30 min at 40 °C. Centrifugation between each extraction: 10 min, 4 °C, 3824× g.	qNMR Dried samples + 1.5 mL methanol/D_2O 3:1 (v/v) + analysis standard: TMSP 0.1 mg/mL. Homogenization and centrifugation at 20,000× g and 4 °C for 10 min. Supernatant was collected in a 5 mm NMR tube. $_1$H and $_{13}$C assignments for linustatin, neolinustatin, and related compounds.	149.4 µg/g	249.1 µg/g	95.41 ± 1.93%	<6%	2017	[61]
	Neolinustatin			230.8 µg/g	384.6 µg/g				
	Linamarin			7.1 µg/g	11.8 µg/g				
	Lotaustralin			7.6 µg/g	12.7 µg/g				
	Epilotaustralin			6.3 µg/g	10.6 µg/g				
	Amygdalin			193.8 µg/g	323.0 µg/g				
Linforce® (flaxseed coated with two herbal extracts—*Senna alexandrina mill* and *Frangula alnus*) and flaxseed	Linustatin	UAE: ultrasonic bath with 160 W; sonication cycle: 30 s ON, 10 s OFF at room temperature. Centrifugation: with conical rotor at 7000 rpm at room temperature. Cryogrounding: with a freezer/mill; duration time: 9 min (3 × 3 min); frequency: 10 impacts/s; stored at −20 °C until use. Three-step preparation method: 1. Aqueous methanol UAE (methanol/H_2O (75/25, v/v)), resulting supernatant I (after centrifugation) and residue; 2. Residue submitted to alkaline UAE (0.08 M NaOH/methanol (25/75, v/v)), resulting supernatant II (after centrifugation); the combined extract solutions were hydrolyzed by 0.02 M NaOH, acidified (pH 2–3), diluted, and filtered.	UHPLC/ESI-HRMS Analytical column: BEH C_{18} column (2.1 × 100 mm, 1.7 um, 130 Å). ** Mobile phase A: Milli-Q water/formic acid (99.9/0.1, v/v), containing also 50 µM NaCl. Mobile phase B: acetonitrile/formic acid (99.9/0.1, v/v), containing also 50 µM NaCl. Gradient mode; flow rate: 0.2 mL/min. Temperature of the auto sampler: 6 °C. Volume of injection: 5 µL. ESI in positive mode; capillary potential: 4000 V; end-plate offset: 500 V; nebulizing gas: nitrogen, pressure at 276 K Pa; drying gas: nitrogen at 9.0 L/min; temperature: 200 °C.	2.0 µg/g	6.6 µg/g	92.3–102.5%	Repeatability precision: 1.1–6.9%. Intermediate precision: 2.4–6.6%.	2019	[60]
	Neolinustatin								

Table 1. *Cont.*

Samples	Cyanogenic Compounds	Extraction	Analytical Technique	LOD	LOQ	Recovery	Precision (RSD)	Year	Ref.
Field-grown flaxseed from 5 different cultivars and chamber-grown flaxseed from other 5 different cultivars	Linustatin	Derivatization of CNGs/preparation of TMS derivatives: Reaction: the defatted seed powder + HMDS + TMCS + IMD in a microcentrifuge tube for 15 min at 50 °C in a sonicating water bath. Centrifugation: 20,817× *g* for 5 min. Preparation for LC-MS/MS: an aliquot of the silylated extract + mobile phase was put in a microcentrifuge tube. Centrifugation: 14,000 rpm for 5 min. The diluted extract was transferred to a glass vial.	For quantification: GC–Flame ionization detector (FID) Analytical column: a capillary column (50% phenyl methylpolysiloxane, 30 m × 0.32 mm, 0.25 mm film thickness). Injection volume: 1 µL. Temperature gradient: increased from 190 °C to 280 °C at 50 °C/min. Held for 2.2 min, for an analysis of 4 min. Carrier gas: hydrogen. Flow rate: 5 mL/min. Injector temperature: 275 °C. Split ratio: 50:1. FID temperature: 300 °C. Nitrogen (make-up gas) + hydrogen + air flow rates: 30, 40, 430 mL/min, respectively. For characterization: HPLC-MS/MS Analytical column: 50 mm × 2.0 mm Fast Gradient RP-18e. Mobile phase: 100% acetonitrile. Injection volume: 3 µL. Flow rate: 0.4 mL/min. Temperature: ambient temperature. MS analysis: micrOTOF-Q II hybrid quadrupole time-of-flight MS/MS with ESI ion source operated in an MRM mode.	6.43 µg/mL	19.50 µg/mL	In 10 mg matrix: 79.9–107.4%. In 20 mg matrix: 94.2–112.7%.	<15%	2019	[65]
	Neolinustatin			4.72 µg/mL	14.31 µg/mL				
Flaxseed purchased from a supermarket	Linamarin	The samples were ground, dissolved in 80% methanol aqueous solution, placed in an ultrasonic bath, and centrifuged; this extraction was repeated twice. Solid-Phase Extraction (clean-up) Solid-phase extraction column: Prime HBL. Washing solution: 5 mL of water and 2 mL of 10% methanol aqueous solution. Elution solution: methanol/acetonitrile (30:70, *v*/*v*). The elution was repeated 3 times. Eluate was collected and evaporated at 40 °C to dryness under nitrogen. The resultant residue was redissolved in 10% methanol aqueous solution.	UHPLC-QqQ-MS/MS Analytical column: C_{18} RRHD (2.1 mm × 50 mm, 1.8 µm). *** Mobile phase A: water with 0.1% (*v*/*v*) formic acid. Mobile phase B: acetonitrile. Gradients used: isocratic and linear. Flow rate: 0.3 mL/min. Injection volume: 10 µL. Column temperature: 35 °C. MS analysis: QqQ MS equipped with ESI.	5 ng/g	20 ng/g	89.88–125.69%	Intra-day: <10.38% Inter-day: <11.33%	2020	[59]
	Lotaustralin			1 ng/g	5 ng/g				
	Linustatin			5 ng/g	20 ng/g				
	Neolinustatin			5 ng/g	20 ng/g				
	Taxiphyllin			5 ng/g	20 ng/g				
	Amygdalin			2 ng/g	10 ng/g				
	Dhurrin			2.5 ng/g	10 ng/g				
	Prunasin			25 ng/g	100 ng/g				

Table 1. *Cont.*

Samples	Cyanogenic Compounds	Extraction	Analytical Technique	LOD	LOQ	Recovery	Precision (RSD)	Year	Ref.
9 cold-pressed flaxseed oil from "Jingdong Mall"	Linustatin	Solid-Phase Extraction 1. SPE column filled with 120 g of cigarette filter fiber conditioned with 5% (*v*/*v*) isopropanol/n-hexane solution. 2. Loaded with the sample oil diluted in 5% (*v*/*v*) isopropanol/n-hexane solution. 3. Washed with 5% (*v*/*v*) isopropanol/n-hexane solution. 4. Desorption of the CNGs with methanol. 5. Blow-drying under nitrogen of the eluent, redissolution in 30% methanol aqueous solution.	UHPLC-MS/MS Analytical column: HSS T3 column (2.1 mm × 100 mm, 1.7 µm) **** Mobile phase A: water; Mobile phase B: acetonitrile. Temperature: 40 °C. Injection volume: 10 µL. Flow rate: 0.4 mL/min. Gradient mode. Detection with an ESI in the negative mode under selective detection mode (SDM). Voltage: 2800 V. Capillary temperature: 350 °C. Dry temperature: 300 °C.	0.23 pg/g	0.76 pg/g	113–133%	0.8–20.5%	2022	[62]
	Neolinustatin			0.13 pg/g	0.44 pg/g				
	Linamarin			0.43 pg/g	1.43 pg/g				
	Lotaustralin			0.44 pg/g	1.48 pg/g				
Flaxseed, flaxseed cake, and products containing these ingredients	Amygdalin	Extraction solvent: 1% formic acid in methanol/water (25/75) (*v*/*v*). Extraction: 30 min on a rotary tumbler. Centrifugation: 10 min at 3000 rpm. The supernatant is transferred to a filter vial and diluted.	UHPLC-MS/MS Analytical column: BEH C_{18} (100 × 2.1 mm, 1.7 µm). Mobile phase A: 0.1% formic acid in water. Mobile phase B: methanol (acetonitrile can also be used in small retention times). Column temp.: 50 °C. Flow rate: 0.4 mL/min. Injection volume: 2–5 µL. Gradient mode. ESI operated in positive mode; performing MRM: capillary voltage: 30 kV; cone voltage: 30 V; source temperature: 150 °C; desolvation temperature: 600 °C; cone gas flow: 150 L/h; desolvation gas flow: 800 L/h; CID gas: argon (0.0043 mbar); solvent discard: 0–2 and 10–11.5 min.	N.A.	1 mg HCN eq./kg	N.A.	N.A.	2023	[66]
	Linamarin								
	Linustatin								
	Lotaustralin								
	Neolinustatin								
	Prunasin								

BEH: ethylene bridged hybrid; CNG: cyanogenic glycoside; D_2O: deuterium oxide; ESI: electrospray ionization; H_2PO_4: phosphoric acid; HCN: hydrogen cyanide; HLB: hydrophilic–lipophilic balance; HMDS: hexamethyldisilazane; HPLC: high-performance liquid chromatography; HRMS: high-resolution mass spectrometry; HSS: high-strength silica; IMD: methylimidazole; LOD: limit of determination; LOQ: limit of quantification; MRM: multiple reaction monitoring; MS/MS: tandem mass spectrometry; N.A.: data not available; NaCl: sodium chloride; NMR: nuclear magnetic resonance; qNMR: quantitative nuclear magnetic resonance; QqQ-MS/MS: triple-quadrupole mass spectrometry; RRHD: rapid resolution high definition; RSD: relative standard deviation; TMCS: trimethylchlorosilane; TMS: trimethylsilyl; TMSP: trimethylsilylpropionoic acid; UHPLC: ultra-high-performance liquid chromatography. ** BEH is trifunctionally bonded to C_{18}. This new technology is versatile, chemical resistant (pH 1–12 and can handle temperatures up to 80 °C), and mechanically stable (can stand high pressures), enhancing retention, specificity, and sensitivity for complex mixtures and a wide variety of compounds [67]. *** This column is stable up to 1200 bar for fast, high-resolution separations of complex samples. It provides good peak shapes for different compounds across a pH of 2–9. The maximum operating temperature is 60 °C [68]. **** High-strength silica particles (made of 100% silica) enhance chromatographic performance by offering increased retention for polar and hydrophobic compounds, providing unique selectivity compared to BEH. It has mechanical strength (pressures up to 1240 bar) and great stability [69].

Table 2. Compilation of methods to determine total HCN in flaxseed or flaxseed products.

Sample	Hydrolysis (for HCN Determination)	Extraction	Analytical Technique	LOD	LOQ	Recovery	Precision (RSD)	Year	Ref.
Flaxseed from a previous study, ethanol-extracted ground flax, flax pressed cake, spent flaxseed, and CO_2-extracted flaxseed.		Extraction of cyanogenic glycosides Extraction solvent: 70% ethanol aqueous solution. Extraction conditions: in a sonic water bath for 30 min at 30 °C. Filtration: with a 0.45 μm filter through glass wool. Extraction of the crude enzyme 1. From ground flaxseed with cold acetone (−20 °C) in a blender for 1 min. 2. Filtration under vacuum through a Whatman No. 1 paper and re-extraction of the residue 3 more times with acetone. 3. Solvent removal on a desiccator under vacuum at 4 °C.	Using a barbituric acid–pyridine reagent: 1. The flaxseed extract was transferred to a stoppered test tube and evaporated with a nitrogen stream. 2. Adding sodium acetate buffer 0.1 M (pH 6) and incubation of the solution with the crude enzyme extract for 1 h at 30 °C. 3. The reaction ends by adding 0.2 M NaOH, then it stands for 5 min at room temperature, and is neutralized with 0.2 M HCl. 4. Chloramine T 0.5% (*w*/*v*) + buffered extract + barbituric acid–pyridine reagent in a test tube that stands at room temperature for 4 min (pink complex). 5. Absorbance measured at 585 nm (using a blank solution). 6. The calibration curve was traced using potassium cyanide as a reference standard (concentration range of 0.1–3.5 μg of HCN).	N.A.	N.A.	94 ± 3%	36.27%	1996 *	[63,64]
	–	4. Extract + cold sodium acetate buffer (pH 6) in a blender for 1 min. 5. Centrifugation: 12,000× *g* for 2 h. 6. The supernatant was put in a fume hood for 3 h at room temperature to allow hydrolysis of any residual cyanogenic glycosides to HCN. 7. The extract was freeze-dried and stored.	Using pyridine–pyrazolone reagent: 1. The flaxseed extract was transferred to a stoppered test tube and evaporated with a nitrogen stream. 2. Adding sodium phosphate buffer 0.1 M (pH 6) and incubation of the solution with the crude enzyme extract for 1 h at 30 °C. 3. The reaction ends by adding 0.1 M NaOH and 0.1 M sodium phosphate buffer (pH 6), to ensure total decomposition of the hydroxynitrites that result from enzymatic activity. 4. Chloramine T 0.5% (*w*/*v*) + buffered extract in a stoppered test tube, mixed, and put in an ice-water bath for 5 min. 5. Adding the pyridine–pyrazolone reagent to the mixture and let the tube stand for 1 h at room temperature (blue complex). 6. Absorbance measured at 620 nm (using a blank solution). 7. The calibration curve was traced using potassium cyanide as a reference standard (concentration range 0.1–3.5 μg of HCN).	N.A.	N.A.		37.37%		[63,64]

Table 2. *Cont.*

Sample	Hydrolysis (for HCN Determination)	Extraction	Analytical Technique	LOD	LOQ	Recovery	Precision (RSD)	Year	Ref.
Edible plants from 14 generas including *Linum* (flaxseed) from both South and North Korea, and China.	Acid Hydrolysis (50 min) 1. Centrifugation: sample + 0.1 M phosphoric acid for 20 min at 8000 rpm. 2. The supernatant + 4 M sulfuric acid added in a tightly capped vial. 3. Start hydrolysis with heating up to 100 °C. 4. The reaction mixture is cooled in ice.	Distillation Performed in a Micro-Dist tube. 1. The hydrolysis mixture + 0.79 M $MgCl_2$ is heated in the tube for 45 min. 2. Cooling at ambient temperature. 3. Cyanide is collected in the 0.2 M NaOH solution.	Ion Chromatography Analytical column: IonPac AS7 (40 mm × 250 mm, 10 µm particle size). Guard column: IonPac AG7 (40 mm × 50 mm, 10 µm particle size). Detector: ED40 (electrochemical detector), DC Amperometry. Flow rate: 1.0 mL/min (isocratic). Injection volume: 200 µL. Electrode cell: silver working electrode (0.00 V vs. Ag/AgCl reference). Mobile phase: 0.5 M sodium acetate/0.1 M sodium hydroxide/0.5% (*v*/*v*) Ethylenediamine.	0.005 mg/L	N.A.	N.A.	<10%	2013	[70]
3 trademarks of each whole flaxseed type: brown and golden. 3 trademarks of bran of each flaxseed type: brown and golden. * All acquired in main supermarkets or natural products stores in Natal/Rio Grande do Norte.	Acid Hydrolysis (3 h) 1. The sample is transferred into a round bottom flask coupled to a distiller. 2. Closing of the distillation system by dipping the condenser's end in an Erlenmeyer containing 2.5% NaOH solution. 3. Hydrolysis starts by adding distilled water and sulfuric acid 10%.	Distillation 1. Starts when sulfuric acid 10% is added. 2. First distillation: distillate 1 is collected into the Erlenmeyer containing the 2.5% NaOH solution. 3. Second distillation: begins when distilled water and 10% H_2SO_4 are added. The distillate 2 is collected in another Erlenmeyer with 2.5% NaOH solution.	Colorimetric Determination Sample tube: an aliquot of distillate + alkaline picrate 0.5% solution (red-orange compound). Blank tube: distilled water + alkaline picrate 0.5% solution. The tubes were shaken, closed, and put in a water bath at 70 °C for 10 min. Calibration curve: for its construction, a sodium cyanide solution (50 µg/mL) was used to prepare 5 standard solutions corresponding to 50, 100, 150, 200, and 250 µg CN^-. Absorbance was measured at 490 nm.	N.A.	N.A.	N.A.	N.A.	2023	[71]

* These two colorimetric methods were compared to the first HPLC in Table 1. While, in this case, all three methods showed high correlation in measuring cyanogenic compounds, the HPLC method is still the most accurate and sensitive because it can directly distinguish all three cyanogenic compounds. Ag: silver; AgCl: silver chloride; H_2SO_4: sulfuric acid; HCl: hydrochloric acid; HCN: hydrogen cyanide; LOD: limit of determination; LOQ: limit of quantification; $MgCl_2$: magnesium chloride; N.A.: not available; NaOH: sodium hydroxide; RSD: relative standard deviation.

The UHPLC columns used for these two methods were a C18 RRHD (Rapid Resolution High Definition) (1.8 μm, 2.1 mm × 50 mm) by Zhong et al. and an HSS (High-strength Silica) T3 (trifunctional ligand type) column (2.1 mm × 100 mm, I.D., 1.7 μm) by Cai et al. [59,62]. However, among all the LC methods analyzed, the most used analytical column was C18, made of silica (polar) bonded with octadecylsilane, an 18-carbon chain (non-polar). More specifically a C18 (1.7 μm, 100 × 2.1 mm) column with BEH (ethylene bridged hybrid) particle technology.

6.2. Levels of Cyanogenic Glycosides in Flaxseeds

On average, flaxseed contains 250–550 mg of cyanogenic glycosides (CNGs) per 100 g [4]. The Joint FAO/WHO Expert Committee on Food Additives (JECFA) has suggested that acute symptoms of cyanide toxicity, such as vomiting, nausea, headache, mental confusion, hyperpnea, and a decrease in blood pressure, can occur at doses exceeding this range. JECFA has established a safe daily intake of cyanide at 90 μg per kilogram of body weight [71,72], while toxic levels for humans are reported to be between 30 and 100 mg per day [70]. The diary flaxseed consumption of 50 mg showed no adverse effects on human health, which happens because the adult human body is able to detoxify ≤100 mg cyanide/day, and food processing like cooking can reduce the content of these compounds [26].

The EFSA (The European Food Safety Authority) Panel on Contaminants in the Food Chain (CONTAM) issued a scientific opinion, concluding that the acute reference dose (ARfD) for cyanide established at 20 μg per kg of body weight in raw apricot kernels is applicable, as it is considered protective against acute cyanide effects, regardless of the dietary source [73]. Even though setting different ARfDs for different food types is not considered appropriate, this value may be overly conservative due to the lower bioavailability of cyanide from flaxseed, which led to the establishment of a bioavailability factor of 3 [73]. Flaxseed was one of the foods with the highest mean cyanide concentration reported (192.1 mg/kg) in the EFSA database; however, there are no reports in the literature of human poisonings due to the consumption of flaxseed. Using the highest cyanide value reported in the EFSA database (407 mg of CN^-/kg) as a worst-case scenario, the maximum amount of ground flaxseed that can be consumed without exceeding the ARfD ranges from 1.3 g to 14.7 g, depending on body weight. The ARfD would be exceeded in a toddler with a consumption of about 4 g (almost a teaspoon). Due to all uncertainties, some risks remain for adolescents if they consume ground flaxseed [73]. Care is needed with foods like ground flaxseed, particularly for sensitive groups like toddlers and adolescents, and the consumption of intact flaxseed would likely result in much lower cyanide exposures. It was also concluded that the primary contributors to exposure were biscuits, juice or nectar, and pastries and cakes [73]. This report by CONTAM also presents the acute lethal oral dose of cyanide in humans that is estimated to be between 0.5 and 3.5 mg/kg of body weight [73].

The CN- reported by Cho et al. ranged from 261.9 to 345.4 μg/g, which is in the same range as the one reported by Pereira et al., which was 348.40–467.54 μg/g [70,71]. The values for linustatin and neolinustatin registered, respectively, were 1.37 mg/g and 1.55 mg/g according to Zhao et al., 2.0–5.7 mg/g and 0.9–3.9 mg/g according to Shawar et al., and 0.22–2.83 mg/g and 0.29–3.65 mg/g according to Zhong et al. [59,60,65]. The study by Pereira et al. also calculated the equivalent of the suggested safe daily intake, which corresponded to two full tablespoons of whole seeds per day or one full tablespoon of bran per day should be the maximum quantity consumed, which shows the importance of adding the total CN- content in commercial packages to ensure the consumption of safe and nutritious food [71]. Also, in the same study, significant differences were found between whole flaxseed and flaxseed bran, but not between brown and golden types.

The European Union, concerned about the presence of cyanogenic glycosides, has established regulatory limits to ensure food safety (Commission Regulation EU 2023/915 of 25 April 2023). The maximum content of hydrogen cyanide or hydrogen cyanide in cyanogenic glycosides in flaxseed. For unprocessed flaxseeds, whole, crushed, ground, split, and chopped are not placed on the market for the final consumer, and 250 mg/kg is the maximum content allowed to be present. For the ones placed on the market for the final consumer, 150 mg/kg is the higher value of this contaminant that can be present in flaxseed [74].

The Rapid Alert System for Food and Feed (RASFF) is a crucial tool used by the European Union to ensure food safety across its member states. It allows for the rapid exchange of information about potential risks to human health posed by food, feed, and food-contact materials, including alerts [75]. Through RASFF, EU countries, along with the European Commission, can rapidly coordinate responses, such as product recalls or import rejections, to protect consumers. The purpose of this tool is to enhance transparency and consumer protection and maintain high food safety standards across Europe [75]. From a quick search on the RASFF window notification of alerts referring to cyanogenic compounds on flaxseed, five recent notifications were found. In May 2022, there was an alert on increased cyanide content (280 ± 77 mg/kg) in organic flaxseed from Germany, by official controls on the market. Many European organizations were involved, and the measures taken by Germany were withdrawal from the market, informing recipient(s), and destruction [76]. In February 2023, there was an alert on hydrogen cyanide in organic blond flaxseed from Turkey (300 ± 60 mg/kg), based on the company's own test. The recipient(s) were informed (by the Netherlands) and destruction also took place (by Romania) [77]. In April 2024, Denmark notified others of the presence of hydrogen cyanide in flaxseed (320 mg/kg) from Poland and hydrogen cyanide in flaxseed from Belgium (340 mg/kg), both resulting from an official control of the market and involving other European countries on the follow-up [78,79]. The product was withdrawn from the market (by Denmark), and measurements were taken related to the first notification [78]. Lastly, in the same month, Denmark alerted others to the presence of hydrogen cyanide in organic brown flaxseed from India (320 mg/kg) following another official control on the market. The product was then withdrawn from the market [80]. Fortunately, over the past two years, there have been only five notifications, all of which were resolved quickly without any harm to the consumer. The goal is to further reduce the number of such issues and to detect these occurrences as early as possible. Identifying these elevated, undesirable levels is crucial to prevent disastrous events. Therefore, the development and enhancement of detection techniques and tools like communication channels are clearly both useful and essential. Detoxification methods are as valuable to the purposes of protecting the consumer as the other resources referred to before. Some of these methods are described below.

6.3. Detoxification Methods

It is crucial to ensure the removal or reduction of these antinutritional compounds. Cyanogenic glycosides can be significantly reduced through food processing methods, both on a large scale in the food industry and on a smaller scale with simple household techniques. The food processing industry can develop new methods or improve existing ones to effectively control cyanogenic glycoside levels, ensuring the safety and nutritional quality of flaxseed products. In Table 3 there are some studied and detailed methods for these purposes (Figure 3).

The use of microwaves and boiling water in cooking, common daily practices, can be effective allies in removing cyanogenic glycosides from flaxseed [81]. Soaking, although the least effective method, still contributes to safer flaxseed consumption [81].

Table 3. Compilation of detoxification methods to remove cyanogenic compounds in flaxseed or flaxseed products.

Sample	Detoxification Technique	Description	Removal Rate	Study Conclusion	Year	Ref.
Flaxseed, Nei Ya-3 Cultivar, northeast China	Extrusion	Equipment: co-rotating twin-screw extruder; length-to-diameter ratio of 27.9:1; screw diameter of 47 mm; circular die of 5.2 mm; equipped with two heating units. Optimal conditions obtained: temperature: 146.0 °C; feeding rate: 32.7 kg/h; crew speed: 152.5 rpm; moisture content: 12.5%.	91.62%	The predicted value for the removal of HCN from flaxseed was 93.23% and the experimental result was 91.62%, constituting a relative error of only 1.76%. The feeding rate was increased to 60 kg/h to improve productivity. The experimental value was 83.32% (relative error of 1.27%). Both detoxification levels were within the required limits.	2008	[82]
Fresh flaxseed from the Lanxi County of Heilongjiang Province, China	Enzymatic treatment/fermentation	Enzymatic preparation: 12.5% human liver β-glucosidase and 8.9% (*w*/*w*) *Bacillus* sp. cyanide hydratase prepared in the laboratory. Standard fermentation medium: flaxseed power + water + $MgCl_2$ + $MnCl_2$; pH adjusted to 6.3; autoclaved at 115 °C for 1 h to inactivate the endogenous β-glucosidase. The fermentation medium and the enzymatic preparation were mixed and incubated at 46.8 °C for 48 h. After that, the residual cyanide and the CNGs were measured in the samples.	99.3% (degradability)	The flaxseed power treated with the enzymatic preparation retained lignans and fatty acids. Although this method seems efficient, low-cost, and a protector of beneficial nutrients, it is necessary to prove that the detoxified flaxseed is safe for consumption. This method also provides new ways of removing CNGs from other edible plants.	2012	[83]
Flaxseed from 3 Polish high-α-linolenate flax varieties: 1 brown (Szafir variety) and 2 golden seeds (Oliwin and Jantarol varieties).	Solvent extraction method	Defatting previous double cold extraction with hexane. Aqueous extraction: DFF to water ratio of 1:15 (*m*/*v*), under constant stirring with a magnetic stirrer at ambient temperature for 1 h. Centrifugation: for 25 min, 1500× *g*. The supernatant was freeze-dried.	95.7–97.7%	This was a comparative study of flaxseed extracts since they can be applied as food ingredients. Aqueous extracts demonstrated a significant reduction in CNGs but less antioxidant activity. Ethanolic extraction, although it resulted in higher antioxidant activity, also showed a higher CNG content, and if used, ethanolic extracts have to be controlled. For detoxifying purposes, flaxseed processing with water such as extraction, but also soaking and wet autoclaving, can decrease significantly CNG content.	2015	[84]
		Ethanolic extraction: DFF was extracted twice with 60% aqueous ethanol with a flaxseed to solvent ratio of 1:7.5 (*m*/*v*), under constant vigorous shaking using a lab-scale orbital, at ambient temperature for 1 h. Filtration; centrifugation: 20 min, 1500× *g*. Removal of ethanol by evaporation of the supernatant using a rotary vacuum evaporator. The samples were freeze-dried.	30.2–33.6%			
5 samples of flaxseed intended for use in animal feed	Enzymatic hydrolysis followed by evaporation	Mixing, grinding, and extraction (with an acidic aqueous solution): adding one or more cellulose or starch materials (wheat bran and sunflower cake) helps disrupt the cells and facilitates the interaction between β-glycosidases (short-term vapor impregnation for activation) and the CNGs. The maturation: provides sufficient time for the hydrolytic reaction (pH 6 and 38 °C). Extrusion applies pressure and temperature that reduce the HCN and water content. Drying and cooling: decrease the humidity and increase the stability of the final detoxified material (most of the HCN is evaporated during the physical part).	89–92%	The levels of hydrogen cyanide (HCN) in the 5 samples ranged from 86 to 127 mg/kg, which were reduced to <11 mg/kg after decontamination. An additional study on decontaminated samples containing flaxseed from various batches showed an average HCN content of 12 mg/kg. The method proved efficacy, meeting EU requirements for the quality of flaxseed concerning HCN after decontamination, without affecting the nutritional quality or leaving harmful residues. Emissions of HCN into the atmosphere must comply with national regulations.	2017	[85]

Table 3. *Cont.*

Sample	Detoxification Technique	Description	Removal Rate	Study Conclusion	Year	Ref.
Brown and golden flaxseed	Germination	This method was carried out for 5 days in a dark and closed incubator at 25 °C. To avoid microbial growth seeds were washed twice a day and collected daily to measure the changes in root length. The samples were treated and oven-dried at 105 °C to constant weight to determine their content. The results correspond to the percentage of weight loss during drying.	Linustatin: 1309.09 ± 68.63 μg/g to 112.73 ± 7.44 μg/g (≈91.39%). Neolinustatin: 1144.91 ± 22.82 μg/g to 93.89 ± 5.77 μg/g (≈91.80%). Lotaustralin: 1138.46 ± 20.29 μg/g to 72.99 ± 7.26 μg/g (≈93.59%).	Even though germination significantly increased linamarin content, it reduced the other antinutritional compounds. This method also promotes active components, increasing SDG content, flaxseed oil, phenolic and vitamin E content, and thus flaxseed antioxidant activity.	2019	[86]
Flaxseed mixture of brown and golden seed varieties from local market in Beijing	Microwave processing	Samples were dispersed on glass plates. Equipment: household microwave oven. Frequency: 2450 MHz. Output power: 450 W. Heating up time: 12 min.	70.9%	The water-boiling method eliminated the highest CNG content followed by UAD, solvent extraction method, microwave processing, autoclaving, and soaking. However, the analysis of this ranking needs to consider the disadvantages of the different methods. The water-boiling method requires a high temperature that may affect the functional properties of flaxseed since it causes vitamin loss and amino acid and protein denaturation. UAD, in turn, can be an effective method and not affect the beneficial nutrients under controlled conditions.	2020	[81]
	Water-boiling method	DFF was poured into boiling water (100 °C) for 30 min with continuous stirring, then cooled down at room temperature, centrifuged (3000 rpm, 15 min, 25 °C), filtered, and freeze-dried.	94.3%			
	Autoclaving	Equipment: autoclave machine. Temperature: 120 °C. Time: 20 min.	62.6%			
	Ultrasound-assisted detoxification (UAD)	The samples were mixed with distilled water in a solid–liquid ratio of 1:20 (*w*/*v*) and stirred in a water bath for 30 min. Equipment: Sonics Vibracell Probe. Power: 300 W; frequency: 20 kHz; pulsed mode: 10 s ON and 3 s OFF; temperature: 50 °C; time interval: 20 min.	92.1%			
	Soaking	DFF was soaked in distilled water with a continuous change in water. Time interval: 48 h. Samples were submitted to a vacuum filtration and freeze-dried.	48.2%			
	Solvent extraction method	Extraction solvent: methanol/ammonia/water (90/5/5). The mixture was stirred (500 rpm) for 15 min, vacuum filtered, and freeze-dried. The extraction was repeated twice for each sample.	82.4%			

Table 3. *Cont.*

Sample	Detoxification Technique	Description	Removal Rate	Study Conclusion	Year	Ref.
Whole sorrel flaxseed from G.A & Robin Fenton Farm and G&G Edmunds Farm	Bench-scale fermentation	Equipment: square narrow-mouth polypropylene fermentation bottle with a gas trap. Inoculation of a mixed culture of *Lactobacillaceae* (i.e., *Lactobacillus* sp., *Limosilactobacillus* sp., and *Lactiplantibacillus* sp., *Lacticaseibacillus* sp., *Levilactobacillus* sp., and *Lentilactobacillus* sp.) in-house cultured previously isolated from wheat-based thin stillage in a solution containing whole flaxseed + commercial white vinegar (3% acetic acid v/v; pH ≈ 3) + distilled water. Conditions of incubation: water bath at 30 °C for 72 h. Preparation for analysis: the fermentation media was passed through a 12″ 200 µm test sieve, rinsed with tap water, dried overnight at 60 °C, and ground in a coffee grinder.	After 48 h of fermentation, linustatin and neolinustatin were below the detection limits and, consequently, total HCN < 10 mg/kg.	Fermentation of flaxseed using *Lactobacillaceae* successfully removed cyanogenic glycosides and retained beneficial nutritional components such as oil, fatty acids, and SDG. The proposed method can be readily implemented in flaxseed processing based on a pilot-scale feasibility study but lacks further development to produce higher quality flaxseed that can be safely and readily used in food and other product production.	2023	[87]
	Scale-up fermentation	These experiments were conducted at 2 volumes, 4 L and 8 L: sorrel flaxseed and bacterial inoculant. Fermentation took place in a mechanical oven incubator at 30 °C for 72 h. Flaxseed was degummed after being rinsed with tap water and dried, spread in silicone sheets, using a 160 L commercial-grade food dehydrator at 60 °C.	Linustatin: 2659.3 ± 45.7 mg/kg to <500 mg/kg (>81.2%). Neolinustatin: 2761.4 ± 16.8 mg/kg to <500 mg/kg (>81.9%). Total HCN: 351.8 ± 2.5 mg/kg to <10 mg/kg (>97.2%).			

CNG: cyanogenic glycoside; DFF: defatted flaxseed flour; $MgCl_2$: magnesium chloride; $MnCl_2$: manganese chloride; SDG: secoisolariciresinol diglycoside; UAD: ultrasound-assisted detoxification.

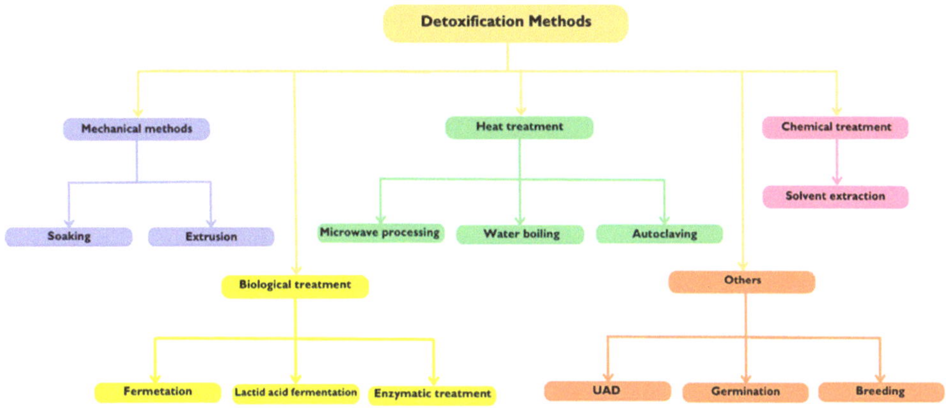

Figure 3. Detoxification methods to reduce the content of flaxseeds in cyanogenic glycosides. UAD: ultrasound-assisted detoxification.

Another mechanical method is extrusion and it also showed a desirable removal rate [82].

Solvent extraction is another viable option, provided the right solvent or mixture is used, though it may also remove valuable constituents [81,84]. Heat treatments, such as water boiling (with the highest removal rate of the physical methods) and autoclaving, can cause protein denaturation and vitamin loss. To mitigate these drawbacks, new techniques have been studied and developed, such as ultrasound-assisted extraction [81], enzymatic treatment, fermentation [83,87], and germination [86]. These methods show promising removal rates and potential for large-scale application. However, some of these methods will require biological control [87].

Instead of processing flaxseed to reduce these compound levels, it is possible to produce seeds with an already low content of cyanogenic glycosides through breeding. A study by Russo & Reggiani on this matter focused specifically on the respective content of these antinutrients in 21 varieties of flaxseed that belonged to 3 different groups of productive attitude: oil, intermediate, and fiber. The intermediate group showed the lowest linustatin content, with the variety Festival standing out for having the lowest total CNG content (0.74 g/kg CN-) and the lowest linustatin content (0.28 g/kg CN-) among all varieties. In the fiber group, the lowest neolinustatin content was observed compared to the other groups, along with varieties that showed low levels of CNG content. In the oil group, the variety Ita269 had the highest CNG content (1.60 g/kg CN-), while the variety Ventimiglia had the lowest neolinustatin content of all varieties. The variability in CNG content across these groups can be explored in breeding programs aimed at reducing CNG levels in flaxseed varieties [88]. However, there are other studies that, apart from evaluating the lower levels of these compounds, also evaluated the differences between compositions to choose the variety that is safer and has more nutritional value. A study by Pavelek et al. evaluated the results of flaxseed breeding in the Czech Republic, which are two recently released Czech varieties and a new breeding line registered for growing that year. The first ones were compared to a standard Dutch variety but had different fatty acid compositions. The last one represents a new quality type given by the content of both linoleic and linolenic fatty acids and the low content of cyanogenic glycosides [89]. A study by Salem et al. applied a comprehensive metabolic and lipidomics approach to the selection of flaxseed varieties. This study evaluated six flaxseed cultivars from different localities in Egypt, identifying Sakha 6 as the most promising. These yellow-colored seeds showed a high content of key amino acids, the highest of two essential PUFAs, α-linolenic acid and linoleic acid, and medicinally important secondary (like riboflavin) metabolites content was higher in this cultivar, while presenting low levels of undesirable antinutrients, such as cyanogenic glycosides. Both the results from breeding and this more recent study suggest that specific flaxseed cultivars are especially beneficial for human consumption and provide valuable data for selecting and developing flaxseed varieties with more quality in the future [90].

Each method has its advantages and disadvantages and may require ongoing refinement. However, it is essential to choose techniques that maximize the removal of cyanogenic glycosides while preserving the beneficial and nutritional components of flaxseed, ensuring its safe consumption.

7. Conclusions

Flaxseed stands out as a highly beneficial dietary component, being a rich source of omega-3 fatty acids, lignans, proteins, and fiber, which play significant roles in cardiovascular health, by reducing blood pressure and positively impacting the lipid profile, cancer prevention, through lignans that exhibit antioxidant properties, and anticancer effects, improving tumor biological markers, diabetes management, by enhancing glycemic control and insulin sensitivity, and gastrointestinal health, working as a prebiotic or increasing the fiber intake, which is very important in managing constipation. Not only does flaxseed help prevent various health issues, but it also enhances the quality of life for healthy individuals. While incorporating flaxseed into the diet has demonstrated promising benefits in these areas, it is not sufficient on its own. With the increasing interest in this nutraceutical, it is crucial to continually explore and deepen our understanding of its health impacts.

However, despite its benefits, this article also underlines potential risks associated with flaxseed consumption. The presence of phytic acid and trypsin inhibitors, but, specially, of cyanogenic glycosides, poses challenges that must be addressed through proper processing and controlled intake. Advances in analytical techniques have significantly improved the extraction, detection, and quantification of these last antinutrients. Techniques such as ultra-high-performance liquid chromatography–tandem mass spectrometry (UHPLC-MS/MS) and quantitative nuclear magnetic resonance (qNMR) have demonstrated high sensitivity, precision, and reliability in identifying and quantifying cyanogenic glycosides.

To ensure consumer safety, a variety of detoxification methods have been explored, ranging from mechanical and biological treatments to heat and chemical approaches. Additionally, innovative techniques such as ultrasound-assisted detoxification and advanced breeding strategies have shown promising removal rates for contaminants like cyanogenic glycosides. Notably, biological or enzymatic treatments present significant potential for large-scale application while minimizing nutrient loss often associated with heat treatments, maintaining the overall nutritional value of flaxseed.

In conclusion, ongoing research, coupled with technological advancements and genetic improvements, is vital for enhancing detection and processing methods for contaminants. Future studies focused on the complex interactions between flaxseed components and various medications will be essential for ensuring their safe integration into diverse dietary regimens. Equally important is the investigation of recommended dosages, which must be tailored according to individual factors such as specific health conditions, medical history, age, and gender. As science continues to advance, flaxseed can be confidently embraced not only as a nutritious and health-promoting addition to diets but also as a secure option for consumers seeking to optimize their overall well-being. By harnessing these innovations and insights, we can unlock the full potential of flaxseed while prioritizing safety and efficacy in its consumption.

Author Contributions: Conceptualization, A.S.S. and S.D.; methodology, S.D. and M.A.S.; software, S.D.; validation, S.D.; formal analysis, S.D.; investigation, S.D. and A.S.S.; resources, A.S.S., S.D. and M.A.S.; data curation, S.D.; writing-original draft preparation, S.D.; writing-review and editing, A.S.S. and M.A.S.; visualization, A.S.S. and S.D.; supervision, A.S.S.; funding acquisition, A.S.S. All authors have read and agreed to the published version of the manuscript.

Funding: This work was financially supported by the research project ValICET (PRIMA/0001/2020, http://doi.org/10.54499/PRIMA/0001/2020)%E2%80%94Valorise foods and Improve Competitiveness through Emerging Technologies applied to food by-products within the circular economy framework (section 2 PRIMA project) funded in Portugal by the Foundation for Science and Technology (FCT). This work received financial support from FCT/MCTES (UIDB/00211/2020, DOI 10.54499/UIDB/00211/2020) through national funds.

Institutional Review Board Statement: Not applicable.

Informed Consent Statement: Not applicable.

Data Availability Statement: Data sharing is not applicable to this article.

Conflicts of Interest: The authors declare no conflict of interest.

References

1. Hall, C., 3rd; Tulbek, M.C.; Xu, Y. Flaxseed. *Adv. Food Nutr. Res.* **2006**, *51*, 1–97. [PubMed]
2. Parikh, M.; Maddaford, T.G.; Austria, J.A.; Aliani, M.; Netticadan, T.; Pierce, G.N. Dietary Flaxseed as a Strategy for Improving Human Health. *Nutrients* **2019**, *11*, 1171. [CrossRef] [PubMed]
3. Canadian Food Inspection Agency The Biology of *Linum usitatissimum* L. (Flax). Available online: https://inspection.canada.ca/en/plant-varieties/plants-novel-traits/applicants/directive-94-08/biology-documents/linum-usitatissimum-flax (accessed on 28 February 2024).
4. Kauser, S.; Hussain, A.; Ashraf, S.; Fatima, G.; Ambreen; Javaria, S.; Abideen, Z.U.; Kabir, K.; Yaqub, S.; Akram, S.; et al. Flaxseed (*Linum usitatissimum*); Phytochemistry, Pharmacological Characteristics and Functional Food Applications. *Food Chem. Adv.* **2024**, *4*, 100573. [CrossRef]
5. Fu, Y.-B. Genetic Evidence for Early Flax Domestication with Capsular Dehiscence. *Genet. Resour. Crop Evol.* **2011**, *58*, 1119–1128. [CrossRef]
6. Cabi Digital Library. *Linum usitatissimum (Flax). CABI Compendium*; CABI Head Office: Wallingford, UK, 2019. [CrossRef]
7. FAOSTAT. Available online: https://www.fao.org/faostat/en/#data/QCL/visualize (accessed on 5 August 2024).
8. Olombrada, E.; Mesias, M.; Morales, F.J. Risk/Benefits of the Use of Chia, Quinoa, Sesame and Flax Seeds in Bakery Products. An Update Review. *Food Rev. Int.* **2024**, *40*, 1047–1068. [CrossRef]
9. Nasiri, F.; Mohtarami, F.; Esmaiili, M.; Pirsa, S. Production of Gluten-Free Biscuits with Inulin and Flaxseed Powder: Investigation of Physicochemical Properties and Formulation Optimization. *Biomass Convers. Biorefin.* **2024**, *14*, 21443–21459. [CrossRef]
10. Epaminondas, P.S.; Araújo, K.L.G.V.; de Souza, A.L.; Silva, M.C.D.; Queiroz, N.; Souza, A.L.; Soledade, L.E.B.; Santos, I.M.G.; Souza, A.G. Influence of Toasting on the Nutritious and Thermal Properties of Flaxseed. *J. Therm. Anal. Calorim.* **2011**, *106*, 551–555. [CrossRef]
11. INSA_pt. Available online: https://jspapp.test.insa.foodcase-services.com/foodcomp/food?23745 (accessed on 2 August 2024).
12. Melelli, A.; Durand, S.; Alvarado, C.; Kervoëlen, A.; Foucat, L.; Grégoire, M.; Arnould, O.; Falourd, X.; Callebert, F.; Ouagne, P.; et al. Anticipating Global Warming Effects: A Comprehensive Study of Drought Impact of Both Flax Plants and Fibres. *Ind. Crops Prod.* **2022**, *184*, 115011. [CrossRef]
13. Zare, S.; Mirlohi, A.; Sabzalian, M.R.; Saeidi, G.; Koçak, M.Z.; Hano, C. Water Stress and Seed Color Interacting to Impact Seed and Oil Yield, Protein, Mucilage, and Secoisolariciresinol Diglucoside Content in Cultivated Flax (*Linum usitatissimum* L.). *Plants* **2023**, *12*, 1632. [CrossRef]
14. Yaqoob, N.; Bhatti, I.A.; Anwar, F.; Mushtaq, M.; Artz, W. Variation in Physico-Chemical/Analytical Characteristics of Oil among Different Flaxseed (*Linum usittatissimum* L.) Cultivars. *Ital. J. Food Sci.* **2016**, *28*, 83–89.
15. Qiu, C.; Wang, H.; Guo, Y.; Long, S.; Wang, Y.; Abbasi, A.M.; Guo, X.; Jarvis, D.I. Comparison of Fatty Acid Composition, Phytochemical Profile and Antioxidant Activity in Four Flax (*Linum usitatissimum* L.) Varieties. *Oil Crop Sci.* **2020**, *5*, 136–141. [CrossRef]
16. Baker, E.J.; Miles, E.A.; Burdge, G.C.; Yaqoob, P.; Calder, P.C. Metabolism and Functional Effects of Plant-Derived Omega-3 Fatty Acids in Humans. *Prog. Lipid Res.* **2016**, *64*, 30–56. [CrossRef] [PubMed]

17. Noreen, S.; Tufail, T.; Bader Ul Ain, H.; Ali, A.; Aadil, R.M.; Nemat, A.; Manzoor, M.F. Antioxidant Activity and Phytochemical Analysis of Fennel Seeds and Flaxseed. *Food Sci. Nutr.* **2023**, *11*, 1309–1317. [CrossRef] [PubMed]
18. Martinchik, A.N.; Baturin, A.K.; Zubtsov, V.V.; Molofeev, V.I. Nutritional value and functional properties of flaxseed. *Vopr. Pitan.* **2012**, *81*, 4–10. [PubMed]
19. Plaha, N.S.; Awasthi, S.; Sharma, A.; Kaushik, N. Distribution, Biosynthesis and Therapeutic Potential of Lignans. *3 Biotech* **2022**, *12*, 255. [CrossRef]
20. Goyal, A.; Sharma, V.; Upadhyay, N.; Gill, S.; Sihag, M. Flax and Flaxseed Oil: An Ancient Medicine & Modern Functional Food. *J. Food Sci. Technol.* **2014**, *51*, 1633–1653. [PubMed]
21. Wang, Y.; Fofana, B.; Roy, M.; Ghose, K.; Yao, X.-H.; Nixon, M.-S.; Nair, S.; Nyomba, G.B.L. Flaxseed Lignan Secoisolariciresinol Diglucoside Improves Insulin Sensitivity through Upregulation of GLUT4 Expression in Diet-Induced Obese Mice. *J. Funct. Foods* **2015**, *18*, 1–9. [CrossRef]
22. Prasad, K. Antihypertensive Activity of Secoisolariciresinol Diglucoside (SDG) Isolated from Flaxseed: Role of Guanylate Cyclase. *Int. J. Angiol.* **2011**, *13*, 7–14. [CrossRef]
23. Bekhit, A.E.-D.A.; Shavandi, A.; Jodjaja, T.; Birch, J.; Teh, S.; Mohamed Ahmed, I.A.; Al-Juhaimi, F.Y.; Saeedi, P.; Bekhit, A.A. Flaxseed: Composition, Detoxification, Utilization, and Opportunities. *Biocatal. Agric. Biotechnol.* **2018**, *13*, 129–152. [CrossRef]
24. Dave Oomah, B.; Mazza, G.; Kenaschuk, E.O. Flavonoid Content of Flaxseed. Influence of Cultivar and Environment. *Euphytica* **1996**, *90*, 163–167. [CrossRef]
25. Associação Portuguesa de Nutrição—Dislipidemias: Caracterização e Tratamento Nutricional. Available online: https://www.apn.org.pt/documentos/ebooks/Ebook_Dislipidemias_CaracterizacaoETratamentoNutricional.pdf (accessed on 6 March 2024).
26. Parikh, M.; Netticadan, T.; Pierce, G.N. Flaxseed: Its Bioactive Components and Their Cardiovascular Benefits. *Am. J. Physiol. Heart Circ. Physiol.* **2018**, *314*, H146–H159. [CrossRef] [PubMed]
27. Caligiuri, S.P.B.; Parikh, M.; Stamenkovic, A.; Pierce, G.N.; Aukema, H.M. Dietary Modulation of Oxylipins in Cardiovascular Disease and Aging. *Am. J. Physiol. Heart Circ. Physiol.* **2017**, *313*, H903–H918. [CrossRef]
28. Rodriguez-Leyva, D.; Weighell, W.; Edel, A.L.; LaVallee, R.; Dibrov, E.; Pinneker, R.; Maddaford, T.G.; Ramjiawan, B.; Aliani, M.; Guzman, R.; et al. Potent Antihypertensive Action of Dietary Flaxseed in Hypertensive Patients. *Hypertension* **2013**, *62*, 1081–1089. [CrossRef] [PubMed]
29. Edel, A.L.; Rodriguez-Leyva, D.; Maddaford, T.G.; Caligiuri, S.P.; Austria, J.A.; Weighell, W.; Guzman, R.; Aliani, M.; Pierce, G.N. Dietary Flaxseed Independently Lowers Circulating Cholesterol and Lowers It beyond the Effects of Cholesterol-Lowering Medications Alone in Patients with Peripheral Artery Disease. *J. Nutr.* **2015**, *145*, 749–757. [CrossRef] [PubMed]
30. Pan, A.; Yu, D.; Demark-Wahnefried, W.; Franco, O.H.; Lin, X. Meta-Analysis of the Effects of Flaxseed Interventions on Blood Lipids. *Am. J. Clin. Nutr.* **2009**, *90*, 288–297. [CrossRef] [PubMed]
31. Dupasquier, C.M.C.; Dibrov, E.; Kneesh, A.L.; Cheung, P.K.M.; Lee, K.G.Y.; Alexander, H.K.; Yeganeh, B.K.; Moghadasian, M.H.; Pierce, G.N. Dietary Flaxseed Inhibits Atherosclerosis in the LDL Receptor-Deficient Mouse in Part through Antiproliferative and Anti-Inflammatory Actions. *Am. J. Physiol. Heart Circ. Physiol.* **2007**, *293*, H2394–H2402. [CrossRef]
32. Francis, A.A.; Deniset, J.F.; Austria, J.A.; LaValleé, R.K.; Maddaford, G.G.; Hedley, T.E.; Dibrov, E.; Pierce, G.N. Effects of Dietary Flaxseed on Atherosclerotic Plaque Regression. *Am. J. Physiol. Heart Circ. Physiol.* **2013**, *304*, H1743–H1751. [CrossRef] [PubMed]
33. Dupasquier, C.M.C.; Weber, A.-M.; Ander, B.P.; Rampersad, P.P.; Steigerwald, S.; Wigle, J.T.; Mitchell, R.W.; Kroeger, E.A.; Gilchrist, J.S.C.; Moghadasian, M.M.; et al. Effects of Dietary Flaxseed on Vascular Contractile Function and Atherosclerosis during Prolonged Hypercholesterolemia in Rabbits. *Am. J. Physiol. Heart Circ. Physiol.* **2006**, *291*, H2987–H2996. [CrossRef]
34. Zhang, W.; Wang, X.; Liu, Y.; Tian, H.; Flickinger, B.; Empie, M.W.; Sun, S.Z. Dietary Flaxseed Lignan Extract Lowers Plasma Cholesterol and Glucose Concentrations in Hypercholesterolaemic Subjects. *Br. J. Nutr.* **2008**, *99*, 1301–1309. [CrossRef]
35. Almario, R.U.; Karakas, S.E. Lignan Content of the Flaxseed Influences Its Biological Effects in Healthy Men and Women. *J. Am. Coll. Nutr.* **2013**, *32*, 194–199. [CrossRef]
36. Holy, E.W.; Forestier, M.; Richter, E.K.; Akhmedov, A.; Leiber, F.; Camici, G.G.; Mocharla, P.; Lüscher, T.F.; Beer, J.H.; Tanner, F.C. Dietary α-Linolenic Acid Inhibits Arterial Thrombus Formation, Tissue Factor Expression, and Platelet Activation. *Arterioscler. Thromb. Vasc. Biol.* **2011**, *31*, 1772–1780. [CrossRef] [PubMed]
37. Ander, B.P.; Weber, A.R.; Rampersad, P.P.; Gilchrist, J.S.C.; Pierce, G.N.; Lukas, A. Dietary Flaxseed Protects against Ventricular Fibrillation Induced by Ischemia-Reperfusion in Normal and Hypercholesterolemic Rabbits. *J. Nutr.* **2004**, *134*, 3250–3256. [CrossRef] [PubMed]
38. Xie, N.; Zhang, W.; Li, J.; Liang, H.; Zhou, H.; Duan, W.; Xu, X.; Yu, S.; Zhang, H.; Yi, D. α-Linolenic Acid Intake Attenuates Myocardial Ischemia/Reperfusion Injury through Anti-Inflammatory and Anti-Oxidative Stress Effects in Diabetic but Not Normal Rats. *Arch. Med. Res.* **2011**, *42*, 171–181. [CrossRef] [PubMed]

39. Endo, J.; Arita, M. Cardioprotective Mechanism of Omega-3 Polyunsaturated Fatty Acids. *J. Cardiol.* **2016**, *67*, 22–27. [CrossRef] [PubMed]
40. Kavyani, Z.; Pourfarziani, P.; Mohamad Jafari Kakhki, A.; Sedgh Ahrabi, S.; Hossein Moridpour, A.; Mollaghasemi, N.; Musazadeh, V.; Hossein Faghfouri, A. The Effect of Flaxseed Supplementation on Glycemic Control in Adults: An Updated Systematic Review and Meta-Analysis of Randomized Controlled Trials. *J. Funct. Foods* **2023**, *110*, 105816. [CrossRef]
41. Zhu, L.; Sha, L.; Li, K.; Wang, Z.; Wang, T.; Li, Y.; Liu, P.; Dong, X.; Dong, Y.; Zhang, X.; et al. Dietary Flaxseed Oil Rich in Omega-3 Suppresses Severity of Type 2 Diabetes Mellitus via Anti-Inflammation and Modulating Gut Microbiota in Rats. *Lipids Health Dis.* **2020**, *19*, 20. [CrossRef] [PubMed]
42. Xi, H.; Zhou, W.; Sohaib, M.; Niu, Y.; Zhu, R.; Guo, Y.; Wang, S.; Mao, J.; Wang, X.; Guo, L. Flaxseed Supplementation Significantly Reduces Hemoglobin A1c in Patients with Type 2 Diabetes Mellitus: A Systematic Review and Meta-Analysis. *Nutr. Res.* **2023**, *110*, 23–32. [CrossRef]
43. Calado, A.; Neves, P.M.; Santos, T.; Ravasco, P. The Effect of Flaxseed in Breast Cancer: A Literature Review. *Front. Nutr.* **2018**, *5*, 4. [CrossRef]
44. Haggans, C.J.; Hutchins, A.M.; Olson, B.A.; Thomas, W.; Martini, M.C.; Slavin, J.L. Effect of Flaxseed Consumption on Urinary Estrogen Metabolites in Postmenopausal Women. *Nutr. Cancer* **1999**, *33*, 188–195. [CrossRef]
45. Thompson, L.U.; Chen, J.M.; Li, T.; Strasser-Weippl, K.; Goss, P.E. Dietary Flaxseed Alters Tumor Biological Markers in Postmenopausal Breast Cancer. *Clin. Cancer Res.* **2005**, *11*, 3828–3835. [CrossRef]
46. McCann, S.E.; Edge, S.B.; Hicks, D.G.; Thompson, L.U.; Morrison, C.D.; Fetterly, G.; Andrews, C.; Clark, K.; Wilton, J.; Kulkarni, S. A Pilot Study Comparing the Effect of Flaxseed, Aromatase Inhibitor, and the Combination on Breast Tumor Biomarkers. *Nutr. Cancer* **2014**, *66*, 566–575. [CrossRef]
47. Mueed, A.; Shibli, S.; Korma, S.A.; Madjirebaye, P.; Esatbeyoglu, T.; Deng, Z. Flaxseed Bioactive Compounds: Chemical Composition, Functional Properties, Food Applications and Health Benefits-Related Gut Microbes. *Foods* **2022**, *11*, 3307. [CrossRef] [PubMed]
48. Luo, J.; Li, Y.; Mai, Y.; Gao, L.; Ou, S.; Wang, Y.; Liu, L.; Peng, X. Flaxseed Gum Reduces Body Weight by Regulating Gut Microbiota. *J. Funct. Foods* **2018**, *47*, 136–142. [CrossRef]
49. Ma, J.; Sun, J.; Bai, H.; Ma, H.; Wang, K.; Wang, J.; Yu, X.; Pan, Y.; Yao, J. Influence of Flax Seeds on the Gut Microbiota of Elderly Patients with Constipation. *J. Multidiscip. Healthc.* **2022**, *15*, 2407–2418. [CrossRef] [PubMed]
50. Austria, J.A.; Richard, M.N.; Chahine, M.N.; Edel, A.L.; Malcolmson, L.J.; Dupasquier, C.M.C.; Pierce, G.N. Bioavailability of Alpha-Linolenic Acid in Subjects after Ingestion of Three Different Forms of Flaxseed. *J. Am. Coll. Nutr.* **2008**, *27*, 214–221. [CrossRef] [PubMed]
51. Coulibaly, A.; Kouakou, B.; Chen, J. Phytic Acid in Cereal Grains: Structure, Healthy or Harmful Ways to Reduce Phytic Acid in Cereal Grains and Their Effects on Nutritional Quality. *Am. J. Plant Nutr. Fertil. Technol.* **2010**, *1*, 1–22. [CrossRef]
52. Nissar, J.; Ahad, T.; Naik, H.R.; Hussain, S.K. A Review Phytic Acid: As Antinutrient or Nutraceutical. Available online: https://www.phytojournal.com/archives/2017/vol6issue6/PartV/6-6-208-319.pdf (accessed on 15 May 2024).
53. Khare, B.; Sangwan, V.; Rani, V. Influence of Sprouting on Proximate Composition, Dietary Fiber, Nutrient Availability, Antinutrient, and Antioxidant Activity of Flaxseed Varieties. *J. Food Process. Preserv.* **2021**, *45*, e15344. [CrossRef]
54. Avilés-Gaxiola, S.; Chuck-Hernández, C.; Serna Saldívar, S.O. Inactivation Methods of Trypsin Inhibitor in Legumes: A Review. *J. Food Sci.* **2018**, *83*, 17–29. [CrossRef]
55. Liener, I.E. Trypsin Inhibitors: Concern for Human Nutrition or Not? *J. Nutr.* **1986**, *116*, 920–923. [CrossRef]
56. Dzuvor, C.K.O.; Taylor, J.T.; Acquah, C.; Pan, S.; Agyei, D. Bioprocessing of Functional Ingredients from Flaxseed. *Molecules* **2018**, *23*, 2444. [CrossRef]
57. Vetter, J. Plant Cyanogenic Glycosides. In *Plant Toxins*; Springer: Dordrecht, The Netherlands, 2017; pp. 287–317. ISBN 9789400764637.
58. Kajla, P.; Sharma, A.; Sood, D.R. Flaxseed-a Potential Functional Food Source. *J. Food Sci. Technol.* **2015**, *52*, 1857–1871. [CrossRef] [PubMed]
59. Zhong, Y.; Xu, T.; Chen, Q.; Li, K.; Zhang, Z.; Song, H.; Wang, M.; Wu, X.; Lu, B. Development and Validation of Eight Cyanogenic Glucosides via Ultra-High-Performance Liquid Chromatography-Tandem Mass Spectrometry in Agri-Food. *Food Chem.* **2020**, *331*, 127305. [CrossRef] [PubMed]
60. Zhao, M.; Bergaentzlé, M.; Flieller, A.; Marchioni, E. Development and Validation of an Ultra-High Performance Liquid Chromatography-High Resolution Mass Spectrometry Method for Simultaneous Quantification of Cyanogenic Glycosides and Secoisolariciresinol Diglucoside in Flaxseed (*Linum usitatissimum* L.). *J. Chromatogr. A* **2019**, *1601*, 214–223. [CrossRef] [PubMed]
61. Roulard, R.; Fontaine, J.-X.; Jamali, A.; Cailleu, D.; Tavernier, R.; Guillot, X.; Rhazi, L.; Petit, E.; Molinie, R.; Mesnard, F. Use of qNMR for Speciation of Flaxseeds (*Linum usitatissimum*) and Quantification of Cyanogenic Glycosides. *Anal. Bioanal. Chem.* **2017**, *409*, 7011–7026. [CrossRef] [PubMed]

62. Cai, B.-D.; Wu, J.-Y.; Bai, Y.-L.; Feng, Y.-Q. Highly Sensitive Analysis of Cyanogenic Glycosides in Cold-Pressed Flaxseed Oil by Employing Cigarette Filter Fiber-Based SPE Coupled with Ultra-Performance Liquid Chromatography-Tandem Mass Spectrometry. *Food Chem.* **2022**, *377*, 131962. [CrossRef]
63. Kobaisy, M.; Oomah, B.D.; Mazza, G. Determination of Cyanogenic Glycosides in Flaxseed by Barbituric Acid−Pyridine, Pyridine−Pyrazolone, and High-Performance Liquid Chromatography Methods. *J. Agric. Food Chem.* **1996**, *44*, 3178–3181. [CrossRef]
64. Oomah, B.D.; Mazza, G.; Kenaschuk, E.O. Cyanogenic Compounds in Flaxseed. *J. Agric. Food Chem.* **1992**, *40*, 1346–1348. [CrossRef]
65. Shahwar, D.; Young, L.W.; Shim, Y.Y.; Reaney, M.J.T. Extractive Silylation Method for High Throughput GC Analysis of Flaxseed Cyanogenic Glycosides. *J. Chromatogr. B Analyt. Technol. Biomed. Life Sci.* **2019**, *1132*, 121816. [CrossRef]
66. EURL-MP-Method_010 Cyanogenic Glucosides (CNGs) in Food and Feed by LC-MSMS. WFSR Wageningen University & Research, 2023, Version 1. pp. 1–20. Available online: https://www.wur.nl/en/show/eurlmp-method_010-cyanogenic-glucosides-cngs-in-food-and-feed-by-lc-msms-v1.htm (accessed on 12 July 2024).
67. Water Corporation—Acquity UPLC Columns. Available online: https://imchem.fr/assets/files/Waters/ACQUITY_UPLC_%202012%20720001140en.pdf (accessed on 2 September 2024).
68. Agilent—ZORBAX Part Number 959757-902. Available online: https://www.agilent.com/store/pt_BR/Prod-959757-902/959757-902 (accessed on 2 September 2024).
69. Waters Corporation—How to Choose a Column. Available online: https://www.waters.com/webassets/cms/library/docs/lcHowToChooseAColumn.pdf (accessed on 2 September 2024).
70. Cho, H.-J.; Do, B.-K.; Shim, S.-M.; Kwon, H.; Lee, D.-H.; Nah, A.-H.; Choi, Y.-J.; Lee, S.-Y. Determination of Cyanogenic Compounds in Edible Plants by Ion Chromatography. *Toxicol. Res.* **2013**, *29*, 143–147. [CrossRef]
71. Pereira, M.L.d.S.; de Souza, R.d.C.P.; de Medeiros, J.V.F.; de Brito, G.Q.; Schwarz, A. Determination of Cyanide in Whole Seeds and Brans of Linseed (*Linum usitatissimum Linn*) by Molecular Spectrophotometry. *Braz. J. Pharm. Sci.* **2023**, *59*, e23059. [CrossRef]
72. Nyirenda, K.K. Toxicity Potential of Cyanogenic Glycosides in Edible Plants. In *Medical Toxicology*; Erkekoglu, P., Ogawa, T., Eds.; IntechOpen: London, UK, 2021; ISBN 9781838802776.
73. EFSA Panel on Contaminants in the Food Chain (CONTAM); Schrenk, D.; Bignami, M.; Bodin, L.; Chipman, J.K.; Del Mazo, J.; Grasl-Kraupp, B.; Hogstrand, C.; Hoogenboom, L.R.; Leblanc, J.-C.; et al. Evaluation of the Health Risks Related to the Presence of Cyanogenic Glycosides in Foods Other than Raw Apricot Kernels. *EFSA J.* **2019**, *17*, e05662. [PubMed]
74. Regulation-2023/915-EN-EUR-Lex. Available online: https://eur-lex.europa.eu/eli/reg/2023/915/oj (accessed on 21 July 2024).
75. RASFF. Available online: https://food.ec.europa.eu/safety/rasff_en (accessed on 6 August 2024).
76. European Commission—RASFF Window. Notification 2022.2641: Increased Cyanide Content in Organic Flaxseed from Germany. 2022. Available online: https://webgate.ec.europa.eu/rasff-window/screen/notification/547009 (accessed on 8 August 2024).
77. European Commission—RASFF Window. Notification 2023.0750: Hydrocyanic Acid (HCN) in Organic Blond Flaxseed from Türkiye. 2023. Available online: https://webgate.ec.europa.eu/rasff-window/screen/notification/593771 (accessed on 8 August 2024).
78. European Commission—RASFF Window. Notification 2024.2851: Hydrocyanic Acid in Flaxseed from Poland. 2024. Available online: https://webgate.ec.europa.eu/rasff-window/screen/notification/677295 (accessed on 8 August 2024).
79. European Commission—RASFF Window. Notification 2024.3101: Hydrocyanic Acid in Flaxseed from Belgium. 2024. Available online: https://webgate.ec.europa.eu/rasff-window/screen/notification/679173 (accessed on 8 August 2024).
80. European Commission—RASFF Window. Notification 2024.3166: Hydrocyanic Acid in Organic Brown Flaxseed from India. 2024. Available online: https://webgate.ec.europa.eu/rasff-window/screen/notification/679478 (accessed on 8 August 2024).
81. Safdar, B.; Pang, Z.; Liu, X.; Rashid, M.T.; Jatoi, M.A. Structural and Functional Properties of Raw and Defatted Flaxseed Flour and Degradation of Cynogenic Contents Using Different Processing Methods. *J. Food Process Eng.* **2020**, *43*, e13406. [CrossRef]
82. Wu, M.; Li, D.; Wang, L.-J.; Zhou, Y.-G.; Brooks, M.S.-L.; Chen, X.D.; Mao, Z.-H. Extrusion Detoxification Technique on Flaxseed by Uniform Design Optimization. *Sep. Purif. Technol.* **2008**, *61*, 51–59. [CrossRef]
83. Wu, C.-F.; Xu, X.-M.; Huang, S.-H.; Deng, M.-C.; Feng, A.-J.; Peng, J.; Yuan, J.-P.; Wang, J.-H. An Efficient Fermentation Method for the Degradation of Cyanogenic Glycosides in Flaxseed. *Food Addit. Contam. Part A Chem. Anal. Control Expo. Risk Assess.* **2012**, *29*, 1085–1091. [CrossRef] [PubMed]
84. Waszkowiak, K.; Gliszczyńska-Świgło, A.; Barthet, V.; Skręty, J. Effect of Extraction Method on the Phenolic and Cyanogenic Glucoside Profile of Flaxseed Extracts and Their Antioxidant Capacity. *J. Am. Oil Chem. Soc.* **2015**, *92*, 1609–1619. [CrossRef]
85. EFSA Panel on Contaminants in the Food Chain (CONTAM); Knutsen, H.K.; Alexander, J.; Barregård, L.; Bignami, M.; Brüschweiler, B.; Ceccatelli, S.; Cottrill, B.; Dinovi, M.; Edler, L.; et al. Assessment of a Decontamination Process for Hydrocyanic Acid in Linseed Intended for Use in Animal Feed. *EFSA J.* **2017**, *15*, e05004.
86. Li, X.; Li, J.; Dong, S.; Li, Y.; Wei, L.; Zhao, C.; Li, J.; Liu, X.; Wang, Y. Effects of Germination on Tocopherol, Secoisolarlciresinol Diglucoside, Cyanogenic Glycosides and Antioxidant Activities in Flaxseed (*Linum usitatissimum* L.). *Int. J. Food Sci. Technol.* **2019**, *54*, 2346–2354. [CrossRef]

87. Huang, C.; Tse, T.J.; Purdy, S.K.; Chicilo, F.; Shen, J.; Meda, V.; Reaney, M.J.T. Depletion of Cyanogenic Glycosides in Whole Flaxseed via *Lactobacillaceae* Fermentation. *Food Chem.* **2023**, *403*, 134441. [CrossRef] [PubMed]
88. Russo, R.; Reggiani, R. Variation in the Content of Cyanogenic Glycosides in Flaxseed Meal from Twenty-One Varieties. *Food Nutr. Sci.* **2014**, *5*, 1456–1462. [CrossRef]
89. Pavelek, M.; Tejklová, E.; Bjelková, M. Results of Linseed Breeding in the Czech Republic. *Tag. Ver. Pfl Anzenzüchter Saatgutkaufl Eute Osterr.* **2011**, *61*, 127–129.
90. Salem, M.A.; Ezzat, S.M.; Giavalisco, P.; Sattar, E.A.; El Tanbouly, N. Application of a Comprehensive Metabolomics Approach for the Selection of Flaxseed Varieties with the Highest Nutritional and Medicinal Attributes. *J. Food Drug Anal.* **2021**, *29*, 214–239. [CrossRef] [PubMed]

molecules

Review

Beneficial Effects of Epigallocatechin Gallate in Preventing Skin Photoaging: A Review

Jiaqiang Sun [1,2,3,†], Yuelu Jiang [4,†], Jing Fu [1,2,3,5], Linlin He [1,2,3,*], Xinmiao Guo [1], Hua Ye [1], Cuiyuan Yin [1], Hongbo Li [4] and Heyuan Jiang [5,*]

1 College of Biological Science and Engineering, Shaanxi University of Technology, Hanzhong 723001, China; sunjiaqiang24@163.com (J.S.); fujing@snut.edu.cn (J.F.); 18717215515@163.com (X.G.); yh18716289047@163.com (H.Y.); 15691611707@163.com (C.Y.)
2 Shaanxi Province Key Laboratory of Bio-Resources, Shaanxi University of Technology, Hanzhong 723001, China
3 Qinba Mountain Area Collaborative Innovation Center of Bioresources Comprehensive Development, State Key Laboratory of Biological Resources and Ecological Environment (Incubation), Hanzhong 723001, China
4 College of Food Science and Engineering, Tianjin University of Science and Technology, Tianjin 300457, China; 18969036236@163.com (Y.J.); hljbobo@tust.edu.cn (H.L.)
5 Key Laboratory of Special Economic Animal and Plant Biology and Genetic Breeding, Ministry of Agriculture and Rural Affairs, Tea Research Institute, Chinese Academy of Agricultural Sciences, Hangzhou 310008, China
* Correspondence: helinlin312@163.com (L.H.); jianghy@tricaas.com (H.J.)
† These authors contributed equally to this work.

Abstract: Skin photoaging, primarily caused by ultraviolet (UV) radiation, leads to skin metabolic disorders, which have adverse psychological and physiological effects on individuals. However, traditional medications for repairing skin photoaging cause side effects. Natural bioactive compounds have been shown to prevent and treat skin photoaging with fewer side effects. Epigallocatechin gallate (EGCG), the main substance in tea polyphenols, is a natural bioactive compound with a range of properties. This review summarizes the beneficial effects and mechanisms of EGCG, as well as the application forms of EGCG in repairing photoaged skin. Results indicated that EGCG has repair effects, including improving elasticity, enhancing moisturization, inhibiting damage, and reducing pigmentation of photoaged skin. It has also been demonstrated that EGCG delivery systems, modified EGCG, and combinations with other bioactive substances could be used for repairing photoaged skin due to its poor stability and low bioavailability. EGCG effectively repairs various types of skin damage caused by UV radiation while maintaining normal skin structure and function. It is, therefore, an effective candidate for repairing photoaged skin. These results could provide references for the development and application of EGCG products for the treatment of photoaged skin.

Keywords: ultraviolet (UV); skin photoaging; epigallocatechin gallate (EGCG); beneficial effects

1. Introduction

Skin photoaging refers to the premature aging of human skin resulting from exposure to ultraviolet (UV), artificial, visible, and infrared lights, with UV radiation being the most important factor [1,2]. The skin plays a beneficial role in absorbing UV radiation for the body, facilitating vitamin D synthesis, and maintaining homeostasis of the body [3]. However, the adverse effects of prolonged UV exposure on the skin are well documented. The UV radiation that reaches Earth's surface primarily consists of UVB (290~320 nm) and UVA (320~400 nm) rays, among these, UVA rays penetrate more deeply into the dermis [1,4]. Most UVB is absorbed by the stratum corneum of the epidermis, while a small fraction passes through this layer to affect the upper portion of the dermis [1]. The mechanisms underlying skin photoaging induced by these two different wavelengths may vary. Initially, UVB is absorbed by chromophores in skin cells, leading to DNA photolesions and generating small amounts of reactive oxygen species (ROS). In contrast, UVA predominantly induces oxidative damage by enhancing ROS production, which results in lipid oxidation and subsequently causes indirect DNA damage [5]. Human skin is simultaneously exposed to both wavelengths of UV radiation, which collectively contribute to photoaging processes. Ultimately, these factors lead to loose, less elastic skin that appears dry and flaky with localized pigmentation changes and may even provoke malignant proliferation of skin cells [6]. The probability of human skin photoaging has increased due to ozone layer depletion, a pressing health concern for humanity. Unlike other diseases, skin photoaging imposes considerable social pressure along with a range of physical and psychological effects.

Topical retinoids (tretinoin, tazarotene, adapalene, etc.) effectively reduce the signs of skin aging caused by UV radiation, such as wrinkles, reduced elasticity, and hyperpigmentation [4,7]. High concentrations of α-hydroxy acids are often used in chemical peels to improve photodamaged skin by exfoliating the epidermal stratum corneum and inducing collagen synthesis [8,9]. However, these medications have significant side effects such as skin irritation, redness, peeling, and in severe cases, scarring. Therefore, there remains a demand for the prevention and repair of photoaged skin, which stimulates the search for safe, reliable, and effective drugs [10].

Natural bioactive substances have emerged as one of the main candidates due to fewer side effects and diverse biological activities compared with commonly used medicines for treating photoaged skin. Studies have demonstrated that natural bioactive compounds such as collagen peptide [11], ginsenoside [12], and anthocyanins [13] are beneficial for the prevention and treatment of skin photoaging. Epigallocatechin gallate (EGCG), a primary natural polyphenol derived from green tea, is known for various properties, including anti-inflammatory, antioxidant, anti-tumor, and anti-bacterial [14,15]. An increasing amount of research indicates its potential in repairing photoaged skin.

This review summarizes reports on the beneficial effects of EGCG on photoaged skin, such as improving skin elasticity, enhancing skin moisturization, inhibiting skin damage, and reducing skin hyperpigmentation (Figure 1). It also discusses various application forms of EGCG for repairing photoaged skin due to its poor stability and low bioavailability. These application forms might offer references for developing and applying EGCG products to repair photoaged skin.

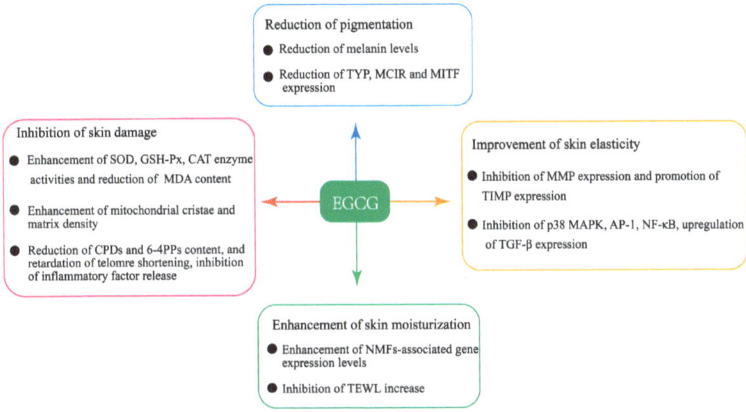

Figure 1. Beneficial effects of EGCG for repairing photoaged skin. TYR, tyrosinase; MC1R, melanocortin 1 receptor; MITF, microphthalmia-associated transcription factor; CPDs, cyclobutane pyrimidine dimers; 6-4PPs, pyrimidine-6,4-pyrimidone photoproducts; CAT, catalase; GSH-Px, glutathione peroxidase; SOD, superoxide dismutase; MDA, malondialdehyde; NMFs, natural moisturizing factors; TEWL, transepidermal water loss.

2. Structural Characterization and Bioavailability of EGCG

EGCG consists of one molecule each of gallic acid and catechin, featuring a chemical structure that includes a benzenediol ring (A) connected to a tetrahydrophan moiety (C), a pyrogallol ring (D), and a galloyl group linked to the B ring. EGCG contains eight phenolic hydroxyl groups (Figure 2) and exhibits strong anti-inflammatory and antioxidant properties, with the B-ring identified as the primary site for antioxidant activity [10]. This property is further enhanced by the trihydroxy structure present in the D-ring [10,16]. According to the Lipinski rule of 5, EGCG has a logP value of less than 5; however, it possesses eight hydrogen bond donors and has a molecular weight of 458.372 g/mol, which is anticipated to result in poor oral bioavailability [17].

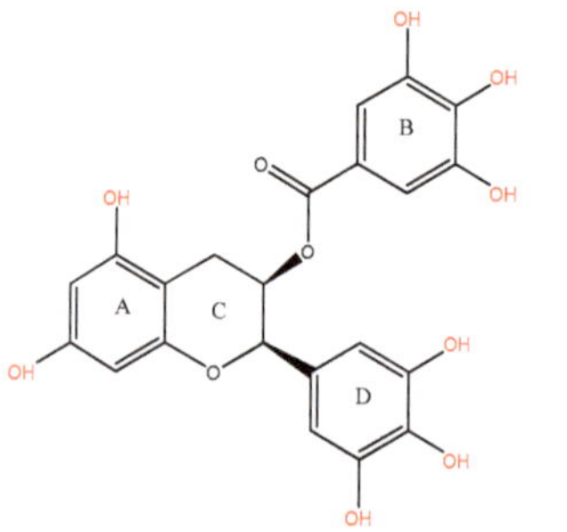

Figure 2. Structure of epigallocatechin gallate (EGCG).

Oral administration is the most commonly used form of EGCG. The plasma EGCG concentration peaks 90 min after oral administration and disappears 12 h later in healthy volunteers [18]. The serum peak concentration reached about 1.4–2.4 h after drinking decaffeinated green tea, with a half-life of 5.0–5.5 h [19]. The metabolokinetic half-life of EGCG in humans appears to be shorter than in rats [20]. After oral administration of EGCG in rats, less than 5% of the dose is absorbed into the bloodstream [21]. The main reasons that limit the utilization of EGCG when taken orally are poor stability and low absorption efficiency of EGCG in vivo. EGCG is very unstable in the internal digestive system, such as the oral cavity, stomach, and intestine; especially under the alkaline condition of the intestine, free EGCG retention is rapidly reduced [22]. At the same time, numerous unfavorable metabolisms of EGCG also occur in the gastrointestinal tract and liver [23]. Catalytic reactions by enzymes such as catechol-O-methyl transferase (COMT), sulfatases, and β-glucuronidases located in the intestine, liver, or target organs can alter the structure of EGCG, resulting in reduced stability [23]. In addition, intestinal microorganisms are also involved in the metabolism and degradation of EGCG [22]. EGCG is completely metabolized in the porcine cecum within 4–8 h [24]. In a clinical study, after the repeated drinking of black tea, the EGCG content in volunteers' blood, urine, and feces was only 0.14% [25]. This result suggests that EGCG undergoes degradation and adverse metabolism in the digestive system and is, therefore, unstable in the body.

Another important reason for limiting the utilization of EGCG when taken orally is the low absorption efficiency, most of which does not enter the blood but enters the excretory system through bile and is excreted by feces [22]. At the same time, regulated by the efflux pump of multidrug resistance-associated protein (MRP), EGCG re-enters the intestine from the blood through efflux, thereby reducing the absorption of EGCG by body cells [22,26]. Based on the above two main reasons, only a small part of EGCG is absorbed by the body after oral administration, thus limiting its biological function.

In addition to oral administration, topical use is an application form of EGCG. Topical formulations appear to be more effective in delivering EGCG than oral administration [27], as the gastrointestinal environment and in vivo metabolism do not affect the stability of topical formulations. However, topical formulations have also shown low bioavailability, mainly due to the low efficiency of transdermal administration and automatic photodegradation of EGCG [28]. The penetration of EGCG into the skin stratum corneum is a key obstacle due to its high molecular weight [29]. Natural sunlight exposure can cause photodegradation of EGCG in the environment, greatly reducing its content and activity, which limits its clinical application for external use [30]. Additionally, factors such as antioxidants, temperature, and pH can also affect the stability of EGCG in topical preparations [28,30].

3. Beneficial Effects of EGCG in Repairing Photoaged Skin

Studies have demonstrated the reparative effects of EGCG on UV-induced photoaged skin. The beneficial effects of EGCG on photoaged skin primarily include improvements in elasticity, enhanced moisturization, inhibition of damage, and reduction in pigmentation. The main molecular targets of EGCG for repairing photoaged skin are outlined in Table 1.

Table 1. Molecular targets of EGCG for repairing photoaged skin.

Compounds	Model	Treatments	Effects	Ref.
EGCG	HSF cells	EGCG (1–20 μM) pretreatment for 45 min, TNF-α (20 ng/mL) induction for 15 min	↓ MMP-1, ERK ↓ MEK, Src	[31]
		EGCG was applied to the medium for 24 h, and HSF cells were exposed to UVA radiation	↓ MMPs, MDA ↑ TGF-β, TIMP-1, SOD, GSH-Px, CAT	[32]
		Tea extract (0.5 mg/mL) was applied to the culture medium, and HSF cells were exposed to UVB radiation (0–20 mJ/cm^2)	↓ MMP-2, MMP-9 ↑ Cell viability	[33]
	HDF cells	10 mg/mL EGCG treated HDF cells exposed to UVA (10 mW/cm^2)	↑ Collagen deposition, insoluble elastin ↓ MDA	[34]
	Artificial skin	EGCG (0.01 mM) was applied to the medium for 6 h and exposed to UVA (20 J/cm^2) radiation	↓ MMPs ↑ TIMP-1	[35]
EGCG–chitosan nanoparticles	HaCaT cells	EGCG nanoformulations were pretreated for 24 h and exposed to UVB (40 mJ/cm^2) radiation for 4 h	↓ CPDs, 6-4PPs	[36]
EGCG	HaCaT cells B16F10 cells	EGCG was used to pretreat HaCaT cells for 30 min, UVB (30 mJ/cm^2) radiation, B16F10 cells were co-treated with αMSH and EGCG (0–100 μM) for 48 h	↑ NMFs, HYAL	[37]
	HaCaT cells	Exposure to UVB (30/60/90 mJ/cm^2) radiation followed by application of EGCG (200 μg/mL)	↓ IL-6, TNF-α, P53 ↓ P21, C-Fos	[38]
	Zebrafish HSF cells	0–100 μM EGCG used to treat zebrafish, 50 μM EGCG used to treat HSF cells for 72 h followed by exposure to UVR (UVA (10 J/cm^2) and UVB (30 mJ/cm^2))	↓ ROS, p38 MAPK, NF-κB, AP-1 ↑ SOD ↓ IL-1α, IL-6, TNF-α, MMP-1	[39]
	B16F10 cells	EGCG (100 μg/mL) was applied to the medium, and the experimental group was treated with the addition of α-MSH	↓ TYR, α-MSH ↓ MC1R, MITF	[40]

Table 1. *Cont.*

Compounds	Model	Treatments	Effects	Ref.
EGCG–collagen	Mice	Hairless back treated topically with EGCG–collagen (0.5 mL, 20 mg mL^{-1}) followed by exposure to UVB (80 μW/cm^2)	↑ SOD, GSH-Px ↑ Epidermal thickness, HYP ↓ MDA, TNF-α, MMP-1	[41]
CMC-Na GTPs	Mice	Hairless mice were exposed to UVB (540 mJ/cm^2) radiation by applying CMC-Na GTPs externally on the back for 30 min	↓ Inflammatory cell infiltration ↑ Nrf2, GSH-Px, CAT, SOD ↓ Bach1, MDA	[42]
EGCG	Mice	Hairless mice topically coated with EGCG preparation (20 μL/cm^2) and exposed to UVB (fluence rate 1.7 μmol/m^2 s) 45 min	↓ Melanosomes ↑ Mitochondrial structure ↑ Collagen, elastic fibers	[43]
EGCG nano-transfersomes	HaCaT cells	Application of EGCG (10 mg/mL) and UVB (60 J/m^2) radiation	↓ MDA, ROS, MMP-2, MMP-9	[44]
Glc-EGCG	Human skin cell models	Exposure to UVA (3 J/cm^2) and UVB (50 mJ/cm^2), application of EGCG (0–50 μM) to skin cells	↓ ROS, IL-6, IL-8, IL-1	[45]
Tea extracts	HaCaT cells HDF cells	The cells were incubated with tea extract (50 μg/mL), exposed to UVA (1 J/cm^2) and UVB (30 mJ/cm^2), and incubated for 24 h	↑ Hyaluronic acid, collagen ↓ IL6, IL8, MMP1, MMP9	[46]
GTE	Healthy white	50 healthy white (Fitzpatrick skin type I–II) adults aged 18–65 years were randomized to a combination of GTE 540 mg plus vitamin C 50 mg or to placebo twice daily for 12 weeks, exposure to UVR mimicking sunlight (290–400 nm; 5% UVB, 95% UVA)	↑ Fibulin-5	[47]
GTE	Female adults	60 female volunteers consumed green tea polyphenols providing 1402 mg total catechins/d (50% EGCG) for 12 weeks; individual MED was determined for each participant. Irradiation was applied to dorsal skin, with 1.25 MEDs using a blue-light solar simulator	↓ Erythema ↑ Skin elasticity, roughness, scaling, density, and water homeostasis	[48]

Abbreviations: ↑ denotes increase; ↓ denotes decrease; GTE, green tea extract; Glc-EGCG, glucosyl form of EGCG; CMC-Na GTPs, carboxymethyl cellulose sodium green tea polyphenols; HSF cells, human skin fibroblasts; HDF cells, human dermal fibroblasts; HaCaT cells, human keratinocytes; B16F10 cells, B16F10 mouse melanoma cells; MED, minimal erythemal dose; IL-6/8/1α, interleukin-6/8/1α; TNF-α, tumor necrosis factor-α; TGF-β, transforming growth factor-β; EGFR, epidermal growth factor receptor; ERK, extracellular signal-regulated kinase; MEK, mitogen-activated protein extracellular kinase; p38 MAPK, p38 mitogen-activated protein kinase; NF-κB, nuclear transcription factor kappa B; AP-1, activator protein 1; Nrf2, nuclear factor erythroid 2-related factor-2; Bach1, BTB and CNC homology 1; MMPs, matrix metalloproteinases; MMP-1/2/9, matrix metalloproteinases 1/2/9; TIMP-1, tissue inhibitor of metalloproteinase-1; SOD, superoxide dismutase; GSH-Px, glutathione peroxidase; CAT, catalase; MDA, malondialdehyde; CPDs, cyclobutane pyrimidine dimers; 6-4PPs, pyrimidine-6,4-pyrimidone photoproducts; HYP, hydroxyproline; NMFs, natural moisturizing factors; HYAL, hyaluronidase; TYR, tyrosinase; α-MSH, α-melanocyte-stimulating hormone; MC1R, melanocortin 1 receptor; MITF, microphthalmia-associated transcription factor.

3.1. EGCG Improves the Elasticity of Photoaged Skin

UV radiation can accelerate the degradation of skin collagen, leading to a reduction in skin elasticity. UV radiation and the resulting ROS co-promote epidermal growth factor receptor (EGFR) [49], which in turn activates mitogen-activated protein kinases (MAPK) signaling pathways composed of extracellular signal-regulated kinase (ERK), C-Jun N-terminal kinase (JNK), and P38 [4,50]. The activation of JNK and P38 pathways increases the expression of the *C-Jun* gene, while activated ERK enters the nucleus to stimulate the expression of the *C-Fos* gene [50,51]. Consequently, MAPK pathway activation leads to the activation of activator protein 1 (AP-1), composed of C-Jun and C-Fos, directly promoting matrix metalloproteinases (MMPs) and cathepsin synthesis [52]. Additionally, UV light can activate the nuclear transcription factor-κB (NF-κB) signaling pathway and enhance MMP-related gene expression. Normally, NF-κB combines with inhibitive κB (IκB) to confine NF-κB within the cytoplasm [50]. However, in response to UV irradiation, the IκB kinase phosphorylates and degrades IκB with increased ROS levels in vivo, allowing the translocation of NF-κB to the nucleus, leading to increased MMP transcription [50]. Furthermore, tumor necrosis factor-α (TNF-α) also induces MMP expression [53,54]. Finally, the increase in MMPs and cathepsin caused by UV radiation increases the degradation of skin collagen [51], which gradually loosens the skin collagen network layer and decreases skin elasticity.

UV radiation can also hinder the synthesis of skin collagen, leading to a decrease in skin elasticity. Transforming growth factor-β (TGF-β) plays a crucial role in regulating collagen synthesis by binding to specific receptors, such as TGF-βI receptor (TβRI) and TGF-βII receptor (TβRII). The binding of TGF-β binds to TβRII activates the intrinsic serine/threonine kinase activity of TβRI and subsequently phosphorylates the transcription factors Smad2 and Smad3, thereby promoting collagen synthesis [55]. Following exposure to UV light, the expression of TβRII is reduced while the expression of inhibitory factor Smad7 is increased. This results in inhibition of activation of transcription factors Smad 2 and Smad 3, consequently reducing collagen synthesis [56]. UV radiation activates AP-1, which in turn inhibits the TGF-β pathway and reduces collagen synthesis [51]. Furthermore, c-Jun can bind to Smad3 to form a complex that blocks Smad3 from binding to specific cis-acting progenitors and reduces collagen synthesis [57]. Ultimately, decreased collagen synthesis due to UV radiation leads to a reduction in skin elasticity.

EGCG has been found to inhibit the increase in collagen, elastin degradation, and the decrease in collagen synthesis caused by UV radiation, thus preventing the reduction in skin elasticity resulting from UV exposure (Figure 3). EGCG reduces the expression of MMPs in UV-irradiated human skin fibroblast (HSF), human dermal fibroblast (HDF), human keratinocytes (HaCaT) cells, artificial skin, zebrafish, and hairless mouse dorsal skin [32,33,35,39,41,44,46]. Furthermore, treatment with EGCG resulted in an increased expression of TGF-β in UV-irradiated HSF cells, as well as elevated levels of tissue metalloproteinase inhibitors (TIMPs) in both UV-irradiated HSF cells and artificial skin [32,35]. Moreover, EGCG inhibited the expression of p38 MAPK, AP-1, NF-κB, and C-Fos transcription in UV-irradiated zebrafish and HSF cells [39], reducing collagen degradation. At present, it remains unclear whether the inhibitory effect of EGCG on AP-1 is attributed to JNK activation or ERK activation. It has been reported that the inhibition of AP-1 by EGCG is due to its suppression of JNK activation and not related to ERK activation [58]; others have suggested that it is due to its inhibitory effect on the phosphorylation of ERK and mitogen-activated protein extracellular kinase (MEK) and Src [31].

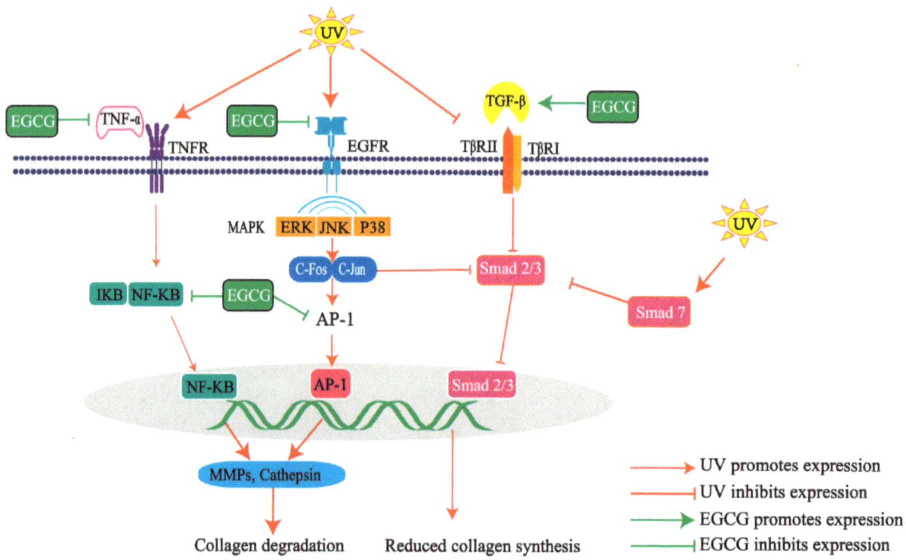

Figure 3. Mechanisms of UV damage and EGCG repair of collagen network layers in photoaged skin. EGCG inhibits activated protein-1 (AP-1) and nuclear transcription factor-κB (NF-κB), contributing to decreased expression of matrix metalloproteinases (MMPs) and increased collagen synthesis through the epidermal growth factor receptor (EGFR), mitogen-activated protein kinase family (MAPK) signaling pathway, tumor necrosis factor-α (TNF-α), and transcriptional growth factor-β (TGF-β). AP-1 also inhibits collagen synthesis by inhibiting the transcription factors Smad 2 and Smad 3. ERK, extracellular signal-regulated kinase; p38, p38 mitogen-activated protein kinase; JNK, C-Jun N-terminal kinase; TβRI/II, transforming growth factor-β type I/II receptor; TNFR, tumor necrosis factor receptor.

The results of animal studies showed that local application of an EGCG preparation could effectively reduce the epidermal layer thickness of the dorsal skin of hairless mice irradiated by UVB, improve the tightness of skin collagen fibers, reduce the formation of cavities, and increase collagen content [41,43]. Clinical trials have also demonstrated the therapeutic potential of EGCG in addressing skin photoaging caused by UV exposure. A randomized controlled clinical trial involving 50 healthy white adults aged 18–65 revealed that oral green tea extract (GTE) and vitamin C could mitigate the degradation of skin elastin fibers due to UV irradiation, thereby preserving skin elasticity [47]. In GTE, EGCG is present in higher amounts, and it is suggested that EGCG may possess similar therapeutic potential. It is important to note that the exact components within GTE responsible for inhibiting elastic fiber degradation remain unclear, particularly due to the addition of vitamin C and the complex composition of GTE used in the experiment. While previous studies have not found evidence supporting the ability of vitamin C to promote collagen synthesis, it can be inferred that any observed improvement in skin condition among subjects is attributable to the beneficial properties of GTE. However, this does not discount a potential role for vitamin C. These experiments confirm that EGCG may hold promise in increasing collagen and elastin content in the skin following UV irradiation, thus aiding in repairing ECM degradation associated with skin photoaging and ultimately effectively preventing and addressing increased skin sagging and wrinkles.

3.2. EGCG Enhances Skin Moisturization

In addition to reducing the elasticity of photoaged skin, UV exposure also reduces skin moisture, resulting in skin dryness and peeling of epidermis. Natural moisturizing factors (NMFs) composed of hyaluronic acid (HA) and filaggrin affect the moisturizing barrier of the skin, with HA being a key regulator of skin moisture levels. Acute UV irradiation can activate *HYAL* gene and increase its expression level, leading to increased HA degradation [37], while chronic UV irradiation can downregulate HA synthesis [59]. Whether promoting HA degradation or inhibiting its synthesis, UV irradiation will eventually reduce the NMF content, destroying the moisturizing barrier and decreasing skin moisture.

Cell experiments have demonstrated that EGCG can reduce *HYAL* gene expression and HA degradation in UVB-irradiated HaCaT cells and increase the expression of NMF-related genes in cells, thereby maintaining the skin water barrier [37]. Animal experiments have shown that dietary supplementation of EGCG inhibits the increase in rat transepidermal water loss (TEWL) caused by UVB irradiation, thereby reducing the skin water loss of rats [60]. Clinical trials have indicated that EGCG improves UV-damaged skin moisture. In a clinical trial involving 45 women aged 40–65 whose back skins were exposed to blue light simulation while orally consuming GTE containing 50% EGCG, they had improved skin elasticity, hydration, and structure compared to the control group. This result suggests that EGCG may enhance women's skin elasticity, hydration, and structure while alleviating skin photoaging effects [48].

3.3. EGCG Inhibits Photoaged Skin Damage

EGCG can repair UV-radiation-induced photoaged skin damage primarily by reducing oxidative stress, DNA damage, and inflammation and inhibiting mitochondrial dysfunction.

3.3.1. EGCG Reduces Oxidative Stress

Under the stimulation of UV radiation, the body undergoes oxidative stress and activates its antioxidant system. During this process, nuclear factor erythroid 2-related factor 2 (Nrf2), an essential transcription factor located in the cytoplasm responsible for activating the antioxidant system, translocates into the nucleus to promote the expression of antioxidant enzymes such as heme oxygenase-1 (HO-1), thereby mitigating oxidative stress [61–63]. However, UV radiation induces DNA damage and downregulates Nrf2 expression through transcriptional inhibition, which leads to reduced Nrf2 activation; this makes it challenging to inhibit UV toxicity [62]. Additionally, BTB and CNC homolog 1 (Bach1) compete with Nrf2 by inhibiting the synthesis and expression of antioxidant enzymes within the nucleus [64].

EGCG is a potent antioxidant that reduces elevated ROS levels caused by UV irradiation, effectively inhibiting decreased activity of antioxidant enzymes and increased production of peroxide products. This ultimately reduces oxidative stress and prevents skin photoaging (Figure 4). Following UV irradiation, EGCG has been shown to increase superoxide dismutase (SOD) and glutathione peroxidase (GSH-Px) activities in HSF and HDF cells while inhibiting malondialdehyde (MDA) content increase [32,34,39]. Furthermore, animal experiments have demonstrated antioxidant properties associated with EGCG. Topical preparations containing EGCG or green tea polyphenols (GTPs) have been found to reduce hydrogen peroxide and MDA levels in skin tissue from hairless mice exposed to UVB radiation while also increasing SOD, catalase (CAT), and GSH-Px enzyme activities [41,42].

Additionally, GTPs promote Nrf2 accumulation within the nucleus while facilitating Bach1 export from the nucleus, subsequently resulting in Bach1 degradation upon entry into the cytoplasm, which slows down its inhibitory effect on Nrf2, leading to increased antioxidase expression [42]. EGCG is a primary component within GTPs, consistent with findings suggesting its potential impact on the Nrf2 pathway [61]. However, the authors did not provide detailed information regarding GTP content or specific components [42].

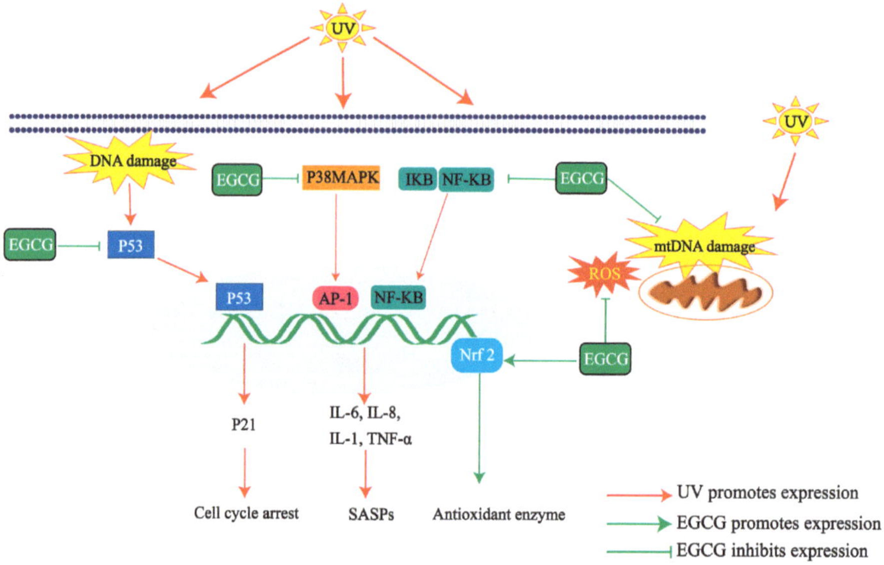

Figure 4. Mechanisms of UV damage and EGCG repair in photoaged skin. EGCG inhibits the release of pro-inflammatory mediators such as tumor necrosis factor α (TNF-α) and interleukin-6/8/1 (IL-6/8/1) through p38 mitogen-activated protein kinase (P38MAPK) and nuclear transcription factor kappa B (NF-κB) and inhibits senescent cells from exhibiting senescence-associated secretory phenotypes (SASPs). EGCG avoids cell cycle arrest by inhibition of DNA damage and expression of P53 and P21. EGCG decelerates mitochondrial dysfunction and maintains the integrity of mitochondrial DNA (mtDNA). EGCG increases the antioxidant enzyme activity and inhibits reactive oxygen species (ROS) content through nuclear factor erythroid 2-related factor 2 (Nrf2).

3.3.2. EGCG Alleviates DNA Damage and Inflammation

Excessive production of ROS due to UV exposure can result in various types of cellular DNA damage, including structural damage, mutations in proto-oncogenes and oncogenes, and even the development of skin cancer. UVB irradiation induces the formation of cyclobutane pyrimidine dimers (CPDs) and pyrimidine-6, 4-pyrimidine photoproducts (6-4PPs), which directly damages DNA and leads to the shortening of telomeres in cell chromosomes [65,66]. On the other hand, UVA indirectly damages DNA through its interaction with cellular chromophores [67].

As photoaged cells proliferate, the number of cells in the G1 phase increases to mitigate the effects of DNA damage caused by UV exposure. This subsequently leads to ECM degradation and ROS production while indirectly stimulating pyruvate signaling and activating a skin inflammatory response [68]. DNA damage and disorders in cell metabolism can lead to increased inflammatory response and expression of inflammatory factors, resulting in inflammatory aging [69]. Inflammatory responses are primarily triggered by the NF-κB and p38MAPK pathways that transmit signals through the neuroendocrine system, ultimately leading to the synthesis and release of a variety of pro-inflammatory mediators such as interleukin, histamine, serotonin, and prostaglandin E2 (PGE2) [68,70]. In particular, PGE2 prevents collagen production and induces the expression of MMPs, thereby affecting the EPK pathway [71]. The release of these mediators increases cellular capillary permeability, leading to infiltration and activation of neutrophils and other phagocytes [72]. At the same time, the increase in inflammatory factors stimulates the overactivation of immune cells, which, together with senescent cells, trigger senescence-associated secretory phenotypes (SASPs), affect the skin microenvironment, and continue to promote cell aging [69].

EGCG reduces DNA damage caused by UV radiation (Figure 4). Pretreatment with a nano-formulated EGCG preparation reduced the increase in CPD and 6-4PP content in HaCaT cells induced by UVB irradiation, thereby slowing down the DNA damage of the cells [36]. Additionally, EGCG can inhibit telomere shortening and cell cycle arrest induced by UVA irradiation [32]. It also reduced the frequency of *HPRT* gene mutation in HSF cells while increasing the senescence and apoptosis of HSF cells induced by UVA irradiation [73]. Morley's study found that drinking 540 mL of green tea could reduce the DNA damage of peripheral blood cells of 10 volunteers aged 30~57 years after UVA irradiation [74].

In addition, EGCG can counteract the skin's inflammatory response during photoaging (Figure 4). Studies have demonstrated that tea polyphenols (TPs) emulsified with sodium carboxymethyl cellulose can effectively inhibit acute inflammatory cell infiltration in the skin of hairless mice irradiated by UVB [42]. Furthermore, EGCG or Gcl-EGCG can inhibit the production of inflammatory factors IL-6, IL-8, and IL-1 [45]. Additionally, EGCG decreased the expression of inflammatory factors TNF-α, IL-1α, and IL-6 in zebrafish and HSF cells induced by UVA and UVB radiation [39]. Tea extract with EGCG as its main component inhibited the UV-induced expression of IL-6 and IL-8 [46]. Adding EGCG to a normal diet increased the minimum erythema dose (MED) of the skin of hairless rats after UV irradiation, reduced the severity of skin sunburn caused by UV radiation, and increased the skin's tolerance to UV radiation [75]. A meta-analysis of recent literature regarding catechins for skin erythema demonstrates that regular intake of green tea catechins (GTCs) reduced skin erythema inflammation caused by UV exposure [76]. In a clinical trial conducted by Rhodes [77] on 16 healthy white volunteers exposed to a UV simulator composed of UVA and UVB radiation sources to induce skin photoaging damage, it was found that oral administration of GTE containing 40.3% EGCG and vitamin C for 12 weeks effectively inhibited UV irradiation-induced skin inflammation and that the level of inflammatory mediator 12-hydroxyeicosatetraenoic acid was reduced while PGE2 did not change. The skin's inflammatory response is influenced by the concentration of catechins in the skin and the type of metabolite after oral supplementation. Individual differences in skin response to EGCG preparations should be acknowledged as they may affect efficacy.

Topical skin application of GTC was more effective than oral administration, possibly due to avoiding the effects of early digestion. Topical application of an EGCG preparation before acute UVA irradiation reduces the formation of sunburned cells and increases leukocyte infiltration in rats [78]. Furthermore, topical application of an EGCG preparation can inhibit UVB-induced leukocyte infiltration in mice [79].

While previous experiments confirmed the potential benefits of EGCG in inhibiting UV-induced inflammatory responses, a clinical study conducted by Farrar et al. [80] showed contradictory results. The study recruited 50 white adults aged 18–65 years for UV exposure using a solar simulator, with the treatment group taking 1080 mg of GTE containing 40% EGCG and 100 mg of vitamin C daily for three months. It was found that this treatment did not reduce UV-induced skin erythema, leukocyte infiltration, and eicosanoid levels compared to the control group. The results indicated that this treatment was ineffective in altering the skin inflammatory response caused by UV exposure, possibly due to individual variations in response to GTE supplementation and gender differences among study subjects.

3.3.3. EGCG Inhibits Mitochondrial Dysfunction

The skin is a highly metabolic organ that relies on the constant renewal of keratinocytes and fibroblasts to maintain the skin barrier. Mitochondrial respiration serves as the energy source for the continuous renewal of skin cells. Mitochondrial dysfunction leads to the aging of skin cells due to insufficient energy, a key characteristic of skin cell aging. Additionally, maintaining the integrity of mitochondrial structure and function improves the maintenance of skin collagen fiber density [81]. ROS, produced by UV radiation, primarily target mitochondrial DNA (mtDNA), leading to mutations due to the lack of protective histones and mechanisms. This ultimately results in mitochondrial dysfunction [82]. In mouse skin cells, UV radiation causes mitochondrial degradation, reducing mitochondrial ridge and matrix density while increasing vacuolation, further contributing to mitochondrial dysfunction. Carnitine acetyltransferase (CRAT) is essential in acetyl-CoA homeostasis within the mitochondrial matrix. The content of CRAT in photoaged skin is reduced, the mitochondrial ROS level of skin cells is increased, and the skin collagen fiber density is reduced after CRAT knockdown [83]. Under oxidative stress, mitochondrial energy requirements vary quantitatively and qualitatively, mainly related to their regulatory roles in mitochondrial autophagy and dynamics. Abnormal regulation of mitochondrial dynamics by UV radiation leads to excessive division of mitochondria, promoting overproduction of ROS and cell apoptosis. The disruption of mitochondria autophagy results in an imbalance in energy homeostasis and cellular dysfunction, ultimately leading to cell aging [84].

EGCG is a mitochondria-targeting molecule with properties that prevent mitochondrial degradation and induce mitochondrial biogenesis [85]. Thus, it may support continuous skin cell renewal by protecting mitochondrial structure and function and reducing mitochondria-mediated apoptosis (Figure 4). Animal experiments have demonstrated that the local application of EGCG or its primary compound, gallocatechin gallate (GCG), effectively preserves the mitochondrial structural integrity of the back skin cells of hairless mice irradiated by UVB. This preservation increases the density of the mitochondrial ridge and matrix while reducing vacuolation, thereby inhibiting mitochondrial dysfunction in skin cells [43,86]. Furthermore, long-term treatment of H_2O_2-induced HDF cells with 12.5 μM EGCG resulted in more complete mtDNA and improved mitochondrial membrane potential [87]. However, studies on mitochondrial function and related enzyme activities in mitochondria are still lacking. Therefore, the mechanisms through which EGCG protects mitochondrial structure and function remain unclear.

EGCG can inhibit the decrease in mitochondrial membrane potential and increase in apoptosis of human lens epithelial cells stimulated by UVB. It can also reduce the expression of Bax (bcl-2 associated X protein), cytochrome c, caspase-3, and caspase-9 in these cells while increasing the expression of Bcl2 (B-cell lymphoma 2). These findings indicate that EGCG can inhibit UVB-induced apoptosis of human lens epithelial cells through a mitochondria-mediated apoptosis signaling pathway [88]. Additionally, EGCG can protect cells through mitochondria-mediated apoptosis by inhibiting the release of cytochrome c and activation of caspases [89]. These mechanisms remain to be further explored in the role of EGCG in repairing UV-induced skin aging.

3.4. EGCG Reduces Pigmentation of Photoaged Skin

Melanin is a biological pigment that determines the color of skin, hair, and eyes based on its content as well as its physical and chemical properties [90]. Human epidermal melanocytes (HEMs) produce melanin at the junction between the epidermis and dermis and play a crucial role in transporting melanin to HaCaT cells [1,90]. While melanin protects against radiation damage to the skin, abnormal local skin pigmentation can lead to uneven skin tone, and excessive accumulation of melanin greatly increases melanoma risk [91].

UV radiation serves as an external factor influencing melanin production within the skin, which promotes HEM proliferation, thereby increasing melanin content and its transport within epidermal layers. Additionally, exposure to UV light acts as a potent inducer for expressing proopiomelanocortin (POMC)-derived peptides such as adrenocorticotropic hormone (ACTH) and α-melanocyte stimulating hormone (α-MSH) [92,93]. ACTH and α-MSH bind to the melanocortin 1 receptor (MC1R), activating a signaling pathway dependent on cyclic adenosine monophosphate (cAMP) [94]. This activation stimulates pathways involved in melanin production by upregulating microphthalmia-associated transcription factor (MITF), which then enhances expression levels of three enzymes: tyrosinase (TYP) and tyrosinase-related proteins TRP1 and TRP2, all essential for melanin synthesis. This process facilitates the conversion of tyrosine into additional melanin [95].

EGCG exhibits potent anti-melanin properties and effectively inhibits the activity of TYP enzyme, owing to its strong biological activity [96,97]. It binds to the active center of TYR and interacts with copper ions and key amino acid residues, thereby reducing TYR activity [98]. EGCG can effectively reduce the melanin content and TYR activity without affecting the activity in B16F10 mouse melanoma cells, and it can also inhibit UVA-induced melanosome formation in B16F10 cells [99]. EGCG does not directly affect TYR activity or inactivate α-MSH nor inhibit α-MSH binding to MC1R for melanin synthesis inhibition. Instead, it downregulates the expression of MC1R, MITF, and TYR-related proteins to inhibit melanin formation [91].

In similar experiments, HaCaT and HEM cells were exposed to UVA and TP treatments. The results showed that TPs reduced melanin content, tyrosinase activity, and the expression of MITF, TRP1, and TRP2. Additionally, TPs directly inhibited α-MSH expression [100]. However, the composition of TPs was not described in this experiment, and the main substances in TPs that inhibit melanin synthesis are unknown. Nevertheless, this study confirmed that EGCG, as the main active component of TPs, may have the potential to inhibit melanin production by HaCaT and HEM cells after UV irradiation and may have a preventive effect on local pigmentation caused by skin photoaging.

4. Application Forms of EGCG in Repairing Photoaged Skin

Due to the instability and low bioavailability of EGCG, high-dose formulations are often used to increase its bioavailability. ADME experiments in rats and dogs have shown that high doses of EGCG (250 mg/kg body weight and 400 mg/kg body weight) can increase bioavailability from 1.6% to 13.9% compared to low doses [101]. Volunteers who took 800 mg of EGCG daily for four weeks experienced a 60% increase in free plasma EGCG [102]. However, it has been observed that high doses of GET (10–29 mg/kg/day) in mice may lead to hepatitis [103,104], indicating potential harm associated with the use of high doses. Therefore, it is essential to explore the use of EGCG with both high bioavailability and beneficial effects on the body, as well as an appropriate application form meeting these requirements, such as effective delivery systems, modified EGCG, and combination with other substances.

An effective delivery system for EGCG can inhibit its degradation and improve its bioavailability by constructing an EGCG delivery carrier, modifying EGCG itself, or combining it with other substances. Nano-drug delivery carriers have been found to prolong the intestinal retention time of EGCG, enhance its absorption efficiency by skin and intestinal mucosa, and even alter its cellular uptake mechanism. Common nano-delivery systems for EGCG mainly consist of nanoparticles supported by proteins, polysaccharides, and lipids (Table 2). Proteins can bind to EGCG and enhance the efficiency of EGCG absorption; commonly used encapsulated proteins include gelatin, lactoglobulin, casein, etc. [104]. Chitosan, a mostly used polysaccharide carrier, exhibits good biocompatibility and bioactivity encapsulated within chitosan capsules; EGCG becomes more stable with a better-sustained release effect in simulated gastroenteric fluid [105]. Encapsulation within liposomes improves the efficiency of EGCG's passage through cuticle barriers while reducing water loss from the cuticle [106], ultimately enhancing oral bioavailability [107] and antioxidant activity [108].

Modifying EGCG enhances its bioavailability and bioactivity by altering the number of phenolic hydroxyl groups and the polarity of the acyl portion of EGCG [109]. These modifications typically involve acetylation, glycosylation, and esterification. The antioxidant capacity of EGCG was partially reduced after both esterification [109] and acetylation [110] modifications, but lipid solubility and stability were improved. The antioxidant properties of glycosylated EGCG did not seem to be reduced, and it was also found that glycosylation of EGCG led to specific selection for HaCaT cells [45].

Furthermore, combining EGCG with other substances, such as hyaluronic acid, olive oil, proteins, and antioxidants, improves its bioavailability and bioactivity. Hyaluronic acid possesses potent antioxidant properties and offers effective protection against HaCaT cells. Olive oil enhances mitochondrial function and reduces skin oxidative damage [111]. Proteins serve as carriers for polyphenols in the gastrointestinal tract [112], while collagen, an essential component of skin tissue, exhibits improved stability when combined with EGCG [113]. Due to its susceptibility to oxidation, various antioxidants such as vitamin C, vitamin E, and alpha-lipoic acid are often added to cosmetic creams and used in clinical trials alongside EGCG. In aqueous solution, vitamin C and α-lipoic acid can stabilize the structure of EGCG and inhibit its photodegradation of EGCG to enhance intestinal absorption efficiency [114,115]. Additionally, it has been observed that dietary preparations inhibiting MRP activation may decrease exocytosis, leading to increased absorption of EGCG in the small intestine [116] while also inhibiting COMT structural modifications, thereby enhancing the transport of EGCG [117].

Table 2. Application forms of EGCG for repairing photoaged skin.

EGCG Forms	Other Drugs	Treatment	Results	Ref.
EGCG liposomes	Hyaluronic acid	Culturing HaCaT cells, constructing EGCG nano-transfers, applying UV irradiation and drug treatment; measuring the efficiency of EGCG liposomes through rat skin	Increasing skin penetration efficiency and EGCG deposition	[44]
	-	Culturing HDF cells, constructing EGCG microsomes by film hydration method, applying UVA irradiation and drug treatment	Extending drug release, improving skin penetration efficiency and EGCG deposition	[118]
	-	Construction of solid lipid nanoparticles (SLNs) and evaluation of permeability and antioxidant properties	Improving skin penetration efficiency and stabilizing antioxidant properties	[106]
EGCG nanoparticles	Olive oil	Cultured HaCaT cells and hairless mice, constructed EGCG polymer nanoparticles, EGCG lipid nanoparticles, and EGCG emulsion nanoparticles, and applied UVB irradiation and drug treatments	Promoting EGCG skin penetration efficiency and stability	[36]
	-	DMBA-induced DNA damage in mouse skin tissues; construction of EGCG-PLGA nanosomes for topical application to mouse interscapular skin of both shoulder blades	Improving bioavailability and drug delivery efficiency	[119]
EGCG nanoethosomes	-	Sucrose ester-stabilized EGCG nano-ethanol bodies stored for 6 months and then applied to mouse skin; UVB irradiation of mouse skin induces damage	Enhancing skin penetration efficiency and stabilizing EGCG nanosomes	[120]
EGCG	Pentagalloyl glucose	Culturing HDF cells, applying UVA irradiation, and combining two drugs	Increasing extracellular matrix deposition of elastin and collagen	[34]
	Collagen	EGCG-modified collagen-forming couplers, cultured hairless mice, imposed UVB irradiation and drug treatment	Improving antioxidant properties, more comprehensive effect of combined use	[41,113]
	Soybean protein	Polymerization of EGCG with amino acid residues of soybean 7S globulin and verification of anti-UVB properties by mouse experiments	Inhibiting of UVB-induced apoptosis, inflammatory factors, and MAPK signaling pathways	[114]

Table 2. *Cont.*

EGCG Forms	Other Drugs	Treatment	Results	Ref.
TP emulsion	-	Hairless mice were cultured, carboxymethyl cellulose emulsified TPs to make an emulsion, and UV irradiation and drug treatments were applied separately	Stabilizing TPs in an aqueous solution and increasing deposition; inhibiting inflammatory cell infiltration skin thickness; increasing oxidative stress levels	[42]
Catechin	Vitamin C	Prepared as capsules for combined use, volunteers take orally	Stabilizing catechins in the gut and improving antioxidant properties	[77,80]
Glc-EGCG	-	Culture of HaCaT cells, HSF cells, melanocytes, endothelial cells with Glc-EGCG; UVA, UVB irradiation damage cells	Improving EGCG solubility, stability, and antioxidant properties	[45]
ZNp-EG	-	EGCG-loaded zein nanoparticles, measurement of relevant properties	Increasing antioxidant and anti-tyrosinase activity and stability	[121]
GTE (42.4% EGCG)	Quercetin and multi-nutrients	Oral administration to rats and determination of pharmacokinetic data	56% increase in bioavailability and enhanced intestinal absorption efficiency	[116]

Abbreviations: Glc-EGCG, glucosyl form of EGCG; ZNp-EG, zein nanoparticles containing EGCG; GTE, green tea extract; TPs, tea polyphenols.

5. Discussion and Perspectives

The skin functions as a protective barrier, maintaining homeostasis for itself and other organs through the cutaneous neuro-immuno-endocrine system [3,92]. Upon exposure to UV radiation, the skin synthesizes various small molecules, including vitamin D_3, POMC, and nitric oxide, which are subsequently released into circulation to exert beneficial effects on the human body [3]. Vitamin D and lumisterol hydroxyderivatives play a crucial role in protecting the body from UV-induced damage by enhancing cellular viability, promoting antioxidant activity, facilitating DNA repair processes, and activating anti-inflammatory mechanisms [122]. Hydroxylumisterols, a photoproduct of pre-vitamin D_3, mitigate photodamage induced by UVB radiation in HaCaT cells [123]. Phototherapy is employed in the treatment of various skin disorders and utilizes a range of light sources, including UV, that have demonstrated therapeutic benefits [93]. However, it is essential not to overlook the detrimental impacts of UV exposure on the skin, such as an increased risk of cancer, autoimmune disorders, and skin photoaging. A rational approach to defining the effects of UV radiation on human health while exploring its beneficial aspects seems imperative for future research aimed at preventing skin photoaging.

Skin photoaging is the consequence of premature aging of skin cells following exposure to UV radiation. Among these, HDF cells appear to be the most dominant cell involved in this process [124,125], followed by HEMs, the primary type of epidermal senescent cells. Fibroblasts proliferate slowly in vivo and are prone to the SASP phenotype, leading to inflammatory responses and ECM degradation that destroys the surrounding microenvironment [125]. The decline in HDF function in photoaged skin is mainly attributed to the breakdown of the dermal collagen network [126,127]. The degradation of collagen in photoaged skin results in the destruction of the collagen network, weakening its mechanical strength and a failure to provide physical support for HDFs, and the reduction in the quantity of HDFs ultimately leads to a decline in its function [128]. By inhibiting the expression of MMPs and promoting the expression of TIMPs [33], EGCG delays the degradation of collagen and elastin in photoaged skin, thereby maintaining the collagen network layer structure, which provides a conducive environment for HDF proliferation and function maintenance. HEMs produce melanin to protect adjacent human epidermal keratinocytes (HEKs). Aging HEMs induce skin cell aging by inhibiting the proliferation of surrounding cells and destroying the proliferative role of basal keratinocytes [124]. As senescent skin cells proliferate and accumulate, they release more cytokines, matrix-modifying enzymes, and other molecules that disrupt skin structure and function [124].

ROS play a dual role in skin photoaging due to UV radiation and as an activator of various "harmful" mechanisms within cells [125]. ROS accumulation threatens many biomolecules and organelles within cells, triggering inflammatory responses and DNA damage. Lipids are crucial in maintaining the body's barrier function and are essential for preserving skin moisture [128]. UV radiation exacerbates lipid oxidation in the skin, disrupting lipid metabolism and promoting skin aging. MDA serves as a marker for assessing the degree of lipid oxidation. EGCG has been shown to reduce MDA levels and maintain lipid homeostasis in the skin by enhancing the activity of antioxidant enzymes within cells [129].

EGCG has been extensively researched for its potent antioxidant properties, although reports of its pro-oxidative effects have also been reported. When EGCG is introduced into cell cultures, it triggers the production of hydrogen peroxide and low levels of ROS [130]. ROS play a dual effect in living organisms, acting as both signaling molecules that influence cell signaling and as potentially "toxic" molecules. It is suggested that the pro-oxidation effect of EGCG may be beneficial, serving as a second messenger within the cell and thereby helping to mitigate damage caused by UV radiation. However, further comprehensive research is needed to substantiate this claim. In addition, EGCG elicits a wide range of metabolic responses within the body, with these metabolites appearing to play a role in realizing EGCG's biological function [130]. Different experimental models, with topical or oral administration of GTP and EGCG, have shown the prevention of UV-induced inflammatory responses, collagen network layer breakdown, and oxidative stress [131]. Continued investigation into the biological activity of EGCG and its metabolites in vivo may help elucidate the molecular mechanisms through which EGCG repairs photoaged skin.

The bioavailability of EGCG is influenced by various factors, especially intestinal absorption. The gut plays a crucial role in the absorption, metabolism, and transformation of EGCG, with intestinal microorganisms playing a key role [132]. Short-term GTE supplementation enhances mice's gut microbiota and cecum/skin metabolome, effectively mitigating their stress response to UV radiation [133]. At the same time, the chemical stability and membrane permeability of EGCG were less than optimal [134]. Furthermore, the skin's cuticle barrier effect and lipophilicity pose challenges for applying EGCG topical preparations [135].

Drug delivery systems, structural modifications, and combinations with other substances have been employed to enhance the bioavailability of EGCG to varying degrees. Notably, drug delivery systems encapsulate EGCG to bypass the effects of early digestive processes on its stability while improving skin permeability. This explains why there is interest in EGCG nano-drug delivery systems. Nevertheless, there are still unresolved issues regarding the application of EGCG in treating photoaged skin that require further research. These include understanding how oral EGCG affects intestinal absorption and the roles played by MRP and COMT; investigating how gut microbes influence absorption, metabolism, and transformation of EGCG; and finding ways to improve or upgrade EGCG nano-drug delivery system for enhanced stability of the nanosystem to reduce the sudden release of EGCG [23].

While EGCG holds great promise for preventing and treating photoaged skin, its poor stability and low bioavailability limit its clinical application. Studies have focused on enhancing the stability and bioavailability of EGCG in recent years. For example, studies have been conducted to develop medications combining EGCG with common UV absorbers to protect human skin from UV radiation [136]. In addition, rheological testing may assist in developing future drug delivery vehicles [137]. Nevertheless, thorough research on how combinations involving related substances affect the stability and bioavailability of EGCG has yet to be conducted. Furthermore, the mechanism underlying interactions between drugs and EGCG in preventing and treating photoaged skin remains unclear; likewise, gender differences and individual variations among subjects undergoing treatment with EGCG in clinical studies warrant ongoing exploration [138].

6. Conclusions

EGCG is an effective candidate for repairing photoaged skin. It effectively addresses various types of skin damage induced by ultraviolet radiation while preserving the normal structure and function of the skin. EGCG can inhibit the expression of MMPs and enhance the expression of TIMPs, thereby delaying the degradation of collagen and elastin in the skin. EGCG can enhance skin moisturization by increasing the expression of NMF-associated genes and inhibiting the increase in TEWL. EGCG also improves antioxidant enzyme activity, inhibits skin inflammation and DNA damage, protects the structural integrity of skin cell mitochondria, and promotes normal mitochondrial function. Furthermore, EGCG also inhibits tyrosinase activity by downregulating the expression of MC1R, MITF, and TYR, thereby inhibiting melanin formation and ultimately alleviating local skin pigmentation. EGCG's stability and bioavailability can be enhanced through EGCG delivery systems, EGCG modification, and combination with other substances. Among these methods, nanocarriers of delivery systems show particular promise. Further exploration of the mechanism of EGCG repair of photoaged skin and improvement of its bioavailability by constructing nanocarriers could provide safe and efficient skin repair drugs.

Author Contributions: Conceptualization, L.H., J.F. and H.J.; writing—original draft preparation, J.S., L.H., J.F., H.J., J.S. and Y.J.; revision and editing, J.S., L.H., C.Y., H.J., H.Y., X.G. and Y.J.; supervision, H.J., H.L. and L.H. All authors have read and agreed to the published version of the manuscript.

Funding: This research was funded by Shaanxi University of Science and Technology Specialized National Key Laboratory of Qinba Biological Resources and Ecological Environment Jointly Constructed by the Ministry of Qinba Province (Grant No. SXZC-2102), National Natural Science Foundation of China (Grant No. 32372766), Key Research and Development Program of Zhejiang Province (Grant No. 2023C02040 and No. 2022C02033-3).

Institutional Review Board Statement: Not applicable.

Informed Consent Statement: Not applicable.

Data Availability Statement: Not applicable.

Acknowledgments: The authors wish to express gratitude to Ching Yuan Hu and Anas Yusuf for their invaluable guidance, which significantly enhanced both the content and linguistic quality of this article.

Conflicts of Interest: The authors declare no conflicts of interest.

References

1. Huang, A.H.; Chien, A.L. Photoaging: A Review of Current Literature. *Curr. Dermatol. Rep.* **2020**, *9*, 22–29. [CrossRef]
2. Furukawa, J.Y.; Martinez, R.M.; Morocho-Jácome, A.L.; Castillo-Gómez, T.S.; Pereda-Contreras, V.J.; Rosado, C.; Velasco, M.V.R.; Baby, A.R. Skin Impacts from Exposure to Ultraviolet, Visible, Infrared, and Artificial Lights—A Review. *J. Cosmet. Laser Ther.* **2021**, *23*, 1–7. [CrossRef]
3. Slominski, R.M.; Chen, J.Y.; Raman, C.; Slominski, A.T. Photo-Neuro-Immuno-Endocrinology: How the ultraviolet radiation regulates the body, brain, and immune system. *Proc. Natl. Acad. Sci. USA* **2024**, *121*, e2308374121. [CrossRef] [PubMed]
4. Tanveer, M.A.; Rashid, H.; Tasduq, S.A. Molecular Basis of Skin Photoaging and Therapeutic Interventions by Plant-Derived Natural Product Ingredients: A Comprehensive Review. *Heliyon* **2023**, *9*, e13580. [CrossRef]
5. D'Orazio, J.; Jarrett, S.; Amaro-Ortiz, A.; Scott, T. UV Radiation and the Skin. *Int. J. Mol. Sci.* **2013**, *14*, 12222–12248. [CrossRef]
6. Gromkowska-Kępka, K.J.; Puścion-Jakubik, A.; Markiewicz-Żukowska, R.; Socha, K. The Impact of Ultraviolet Radiation on Skin Photoaging—Review of in Vitro Studies. *J. Cosmet. Dermatol.* **2021**, *20*, 3427–3431. [CrossRef]
7. Poon, F.; Kang, S.; Chien, A.L. Mechanisms and Treatments of Photoaging. *Photodermatol. Photoimmunol. Photomed.* **2015**, *31*, 65–74. [CrossRef]

8. Jr, J.E.F.; Porumb, S. Chemical peels: Their Place Within the Range of Resurfacing Techniques. *Am. J. Clin. Dermatol.* **2004**, *5*, 179–187. [CrossRef]
9. Helander, S.D. Treatment of Photoaged Skin: Efficacy, Tolerability and Costs of Available Agents. *Drugs Aging* **1996**, *8*, 12–16. [CrossRef]
10. Roh, E.; Kim, J.-E.; Kwon, J.Y.; Park, J.S.; Bode, A.M.; Dong, Z.; Lee, K.W. Molecular Mechanisms of Green Tea Polyphenols with Protective Effects against Skin Photoaging. *Crit. Rev. Food Sci. Nutr.* **2015**, *57*, 1631–1637. [CrossRef]
11. Cho, W.; Park, J.; Lee, M.; Park, S.-H.; Jung, J.; Kim, J.; Eun, S.; Kim, J. Gly-Pro-Val-Gly-Pro-Ser Peptide Fish Collagen Improves Skin Moisture and Wrinkles with Ameliorated the Oxidative Stress and Pro-Inflammatory Factors in Skin Photoaging Mimic Models. *Prev. Nutr. Food Sci.* **2023**, *28*, 50–60. [CrossRef] [PubMed]
12. Liang, J.; Yu, Y.; Wang, B.; Lu, B.; Zhang, J.; Zhang, H.; Ge, P. Ginsenoside Rb1 Attenuates Oxygen-Glucose Depriva-tion-Induced Apoptosis in SH-SY5Y Cells via Protection of Mitochondria and Inhibition of AIF and Cytochrome c Release. *Molecules* **2013**, *18*, 12777–12792. [CrossRef] [PubMed]
13. Bae, J.; Lim, S.S.; Kim, S.J.; Choi, J.; Park, J.; Ju, S.M.; Han, S.J.; Kang, I.; Kang, Y. Bog Blueberry Anthocyanins Alleviate Photoaging in ultraviolet-B Irradiation-induced Human Dermal Fibroblasts. *Mol. Nutr. Food Res.* **2009**, *53*, 726–738. [CrossRef]
14. Prasanth, M.; Sivamaruthi, B.; Chaiyasut, C.; Tencomnao, T. A Review of the Role of Green Tea (*Camellia sinensis*) in Antiphotoaging, Stress Resistance, Neuroprotection, and Autophagy. *Nutrients* **2019**, *11*, 474. [CrossRef]
15. Frasheri, L.; Schielein, M.C.; Tizek, L.; Mikschl, P.; Biedermann, T.; Zink, A. Great Green Tea Ingredient? A Narrative Literature Review on Epigallocatechin Gallate and Its Biophysical Properties for Topical Use in Dermatology. *Phytother. Res.* **2020**, *34*, 2170–2179. [CrossRef]
16. Zhang, S.; Mao, B.; Cui, S.; Zhang, Q.; Zhao, J.; Tang, X.; Chen, W. Absorption, Metabolism, Bioactivity, and Biotransformation of Epigallocatechin Gallate. *Crit. Rev. Food Sci. Nutr.* **2024**, *64*, 6546–6566. [CrossRef]
17. Furniturewalla, A.; Barve, K. Approaches to Overcome Bioavailability Inconsistencies of Epigallocatechin Gallate, a Powerful Anti-Oxidant in Green Tea. *Food Chem. Adv.* **2022**, *1*, 100037. [CrossRef]
18. Fernández, V.A.; Toledano, L.A.; Lozano, N.P.; Tapia, E.N.; Roig, M.D.G.; Fornell, R.D.l.T.; Algar, G. Bioavailability of Epigallocatechin Gallate Administered with Different Nutritional Strategies in Healthy Volunteers. *Antioxidants* **2020**, *9*, 440. [CrossRef] [PubMed]
19. Yang, C.S.; Chen, L.; Lee, M.-J.; Balentine, D.; Kuo, M.C.; Schantz, S.P. Blood and urine levels of tea catechins after ingestion of different amounts of green tea by human volunteers. *Cancer Epidemiol. Biomark. Prev.* **1998**, *7*, 351–354.
20. Miyazawa, T. Absorption, Metabolism and Antioxidative Effects of Tea Catechin in Humans. *BioFactors* **2000**, *13*, 55–59. [CrossRef]
21. Dube, A.; Nicolazzo, J.A.; Larson, I. Chitosan Nanoparticles Enhance the Plasma Exposure of (−)-Epigallocatechin Gallate in Mice through an Enhancement in Intestinal Stability. *Eur. J. Pharm. Sci.* **2011**, *44*, 422–426. [CrossRef] [PubMed]
22. Dai, W.; Ruan, C.; Zhang, Y.; Wang, J.; Han, J.; Shao, Z.; Sun, Y.; Liang, J. Bioavailability enhancement of EGCG by structural modification and nano-delivery: A review. *J. Funct. Foods* **2019**, *65*, 103732. [CrossRef]
23. Matić, S.; Takač, M.J.-M.; Barbarić, M.; Lučić, B.; Gall Trošelj, K.; Stepanić, V. The Influence of In Vivo Metabolic Modifications on ADMET Properties of Green Tea Catechins–In Silico Analysis. *J. Pharm. Sci.* **2018**, *107*, 2957–2964. [CrossRef] [PubMed]
24. Slot, G.V.; Humpf, H.-U. Degradation and Metabolism of Catechin, Epigallocatechin-3-Gallate (EGCG), and Related Compounds by the Intestinal Microbiota in the Pig Cecum Model. *J. Agric. Food Chem.* **2009**, *57*, 8041–8048. [CrossRef]
25. Warden, B.A.; Smith, L.S.; Beecher, G.R.; Balentine, D.A.; Clevidence, B.A. Catechins Are Bioavailable in Men and Women Drinking Black Tea throughout the Day. *J. Nutr.* **2001**, *131*, 1731–1737. [CrossRef]
26. Leslie, E.M.; Deeley, R.G.; Cole, S.P. Toxicological Relevance of the Multidrug Resistance Protein 1, MRP1 (ABCC1) and Related Transporters. *Toxicology* **2001**, *167*, 3–23. [CrossRef]
27. Lambert, J.D.; Kim, D.H.; Zheng, R.; Yang, C.S. Transdermal Delivery of (-)-Epigallocatechin-3-Gallate, a Green Tea Polyphenol, in Mice. *J. Pharm. Pharmacol.* **2006**, *58*, 599–604. [CrossRef]
28. Krupkova, O.; Ferguson, S.J.; Wuertz-Kozak, K. Stability of (−)-Epigallocatechin Gallate and Its Activity in Liquid Formulations and Delivery Systems. *J. Nutr. Biochem.* **2016**, *37*, 1–12. [CrossRef]
29. Scalia, S.; Trotta, V.; Bianchi, A. In Vivo Human Skin Penetration of (–)-Epigallocatechin-3-Gallate from Topical Formulations. *Acta Pharm.* **2014**, *64*, 257–265. [CrossRef]
30. Bianchi, A.; Marchetti, N.; Scalia, S. Photodegradation of (−)-Epigallocatechin-3-Gallate in Topical Cream Formulations and Its Photostabilization. *J. Pharm. Biomed. Anal.* **2011**, *56*, 692–697. [CrossRef]
31. Won, H.-R.; Lee, P.; Oh, S.; Kim, Y.-M. Epigallocatechin-3-Gallate Suppresses the Expression of TNF-α-Induced MMP-1 via MAPK/ERK Signaling Pathways in Human Dermal Fibroblasts. *Biol. Pharm. Bull.* **2021**, *44*, 18–24. [CrossRef] [PubMed]
32. Jia, Y.; Mao, Q.; Yang, J.; Du, N.; Zhu, Y.; Min, W. (–)-Epigallocatechin-3-Gallate Protects Human Skin Fibroblasts from Ul-traviolet a Induced Photoaging. *Clin. Cosmet. Investig. Dermatol.* **2023**, *16*, 149–159. [CrossRef] [PubMed]
33. Abotorabi, Z.; Khorashadizadeh, M.; Arab, M.; Fard, M.H.; Zarban, A. Jujube and Green Tea Extracts Protect Human Fibro-blast Cells against UVB- Mediated Photo Damage and MMP-2 and MMP-9 Production. *Avicenna J. Phytomed.* **2020**, *10*, 287–296.
34. Chowdhury, A.; Nosoudi, N.; Karamched, S.; Parasaram, V.; Vyavahare, N. Polyphenol Treatments Increase Elastin and Collagen Deposition by Human Dermal Fibroblasts; Implications to Improve Skin Health. *J. Dermatol. Sci.* **2021**, *102*, 94–100. [CrossRef] [PubMed]

35. Lee, J.H.; Chung, J.H.; Cho, K.H. The Effects of Epigallocatechin-3-Gallate on Extracellular Matrix Metabolism. *J. Dermatol. Sci.* **2005**, *40*, 195–204. [CrossRef]
36. Salaheldin, T.A.; Adhami, V.M.; Fujioka, K.; Mukhtar, H.; Mousa, S.A. Photochemoprevention of Ultraviolet Beam Radiation-Induced DNA Damage in Keratinocytes by Topical Delivery of Nanoformulated Epigallocatechin-3-Gallate. *Nanomed. Nanotechnol. Biol. Med.* **2022**, *44*, 102580. [CrossRef]
37. Kim, E.; Hwang, K.; Lee, J.; Han, S.; Kim, E.-M.; Park, J.; Cho, J. Skin Protective Effect of Epigallocatechin Gallate. *Int. J. Mol. Sci.* **2018**, *19*, 173. [CrossRef] [PubMed]
38. Luo, D.; Min, W.; Lin, X.-F.; Wu, D.; Xu, Y.; Miao, X. Effect of Epigallocatechingallate on Ultraviolet B-Induced Photo-Damage in Keratinocyte Cell Line. *Am. J. Chin. Med.* **2006**, *34*, 911–922. [CrossRef] [PubMed]
39. Zhang, J.; Xu, Y.; Ruan, X.; Zhang, T.; Zi, M.; Zhang, Q. Photoprotective Effects of Epigallocatechin Gallate on Ultravio-let-Induced Zebrafish and Human Skin Fibroblasts Cells. *Mediat. Inflamm.* **2024**, *2024*, 7887678. [CrossRef]
40. Wang, W.; Di, T.; Wang, W.; Jiang, H. EGCG, GCG, TFDG, or TSA Inhibiting Melanin Synthesis by Downregulating MC1R Expression. *Int. J. Mol. Sci.* **2023**, *24*, 11017. [CrossRef]
41. Fan, Z.; Zhou, Y.; Gan, B.; Li, Y.; Chen, H.; Peng, X.; Zhou, Y. Collagen-EGCG Combination Synergistically Prevents UVB-Induced Skin Photoaging in Nude Mice. *Macromol. Biosci.* **2023**, *23*, 2300251. [CrossRef] [PubMed]
42. Li, H.; Jiang, N.; Liu, Q.; Gao, A.; Zhou, X.; Liang, B.; Li, R.; Li, Z.; Zhu, H. Topical Treatment of Green Tea Polyphenols Emulsified in Carboxymethyl Cellulose Protects against Acute Ultraviolet Light B-Induced Photodamage in Hairless Mice. *Photochem. Photobiol. Sci.* **2016**, *15*, 1264–1271. [CrossRef] [PubMed]
43. Sheng, Y.-Y.; Xiang, J.; Lu, J.-L.; Ye, J.-H.; Chen, Z.-J.; Zhao, J.-W.; Liang, Y.-R.; Zheng, X.-Q. Protective Effects of Gallocatechin Gallate against Ultraviolet B Induced Skin Damages in Hairless Mice. *Sci. Rep.* **2022**, *12*, 1310. [CrossRef] [PubMed]
44. Avadhani, K.S.; Manikkath, J.; Tiwari, M.; Chandrasekhar, M.; Godavarthi, A.; Vidya, S.M.; Hariharapura, R.C.; Kalthur, G.; Udupa, N.; Mutalik, S. Skin Delivery of Epigallocatechin-3-Gallate (EGCG) and Hyaluronic Acid Loaded Nano-Transfersomes for Antioxidant and Anti-Aging Effects in UV Radiation Induced Skin Damage. *Drug Deliv.* **2017**, *24*, 61–74. [CrossRef]
45. Nadim, M.; Auriol, D.; Lamerant-FayeL, N.; Lefèvre, F.; Dubanet, L.; Redziniak, G.; Kieda, C.; Grillon, C. Improvement of Polyphenol Properties upon Glucosylation in a UV-Induced Skin Cell Ageing Model. *Int. J. Cosmet. Sci.* **2014**, *36*, 579–587. [CrossRef]
46. Kanlayavattanakul, M.; Khongkow, M.; Klinngam, W.; Chaikul, P.; Lourith, N.; Chueamchaitrakun, P. Recent Insights into Catechins-Rich Assam Tea Extract for Photoaging and Senescent Ageing. *Sci. Rep.* **2024**, *14*, 2253. [CrossRef]
47. Charoenchon, N.; Rhodes, L.E.; Nicolaou, A.; Williamson, G.; Watson, R.E.B.; Farrar, M.D. Ultraviolet Radiation-Induced Degradation of Dermal Extracellular Matrix and Protection by Green Tea Catechins: A Randomized Controlled Trial. *Clin. Exp. Dermatol.* **2022**, *47*, 1314–1323. [CrossRef]
48. Heinrich, U.; Moore, C.E.; De Spirt, S.; Tronnier, H.; Stahl, W. Green Tea Polyphenols Provide Photoprotection, Increase Microcirculation, and Modulate Skin Properties of Women. *J. Nutr.* **2011**, *141*, 1202–1208. [CrossRef]
49. McCarty, M.F.; Benzvi, C.; Vojdani, A.; Lerner, A. Nutraceutical Strategies for Alleviation of UVB Phototoxicity. *Exp. Dermatol.* **2023**, *32*, 722–730. [CrossRef]
50. Chen, X.; Yang, C.; Jiang, G. Research Progress on Skin Photoaging and Oxidative Stress. *Adv. Dermatol. Allergol.* **2021**, *38*, 931–936. [CrossRef]
51. Freitas-Rodríguez, S.; Folgueras, A.R.; López-Otín, C. The Role of Matrix Metalloproteinases in Aging: Tissue Remodeling and Beyond. *Biochim. Biophys. Acta (BBA) Bioenerg.* **2017**, *1864*, 2015–2025. [CrossRef] [PubMed]
52. Quan, T.; Little, E.; Quan, H.; Qin, Z.; Voorhees, J.J.; Fisher, G.J. Elevated Matrix Metalloproteinases and Collagen Frag-mentation in Photodamaged Human Skin: Impact of Altered Extracellular Matrix Microenvironment on Dermal Fibroblast Function. *J. Investig. Dermatol.* **2013**, *133*, 1362–1366. [CrossRef] [PubMed]
53. Bashir, M.M.; Sharma, M.R.; Werth, V.P. TNF-α Production in the Skin. *Arch. Dermatol. Res.* **2009**, *301*, 87–91. [CrossRef]
54. Ågren, M.S.; Schnabel, R.; Christensen, L.H.; Mirastschijski, U. Tumor Necrosis Factor-α-Accelerated Degradation of Type I Collagen in Human Skin Is Associated with Elevated Matrix Metalloproteinase (MMP)-1 and MMP-3 Ex Vivo. *Eur. J. Cell Biol.* **2014**, *94*, 12–21. [CrossRef]
55. Quan, T.; He, T.; Kang, S.; Voorhees, J.J.; Fisher, G.J. Solar Ultraviolet Irradiation Reduces Collagen in Photoaged Human Skin by Blocking Transforming Growth Factor-β Type II Receptor/Smad Signaling. *Am. J. Pathol.* **2004**, *165*, 741–751. [CrossRef]
56. Chen, B.; Li, R.; Yan, N.; Chen, G.; Qian, W.; Jiang, H.-L.; Ji, C.; Bi, Z.-G. Astragaloside IV Controls Collagen Reduction in Photoaging Skin by Improving Transforming Growth Factor-β/Smad Signaling Suppression and Inhibiting Matrix Metalloproteinase-1. *Mol. Med. Rep.* **2015**, *11*, 3344–3348. [CrossRef] [PubMed]
57. Verrecchia, F.; Pessah, M.; Atfi, A.; Mauviel, A. Tumor Necrosis Factor-α Inhibits Transforming Growth Factor-β /Smad Sig-naling in Human Dermal Fibroblasts via AP-1 Activation. *J. Biol. Chem.* **2000**, *275*, 30226–30231. [CrossRef]
58. Singh, B.N.; Shankar, S.; Srivastava, R.K. Green Tea Catechin, Epigallocatechin-3-Gallate (EGCG): Mechanisms, Perspectives and Clinical Applications. *Biochem. Pharmacol.* **2011**, *82*, 1807–1821. [CrossRef]
59. Šínová, R.; Pavlík, V.; Ondrej, M.; Velebný, V.; Nešporová, K. Hyaluronan: A key player or just a bystander in skin photoaging? *Exp. Dermatol.* **2021**, *31*, 442–458. [CrossRef]
60. Jeon, H.Y.; Kim, J.K.; Seo, D.B.; Cho, S.Y.; Lee, S.J. Beneficial Effect of Dietary Epigallocatechin-3-Gallate on Skin via En-hancement of Antioxidant Capacity in Both Blood and Skin. *Ski. Pharmacol. Physiol.* **2010**, *23*, 283–289. [CrossRef]

61. Zhong, J.L.; Edwards, G.P.; Raval, C.; Li, H.; Tyrrell, R.M. The Role of Nrf2 in Ultraviolet A Mediated Heme Oxygenase 1 Induction in Human Skin Fibroblasts. *Photochem. Photobiol. Sci.* **2009**, *9*, 18–24. [CrossRef] [PubMed]
62. Ikehata, H.; Yamamoto, M. Roles of the KEAP1-NRF2 System in Mammalian Skin Exposed to UV Radiation. *Toxicol. Appl. Pharmacol.* **2018**, *360*, 69–77. [CrossRef] [PubMed]
63. Lee, J.H.; Park, J.; Shin, D.W. The Molecular Mechanism of Polyphenols with Anti-Aging Activity in Aged Human Dermal Fibroblasts. *Molecules* **2022**, *27*, 4351. [CrossRef] [PubMed]
64. Mafra, D.; Alvarenga, L.; Cardozo, L.F.; Stockler-Pinto, M.B.; Nakao, L.S.; Stenvinkel, P.; Shiels, P.G. Inhibiting BTB Domain and CNC Homolog 1 (Bach1) as an Alternative to Increase Nrf2 Activation in Chronic Diseases. *Biochim. Biophys. Acta (BBA)-Gen. Subj.* **2022**, *1866*, 130129. [CrossRef] [PubMed]
65. Mullenders, L.H.F. Solar UV Damage to Cellular DNA: From Mechanisms to Biological Effects. *Photochem. Photobiol. Sci.* **2018**, *17*, 1842–1852. [CrossRef]
66. Son, N.; Cui, Y.; Xi, W. Association Between Telomere Length and Skin Cancer and Aging: A Mendelian Randomization Analysis. *Front. Genet.* **2022**, *13*, 931785. [CrossRef]
67. Brem, R.; Guven, M.; Karran, P. Oxidatively-Generated Damage to DNA and Proteins Mediated by Photosensitized UVA. *Free. Radic. Biol. Med.* **2017**, *107*, 101–109. [CrossRef]
68. Ansary, T.M.; Hossain, M.R.; Kamiya, K.; Komine, M.; Ohtsuki, M. Inflammatory Molecules Associated with Ultraviolet Radiation-Mediated Skin Aging. *Int. J. Mol. Sci.* **2021**, *22*, 3974. [CrossRef]
69. He, X.; Gao, X.; Xie, W. Research Progress in Skin Aging and Immunity. *Int. J. Mol. Sci.* **2024**, *25*, 4101. [CrossRef]
70. Salminen, A.; Kaarniranta, K.; Kauppinen, A. Photoaging: UV Radiation-Induced Inflammation and Immunosuppression Accelerate the Aging Process in the Skin. *Inflamm. Res.* **2022**, *71*, 817–831. [CrossRef]
71. Shim, J.H. Prostaglandin E2 Induces Skin Aging via E-Prostanoid 1 in Normal Human Dermal Fibroblasts. *Int. J. Mol. Sci.* **2019**, *20*, 5555. [CrossRef] [PubMed]
72. Fernandes, A.; Rodrigues, P.M.; Pintado, M.; Tavaria, F.K. A Systematic Review of Natural Products for Skin Applications: Targeting Inflammation, Wound Healing, and Photo-Aging. *Phytomedicine* **2023**, *115*, 154824. [CrossRef]
73. Xu, Y.; Zhu, J.; Zhou, B.; Luo, D. Epigallocatechin-3-Gallate Decreases UVA-Induced HPRT Mutations in Human Skin Fibro-blasts Accompanied by Increased Rates of Senescence and Apoptosis. *Exp. Ther. Med.* **2012**, *3*, 625–630. [CrossRef] [PubMed]
74. Morley, N.; Clifford, T.; Salter, L.; Campbell, S.; Gould, D.; Curnow, A. The Green Tea Polyphenol (−)-Epigallocatechin Gallate and Green Tea Can Protect Human Cellular DNA from Ultraviolet and Visible Radiation-induced Damage. *Photodermatol. Photoimmunol. Photomed.* **2004**, *21*, 15–22. [CrossRef]
75. Jeon, H.Y.; Kim, J.K.; Kim, W.G.; Lee, S.J. Effects of Oral Epigallocatechin Gallate Supplementation on the Minimal Erythema Dose and UV-Induced Skin Damage. *Ski. Pharmacol. Physiol.* **2009**, *22*, 137–141. [CrossRef]
76. Kapoor, M.P.; Sugita, M.; Fukuzawa, Y.; Timm, D.; Ozeki, M.; Okubo, T. Green Tea Catechin Association with Ultraviolet Radiation-Induced Erythema: A Systematic Review and Meta-Analysis. *Molecules* **2021**, *26*, 3702. [CrossRef]
77. Rhodes, L.E.; Darby, G.; Massey, K.A.; Clarke, K.A.; Dew, T.P.; Farrar, M.D.; Bennett, S.; Watson, R.E.B.; Williamson, G.; Nicolaou, A. Oral Green Tea Catechin Metabolites Are Incorporated into Human Skin and Protect against UV Radia-tion-Induced Cutaneous Inflammation in Association with Reduced Production of pro-Inflammatory Eicosanoid 12-Hydroxyeicosatetraenoic Acid. *Br. J. Nutr.* **2013**, *110*, 891–900. [CrossRef] [PubMed]
78. Sevın, A.; Öztaş, P.; Senen, D.; Han, Ü.; Karaman, Ç.; Tarimci, N.; Kartal, M.; Erdoğan, B. Effects of Polyphenols on Skin Damage Due to Ultraviolet A Rays: An Experimental Study on Rats. *J. Eur. Acad. Dermatol. Venereol.* **2007**, *21*, 650–656. [CrossRef]
79. Katiyar, S.K.; Mukhtar, H. Green Tea Polyphenol (−)-Epigallocatechin-3-Gallate Treatment to Mouse Skin Prevents UVB-Induced Infiltration of Leukocytes, Depletion of Antigen-Presenting Cells, and Oxidative Stress. *J. Leukoc. Biol.* **2001**, *69*, 719–726. [CrossRef]
80. Farrar, M.D.; Nicolaou, A.; Clarke, K.A.; Mason, S.; Massey, K.A.; Dew, T.P.; Watson, R.E.; Williamson, G.; Rhodes, L.E. A Randomized Controlled Trial of Green Tea Catechins in Protection against Ultraviolet Radiation–Induced Cutaneous Inflammation. *Am. J. Clin. Nutr.* **2015**, *102*, 608–615. [CrossRef]
81. Martic, I.; Papaccio, F.; Bellei, B.; Cavinato, M. Mitochondrial Dynamics and Metabolism Across Skin Cells: Implications for Skin Homeostasis and Aging. *Front. Physiol.* **2023**, *14*, 1284410. [CrossRef] [PubMed]
82. He, H.; Xiong, L.; Jian, L.; Li, L.; Wu, Y.; Qiao, S. Role of Mitochondria on UV-Induced Skin Damage and Molecular Mechanisms of Active Chemical Compounds Targeting Mitochondria. *J. Photochem. Photobiol. B Biol.* **2022**, *232*, 112464. [CrossRef] [PubMed]
83. Song, M.J.; Park, C.; Kim, H.; Han, S.; Lee, S.H.; Lee, D.H.; Chung, J.H. Carnitine Acetyltransferase Deficiency Mediates Mitochondrial Dysfunction-Induced Cellular Senescence in Dermal Fibroblasts. *Aging Cell* **2023**, *22*, e14000. [CrossRef] [PubMed]
84. Zhang, C.; Gao, X.; Li, M.; Yu, X.; Huang, F.; Wang, Y.; Yan, Y.; Zhang, H.; Shi, Y.; He, X. The Role of Mitochondrial Quality Surveillance in Skin Aging: Focus on Mitochondrial Dynamics, Biogenesis and Mitophagy. *Ageing Res. Rev.* **2023**, *87*, 101917. [CrossRef]
85. Chen, B.; Zhang, W.; Lin, C.; Zhang, L. A Comprehensive Review on Beneficial Effects of Catechins on Secondary Mito-chondrial Diseases. *Int. J. Mol. Sci.* **2022**, *23*, 11569. [CrossRef] [PubMed]
86. Liu, C.; Zheng, X.-Q.; Xiang, L.-P.; Lu, J.-L.; Polito, C.A.; Liang, Y.-R. Protective Effect of (-)-Epigallocatechin Gallate on Ul-traviolet b-Induced Skin Damage in Hairless Mice. *Trop. J. Pharm. Res.* **2016**, *15*, 1183. [CrossRef]

87. Meng, Q.; Velalar, C.N.; Ruan, R. Effects of Epigallocatechin-3-Gallate on Mitochondrial Integrity and Antioxidative Enzyme Activity in the Aging Process of Human Fibroblast. *Free Radic. Biol. Med.* **2008**, *44*, 1032–1041. [CrossRef]
88. Wu, Q.; Song, J.; Gao, Y.; Zou, Y.; Guo, J.; Zhang, X.; Liu, D.; Guo, D.; Bi, H. Epigallocatechin Gallate Enhances Human Lens Epithelial Cell Survival after UVB Irradiation via the Mitochondrial Signaling Pathway. *Mol. Med. Rep.* **2022**, *25*, 1–12. [CrossRef]
89. de Oliveira, M.R.; Daglia, M.; Rastrelli, L.; Nabavi, S.M. Epigallocatechin Gallate and Mitochondria—A Story of Life and Death. *Pharmacol. Res.* **2016**, *104*, 70–85. [CrossRef]
90. Ma, L.-P.; Liu, M.-M.; Liu, F.; Sun, B.; Wang, S.-N.; Chen, J.; Yu, H.-J.; Yan, J.; Tian, M.; Gao, L.; et al. Melatonin Inhibits Senescence-Associated Melanin Pigmentation through the p53-TYR Pathway in Human Primary Melanocytes and the Skin of C57BL/6 J Mice after UVB Irradiation. *J. Mol. Med.* **2023**, *101*, 581–593. [CrossRef]
91. Hong, Y.-H.; Jung, E.Y.; Noh, D.O.; Suh, H.J. Physiological Effects of Formulation Containing Tannase-Converted Green Tea Extract on Skin Care: Physical Stability, Collagenase, Elastase, and Tyrosinase Activities. *Integr. Med. Res.* **2013**, *3*, 25–33. [CrossRef] [PubMed]
92. Slominski, A.T.; Slominski, R.M.; Raman, C.; Chen, J.Y.; Athar, M.; Elmets, C. Neuroendocrine signaling in the skin with a special focus on the epidermal neuropeptides. *Am. J. Physiol. -Cell Physiol.* **2022**, *323*, C1757–C1776. [CrossRef] [PubMed]
93. Slominski, A.T.; Zmijewski, M.A.; Plonka, P.M.; Szaflarski, J.P.; Paus, R. How UV Light Touches the Brain and Endocrine System Through Skin, and Why. *Endocrinology* **2018**, *159*, 1992–2007. [CrossRef] [PubMed]
94. Rougier, A.; Schiller, M.; Brzoska, T.; Böhm, M.; Metze, D.; Scholzen, T.E.; Luger, T.A. Solar-Simulated Ultraviolet Radia-tion-Induced Upregulation of the Melanocortin-1 Receptor, Proopiomelanocortin, and α-Melanocyte-Stimulating Hormone in Human Epidermis In Vivo. *J. Investig. Dermatol.* **2004**, *122*, 468–476. [CrossRef] [PubMed]
95. Pourang, A.; Tisack, A.; Ezekwe, N.; Torres, A.E.; Kohli, I.; Hamzavi, I.H.; Lim, H.W. Effects of Visible Light on Mechanisms of Skin Photoaging. *Photodermatol. Photoimmunol. Photomed.* **2021**, *38*, 191–196. [CrossRef]
96. Mita, S.R.; Husni, P.; Putriana, N.A.; Maharani, R.; Hendrawan, R.P.; Dewi, D.A. A Recent Update on the Potential Use of Catechins in Cosmeceuticals. *Cosmetics* **2024**, *11*, 23. [CrossRef]
97. Chaikul, P.; Sripisut, T.; Chanpirom, S.; Ditthawutthikul, N. Anti-Skin Aging Activities of Green Tea (Camelliasinensis (L) Kuntze) in B16F10 Melanoma Cells and Human Skin Fibroblasts. *Eur. J. Integr. Med.* **2020**, *40*, 101212. [CrossRef]
98. Song, X.; Ni, M.; Zhang, Y.; Zhang, G.; Pan, J.; Gong, D. Comparing the Inhibitory Abilities of Epigallocatechin-3-Gallate and Gallocatechin Gallate against Tyrosinase and Their Combined Effects with Kojic Acid. *Food Chem.* **2021**, *349*, 129172. [CrossRef]
99. Liang, Y.; Kang, S.; Deng, L.; Xiang, L.; Zheng, X. Inhibitory Effects of (-)-Epigallocatechin-3-Gallate on Melanogenesis in Ul-traviolet A-Induced B16 Murine Melanoma Cell. *Trop. J. Pharm. Res.* **2014**, *13*, 1825. [CrossRef]
100. Chen, J.; Li, H.; Liang, B.; Zhu, H. Effects of Tea Polyphenols on UVA-Inducedmelanogenesis via Inhibition of α-MSH-MC1R Signalling pathway. *Adv. Dermatol. Allergol./Postępy Dermatol. I Alergol.* **2022**, *39*, 327–335. [CrossRef]
101. Ullmann, U.; Haller, J.; Decourt, J.; Girault, N.; Girault, J.; Richard-Caudron, A.; Pineau, B.; Weber, P. A Single Ascending Dose Study of Epigallocatechin Gallate in Healthy Volunteers. *J. Int. Med Res.* **2003**, *31*, 88–101. [CrossRef] [PubMed]
102. Chow, H.-H.S.; Cai, Y.; Hakim, I.A.; Crowell, J.A.; Shahi, F.; Brooks, C.A.; Dorr, R.T.; Hara, Y.; Alberts, D.S. Pharmaco-kinetics and Safety of Green Tea Polyphenols after Multiple-Dose Administration of Epigallocatechin Gallate and Polyphenon E in Healthy Individuals. *Clin. Cancer Res.* **2023**, *9*, 3312–3319.
103. Oketch-Rabah, H.A.; Roe, A.L.; Rider, C.V.; Bonkovsky, H.L.; Giancaspro, G.I.; Navarro, V.; Paine, M.F.; Betz, J.M.; Marles, R.J.; Casper, S.; et al. United States Pharmacopeia (USP) Comprehensive Review of the Hepatotoxicity of Green Tea Extracts. *Toxicol. Rep.* **2020**, *7*, 386–402. [CrossRef]
104. Yang, Q.-Q.; Wei, X.-L.; Fang, Y.-P.; Gan, R.-Y.; Wang, M.; Ge, Y.-Y.; Zhang, D.; Cheng, L.-Z.; Corke, H. Nanochemoprevention with Therapeutic Benefits: An Updated Review Focused on Epigallocatechin Gallate Delivery. *Crit. Rev. Food Sci. Nutr.* **2019**, *60*, 1243–1264. [CrossRef]
105. Liang, J.; Yan, H.; Puligundla, P.; Gao, X.; Zhou, Y.; Wan, X. Applications of Chitosan Nanoparticles to Enhance Absorption and Bioavailability of Tea Polyphenols: A Review. *Food Hydrocoll.* **2017**, *69*, 286–292. [CrossRef]
106. Dzulhi, S.; Anwar, E.; Nurhayati, T. Formulation, Characterization and In Vitro Skin Penetration of Green Tea (*Camellia sinensis* L.) Leaves Extract-Loaded Solid Lipid Nanoparticles. *J. Appl. Pharm. Sci.* **2018**, *8*, 57–62. [CrossRef]
107. Ramesh, N.; Mandal, A.K.A. Pharmacokinetic, Toxicokinetic, and Bioavailability Studies of Epigallocatechin-3-Gallate Loaded Solid Lipid Nanoparticle in Rat Model. *Drug Dev. Ind. Pharm.* **2019**, *45*, 1506–1514. [CrossRef] [PubMed]
108. Pereira, A.; Ramalho, M.J.; Silva, R.; Silva, V.; Marques-Oliveira, R.; Silva, A.C.; Pereira, M.C.; Loureiro, J.A. Vine Cane Compounds to Prevent Skin Cells Aging through Solid Lipid Nanoparticles. *Pharmaceutics* **2022**, *14*, 240. [CrossRef]
109. Zhuang, Y.; Quan, W.; Wang, X.; Cheng, Y.; Jiao, Y. Comprehensive Review of EGCG Modification: Esterification Methods and Their Impacts on Biological Activities. *Foods* **2024**, *13*, 1232. [CrossRef]
110. Zhang, J.; Cui, H.; Yin, J.; Wang, Y.; Zhao, Y.; Yu, J.; Engelhardt, U.H. Separation and Antioxidant Activities of New Acety-lated EGCG Compounds. *Sci. Rep.* **2023**, *13*, 20964. [CrossRef]
111. Yazihan, N.; Akdas, S.; Olgar, Y.; Biriken, D.; Turan, B.; Ozkaya, M. Olive Oil Attenuates Oxidative Damage by Improving Mitochondrial Functions in Human Keratinocytes. *J. Funct. Foods* **2020**, *71*, 104008. [CrossRef]
112. Li, Y.; He, D.; Li, B.; Lund, M.N.; Xing, Y.; Wang, Y.; Li, F.; Cao, X.; Liu, Y.; Chen, X.; et al. Engineering Polyphenols with Biological Functions via Polyphenol-Protein Interactions as Additives for Functional Foods. *Trends Food Sci. Technol.* **2021**, *110*, 470–482. [CrossRef]

113. Chotphruethipong, L.; Sukketsiri, W.; Battino, M.; Benjakul, S. Conjugate between Hydrolyzed Collagen from Defatted Seabass Skin and Epigallocatechin Gallate (EGCG): Characteristics, Antioxidant Activity and in Vitro Cellular Bioactivity. *RSC Adv.* **2021**, *11*, 2175–2184. [CrossRef] [PubMed]
114. Messire, G.; Serreau, R.; Berteina-Raboin, S. Antioxidant Effects of Catechins (EGCG), Andrographolide, and Curcuminoids Compounds for Skin Protection, Cosmetics, and Dermatological Uses: An Update. *Antioxidants* **2023**, *12*, 1317. [CrossRef] [PubMed]
115. Scalia, S.; Marchetti, N.; Bianchi, A. Comparative Evaluation of Different Co-Antioxidants on the Photochemical- and Func-tional-Stability of Epigallocatechin-3-Gallate in Topical Creams Exposed to Simulated Sunlight. *Molecules* **2013**, *18*, 574–587. [CrossRef] [PubMed]
116. Kale, A.; Gawande, S.; Kotwal, S.; Netke, S.; Roomi, W.; Ivanov, V.; Niedzwiecki, A.; Rath, M. Studies on the Effects of Oral Administration of Nutrient Mixture, Quercetin and Red Onions on the Bioavailability of Epigallocatechin Gallate from Green Tea Extract. *Phytother. Res.* **2010**, *24*, S48–S55. [CrossRef]
117. Choi, E.-H.; Rha, C.-S.; Balusamy, S.R.; Kim, D.-O.; Shim, S.-M. Impact of Bioconversion of Gallated Catechins and Flavonol Glycosides on Bioaccessibility and Intestinal Cellular Uptake of Catechins. *J. Agric. Food Chem.* **2019**, *67*, 2331–2339. [CrossRef]
118. Li, D.; Martini, N.; Wu, Z.; Chen, S.; Falconer, J.R.; Locke, M.; Zhang, Z.; Wen, J. Niosomal Nanocarriers for Enhanced Dermal Delivery of Epigallocatechin Gallate for Protection against Oxidative Stress of the Skin. *Pharmaceutics* **2022**, *14*, 726. [CrossRef]
119. Srivastava, A.K.; Bhatnagar, P.; Singh, M.; Mishra, S.; Kumar, P.; Shukla, Y.; Gupta, K.C. Synthesis of PLGA Nanoparticles of Tea Polyphenols and Their Strong in Vivo Protective Effect against Chemically Induced DNA Damage. *Int. J. Nanomed.* **2013**, *8*, 1451–1462. [CrossRef]
120. Zhang, W.; Yang, Y.; Lv, T.; Fan, Z.; Xu, Y.; Yin, J.; Liao, B.; Ying, H.; Ravichandran, N.; Du, Q. Sucrose Esters Improve the Colloidal Stability of Nanoethosomal Suspensions of (−)-Epigallocatechin Gallate for Enhancing the Effectiveness against UVB-Induced Skin Damage. *J. Biomed. Mater. Res. Part B Appl. Biomater.* **2016**, *105*, 2416–2425. [CrossRef]
121. Vale, E.P.; Morais, E.d.S.; Tavares, W.d.S.; de Sousa, F.F.O. Epigallocatechin-3-Gallate Loaded-Zein Nanoparticles: Characterization, Stability and Associated Antioxidant, Anti-Tyrosinase and Sun Protection Properties. *J. Mol. Liq.* **2022**, *358*, 119107. [CrossRef]
122. Slominski, A.T.; Kim, T.-K.; Janjetovic, Z.; Slominski, R.M.; Li, W.; Jetten, A.M.; Indra, A.K.; Mason, R.S.; Tuckey, R.C. Biological Effects of CYP11A1-Derived Vitamin D and Lumisterol Metabolites in the Skin. *J. Investig. Dermatol.* **2024**, *144*, 2145–2161. [CrossRef] [PubMed]
123. Chaiprasongsuk, A.; Janjetovic, Z.; Kim, T.-K.; Schwartz, C.J.; Tuckey, R.C.; Tang, E.K.Y.; Raman, C.; Panich, U.; Slominski, A.T. Hydroxylumisterols, Photoproducts of Pre-Vitamin D3, Protect Human Keratinocytes against UVB-Induced Damage. *Int. J. Mol. Sci.* **2020**, *21*, 9374. [CrossRef]
124. Dańczak-Pazdrowska, A.; Gornowicz-Porowska, J.; Polańska, A.; Krajka-Kuźniak, V.; Stawny, M.; Gostyńska, A.; Rubiś, B.; Nourredine, S.; Ashiqueali, S.; Schneider, A.; et al. Cellular Senescence in Skin-Related Research: Targeted Signaling Pathways and Naturally Occurring Therapeutic Agents. *Aging Cell* **2023**, *22*, e13845. [CrossRef] [PubMed]
125. Zhang, J.; Yu, H.; Man, M.; Hu, L. Aging in the Dermis: Fibroblast Senescence and Its Significance. *Aging Cell* **2023**, *23*, e14054. [CrossRef]
126. Quan, T. Human Skin Aging and the Anti-Aging Properties of Retinol. *Biomolecules* **2023**, *13*, 1614. [CrossRef]
127. Fisher, G.J.; Wang, B.; Cui, Y.; Shi, M.; Zhao, Y.; Quan, T.; Voorhees, J.J. Skin Aging from the Perspective of Dermal Fibroblasts: The Interplay between the Adaptation to the Extracellular Matrix Microenvironment and Cell Autonomous Processes. *J. Cell Commun. Signal.* **2023**, *17*, 523–529. [CrossRef]
128. He, X.; Gao, X.; Xie, W. Research Progress in Skin Aging, Metabolism, and Related Products. *Int. J. Mol. Sci.* **2023**, *24*, 15930. [CrossRef] [PubMed]
129. Katiyar, S.K.; Afaq, F.; Perez, A.; Mukhtar, H. Green Tea Polyphenol (-)-Epigallocatechin-3-Gallate Treatment of Human Skin Inhibits Ultraviolet Radiation-Induced Oxidative Stress. *Carcinogenesis* **2001**, *22*, 287–294. [CrossRef]
130. Kim, H.-S.; Quon, M.J.; Kim, J.-A. New Insights into the Mechanisms of Polyphenols beyond Antioxidant Properties; Lessons from the Green Tea Polyphenol, Epigallocatechin 3-Gallate. *Redox Biol.* **2014**, *2*, 187–195. [CrossRef]
131. Katiyar, S.K. Skin Photoprotection by Green Tea: Antioxidant and Immunomodulatory Effects. *Curr. Drug Targets-Immune Endocr. Metab. Disord.* **2003**, *3*, 234–242. [CrossRef] [PubMed]
132. Gan, R.-Y.; Li, H.-B.; Sui, Z.-Q.; Corke, H. Absorption, Metabolism, Anti-Cancer Effect and Molecular Targets of Epigallocat-echin Gallate (EGCG): An Updated Review. *Crit. Rev. Food Sci. Nutr.* **2018**, *58*, 924–941. [CrossRef]
133. Jung, E.S.; Park, J.I.; Park, H.; Holzapfel, W.; Hwang, J.S.; Lee, C.H. Seven-Day Green Tea Supplementation Revamps Gut Microbiome and Caecum/Skin Metabolome in Mice from Stress. *Sci. Rep.* **2019**, *9*, 18418. [CrossRef]
134. Revathy, S.; Satyam, S.; Anupama, B. Chemico-biological aspects of (-)-epigallocatechin-3-gallate (EGCG) to improve its stability, bioavailability and membrane permeability: Current status and future prospects. *Crit. Rev. Food Sci. Nutr.* **2022**, *63*, 10382–10411.
135. Aljuffali, I.A.; Lin, C.-H.; Yang, S.-C.; Alalaiwe, A.; Fang, J.-Y. Nanoencapsulation of Tea Catechins for Enhancing Skin Ab-sorption and Therapeutic Efficacy. *AAPS PharmSciTech* **2022**, *23*, 187. [CrossRef]
136. Chen, X.; Yi, Z.; Chen, G.; Ma, X.; Tong, Q.; Tang, L.; Li, X. Engineered Fabrication of EGCG-UV Absorber Conjugated Nano-assemblies for Antioxidative Sunscreens with Broad-Band Absorption. *Colloids Surfaces B Biointerfaces* **2022**, *220*, 112912. [CrossRef] [PubMed]

137. Da Silva, P.B.; Calixto, G.M.F.; Júnior, J.A.O.; Bombardelli, R.L.; Fonseca-Santos, B.; Rodero, C.F.; Chorilli, M. Structural Features and the Anti-Inflammatory Effect of Green Tea Extract-Loaded Liquid Crystalline Systems Intended for Skin Delivery. *Polymers* **2017**, *9*, 30. [CrossRef]
138. Di Sotto, A.; Gullì, M.; Percaccio, E.; Vitalone, A.; Mazzanti, G.; Di Giacomo, S. Efficacy and Safety of Oral Green Tea Prepa-rations in Skin Ailments: A Systematic Review of Clinical Studies. *Nutrients* **2022**, *14*, 3149. [CrossRef]

molecules

Review

Bactrian Camel Milk: Chemical Composition, Bioactivities, Processing Techniques, and Economic Potential in China

Shamila Seyiti [1,*], Abulimiti Kelimu [2,*] and Gulinaer Yusufu [1]

1 School of Economics and Management, Xinjiang University, Shengli Road 666, Urumqi 830046, China; gulinaeryusufu@xju.edu.cn
2 College of Food Science and Pharmacy, Xinjiang Agricultural University, Nongda East Road 311, Urumqi 830052, China
* Correspondence: shamila@xju.edu.cn (S.S.); abkaram@xjau.edu.cn (A.K.)

Abstract: Bactrian camel (BC) milk has gained increasing attention due to its unique nutritional profile and potential bioactivities. This comprehensive review explores the chemical composition, bioactivities, processing techniques, and economic potential of BC milk in China. The distinctive chemical composition of BC milk, including protein, lipid, carbohydrate, vitamin, and mineral content, is discussed, emphasizing its differences from other mammalian milk. The review highlights the various bioactivities of BC milk, such as anti-inflammatory, antidiabetic, lipid-lowering, and anticancer properties, as well as its modulatory effects on intestinal microbiota. The technological properties of BC milk, focusing on its heat stability, coagulation behavior, and potential for product development, are examined. The review also addresses current processing techniques and their impact on milk quality. Finally, the economic potential and future perspectives of BC milk in China are evaluated. This review provides valuable insights into the multifaceted aspects of BC milk, serving as a foundation for future research and development in this emerging field. The motivation for this review stems from the growing interest in BC milk as a functional food and the need for a comprehensive understanding of its properties, applications, and market potential to guide future research and industry development.

Keywords: Bactrian camel; camel milk; chemical composition; bioactivities; economic potential

1. Introduction

Camels play a wide variety of roles, providing milk, meat, long-distance transportation, and other commodities for people living in extremely arid desert areas. Unlike other common livestock, camels possess an unparalleled ability to thrive and provide services under extreme conditions, such as food and water scarcity and high temperatures. However, with the advancement of science and technology, the social roles of camels have transformed into a new era, continuing to contribute to the livelihood of desert people by providing more value-added products, such as milk [1].

The Camelidae family is characterized by two defined subfamilies: *Camelinae* (Old World camelids) and *Laminae* (New World camelids). Old World camelids, which make up the majority of the camel population, include two domesticated and one wild species: the Dromedary camel (DC, *Camelus dromedarius*, one-humped), the BC (*Camelus bactrianus*, two-humped), and the endangered wild camel (*Camelus ferus*, two-humped). The DC, which inhabits hot and arid areas of the Middle East, Eastern and Northern Africa, Southwest Asia, and Australia, represents one of the most prevalent and widely reared species. On the other hand, BC has been recognized as an umbrella species for the fragile ecosystem of Central Asia and mostly occurs in the cooler desert regions of northwestern China, southern Russia, Mongolia, Kazakhstan, and Asia Minor [2]. It is a numerically inferior species, accounting for 10% of the total camel population [3]. As a lesser-known camel population, *Laminae* (New World camelids) are mainly distributed at high altitudes in South America and are generally divided into two wild species, guanaco (*Lama guanicoe*) and vicuña (*vicugna X*) and two domesticated species, llama (*Lama glama*) and alpaca (*Vicugna pacos*) [4].

In recent decades, the camel population in Africa has increased approximately twofold, while it has decreased gradually in Asia. This is because, in arid countries, extensive camel breeding remains the main agricultural activity due to its cost-effectiveness in terms of feed conversion, longer lactation period, and unique ability to adapt to harsh environments compared to other milk-producing mammals. In addition, as an integral part of their diet, camel milk occupies a pivotal position with respect to its role in the nutritional uptake of nomadic communities, as it is an excellent source of protein, fatty acids, minerals, and vitamins [5]. Camels also enable people in marginalized communities in certain geographical areas to make a living and support their families by selling camel products (milk, meat, and wool) and providing services in social activities such as tourism and leisure. In contrast, owing to the decreased necessity of utilizing camels for traditional purposes, such as transportation and agriculture, the number of camels reared in Asia has decreased gradually [6]. As reported by the Food and Agriculture Organization of the United Nations [7], the world camel population reached about 41.77 million in 2022.

Although camel milk has long been known for its excellent nutritional value and various bioactivities, it is consumed primarily by desert dwellers in its fresh state or after converting it into traditional fermented beverages through spontaneous fermentation. Over the past few decades, research on camel milk has increased, leading to several studies investigating its contents and functions [8,9]. Being widely regarded as a sustainable and promising livestock species for the development of health-beneficial foods or ingredients, camel milk consumption is gradually becoming popular among urban consumers due to China's large population and rapid economic development. With the widening gap between supply and demand, alongside the rising price of camel milk, camel farming has evolved into a highly profitable business. Both the government and the private sector recognize that sustained increases in the prices of camel milk and its products could significantly enhance economic livelihoods and alleviate poverty in rural areas with harsh climates, provided that proper management and utilization are available. Given the economic and nutritional significance of camel milk, there has been a gradual increase in capital investment and large-scale commercial production on modern camel farms. Currently, camel milk products, including pasteurized milk, fermented beverages, and dehydrated products, are readily available on the market. While the individual components, physicochemical characteristics, bioactivities, and technological properties of DC milk have received increasing attention since it is the most widely reared species worldwide, BC, which is the predominantly reared species in Asian countries, is understudied in terms of the composition, bioactive components, and technological aspects of its milk, with very limited reviews available, especially in terms of bioactivities and technological aspects. Moreover, studies have shown that some components (protein, fatty acids, oligosaccharides, etc.) of BC milk are significantly different from those of DC milk [10,11]. From a technological perspective, the distinct composition of BC milk may influence its behavior during processing and product development. Given the unique characteristics of BC milk and the limited understanding of its functional and technological aspects, a comprehensive review of the available research findings is warranted. Such a review would not only shed light on the potential benefits and applications of BC milk but also contribute to the broader understanding of camel milk diversity and its potential for value-addition in the dairy industry.

Hence, the main purpose of this review is to summarize the fragmentary knowledge of BC milk, with an emphasis placed on the camel population and milk yield, composition, bioactivities, processing, and economic potential.

2. Methods

A systematic literature search was carried out to identify pertinent studies on Bactrian camel and camel milk. This comprehensive search involved various databases and official websites to ensure a thorough examination of the existing scientific literature. The sources utilized for identifying relevant research articles and reports included Google Scholar (https://scholar.google.com, accessed on 15 August 2024), Science Direct (www.sciencedirect.com, accessed on 10 August 2024), the American Chemical Society Database (https://pubs.acs.org, accessed on 25 July 2024), and the China National Knowledge Infrastructure (CNKI) Database (www.cnki.net, accessed on 15 August 2024). The search strategy was guided by specific keywords such as "camel milk" and "Bactrian camel milk", chosen to capture a broad range of studies on the topic. The focus was primarily on research conducted in China on Bactrian camel milk in recent years.

The inclusion criteria for the literature encompassed studies that specifically addressed the nutritional, biochemical, or therapeutic properties of Bactrian camel milk, research articles, reviews, and reports published in peer-reviewed journals, and studies available in English or with an English abstract. The search was systematically conducted across the selected databases, and relevant articles were screened based on their titles and abstracts. Full texts of potentially relevant articles were then retrieved and reviewed in detail.

3. Camel Population

Throughout Chinese history, camels have played a crucial role in civilization. Due to their unique adaptations to diverse and extreme environments, camels became the primary means of caravan traffic along the Silk Road, facilitating long-distance migration in desert areas. Beyond their role as pack animals, camels provide economically deprived communities with milk, meat, and wool. Most camels in China are Bactrian, mainly distributed in mountainous and marginal regions, where dairy camel farming is increasingly viewed as a profitable business. According to the National Bureau of Statistics, China's camel population has grown at a compound annual growth rate of 10% from 2013 to 2023, reaching approximately 579,700 camels in 2023 (Figure 1). The majority are found in the provinces of Xinjiang (52.44%), Inner Mongolia (37.62%), Gansu (7.19%), and Qinghai (2.48%) [12]. The "National List of Animal Genetic Resources and Varieties" (2021) characterizes China's camel genetic resources by five defined breeds: Alxa BC and Sonid BC from Inner Mongolia, Tarim BC and Junggar BC from Xinjiang, and Qinghai BC [13].

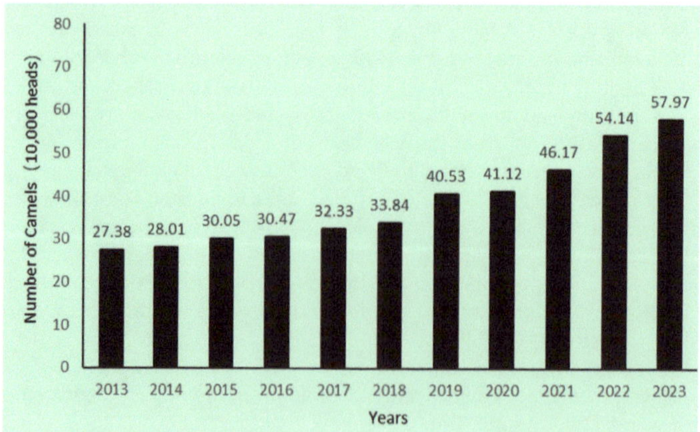

Figure 1. Change in number of camels from 2013 to 2023 (NBS of China 2023).

4. Milk Yield

Generally, in terms of milk production, camels are considered less productive, with BC being even less productive than DC. However, camel milk, especially from BC, is richer in dry matter, including fat, protein, and lactose [14,15]. Nurseitova et al. (2014) reported no significant difference in milk yield between BC (2.9 L/d) and DC (3.2 L/d) reared under the same conditions, although BC milk had higher fat and dry matter content [16]. These conflicting results could be attributed to varying farming conditions [15]. According to a recent study, the milk production of Bactrian camels in China varies between 1470.18 ± 9.75 kg and 978.34 ± 3.80 kg over a period of 300 days [17].

5. Physicochemical Properties of Camel Milk

Camel milk, characterized by its opacity and white color due to its finely dispersed fat, possesses a unique combination of sweetness, sharpness, and occasional saltiness. Agitation of the milk leads to the easy formation of froth. Throughout the lactation cycle, the physicochemical properties of camel milk undergo significant changes. Specifically, the relative density hovers between 1.028 and 1.045; the viscosity ranges from 6.79 to 24.66 mPa·s; the pH varies between 6.35 and 6.79; the acidity falls within 0.17–0.24%; the thörner degree, a measure of titratable acidity, ranges from 19.98 to 20.71; and the conductivity is measured at 0.38–0.57 mS/cm. As the dry matter content decreases during lactation, both the density and viscosity of camel milk gradually decrease, while the other physicochemical properties tend to remain relatively constant [1,18,19]. The composition of camel milk, and consequently its flavor and other attributes, is influenced by several factors, with fodder type and water availability being particularly significant [20]. In terms of shelf life, raw camel milk outperforms bovine milk, remaining stable for an extended period at 30 °C and maintaining its quality for 2–3 days at room temperature without any visible changes. When refrigerated at 4 °C, it can be kept for more than three months without showing any signs of deterioration.

6. Chemical Composition

BC milk, like other animal milks, provides a wide array of nutrients, such as proteins, lipids, lactose, and other essential components, making it a valuable source of nutrition for a balanced diet. However, BC milk is distinct from other mammalian milks in terms of certain chemical components, including individual proteins, fatty acids, and various bioactive ingredients that may offer health-promoting benefits [9,21]. Understanding the nutritional quality of milk is critical for its application and processing. Over the past two decades, considerable efforts have been made to unveil the compositional characteristics of this unique food resource (Table 1). Despite the fact that the milk yield of BC is lower than that of DC, BC milk is richer in protein, fat, and total solids [14,22], which makes it superior to DC milk in terms of nutritional value and suitability for processing. The composition of camel milk is influenced by several factors, including seasonal variations, geographical origins, camelid species, nutritional conditions, breed, stage of lactation, age, and parity [17,19,23–25].

Table 1. Chemical composition of different milk types.

Milk Source	Alxa BC	Sonid BC	Junggar BC	DC	Cow
Total solids (%)	14.63–15.67	14.20 ± 0.01	14.85–17.49	10.80	12.40
Protein (%)	4.31–4.80	4.15 ± 0.15	3.58–3.85	3.20	3.40
Fat (%)	3.38–6.05	4.80 ± 0.14	5.37–7.44	3.10	3.70
Lactose (%)	4.56–6.23	4.90 ± 0.11	4.71–5.70	3.80	4.70
Ash (%)	0.92–0.94	0.81 ± 0.03	0.71–0.91	0.70	0.60
Na (mg/kg)	720	ND	552	474	319
P (mg/kg)	1168	ND	1068	946	901

Table 1. Cont.

Milk Source	Alxa BC	Sonid BC	Junggar BC	DC	Cow
Ca (mg/kg)	1546	1809	1442	1299	1137
K (mg/kg)	1910	1244	1449	1732	1418
Vc (mg/kg)	42.33	ND	28.23	33.00	15.00
V_A (mg/kg)	0.45	ND	1.11	0.27	0.38
V_{B1} (mg/kg)	0.81	ND	0.13	0.48	0.40
V_{B3} (mg/kg)	1.24	ND	ND	0.78	0.80
V_{B5} (mg/kg)	0.50	ND	ND	3.68	3.20
V_{B6} (mg/kg)	0.72	ND	0.50	0.55	0.50
V_E (mg/kg)	2.73	ND	1.43	0.02	1.00
References	[19,26–28]	[27,29]	[18,30–33]	[34–36]	[33–35,37]

Abbreviations: ND, not determined; BC, Bactrian camel; DC, Dromedary camel.

6.1. Camel Milk Protein

Camel milk is increasingly recognized as a valuable source of proteins that exhibit a wide range of abundances and varieties. These proteins can be broadly categorized into three primary groups: casein (CN), whey protein (WP), and milk fat globule membrane (MFGM) proteins. Among these, casein components such as α-CN, β-CN, and κ-CN [38], along with whey proteins such as α-lactalbumin (LA), serum albumin (SA), lactoferrin (LF), immunoglobulins, and peptidoglycan recognition protein, are the most prominent [11]. MFGM proteins also play a significant role in the overall protein composition of camel milk [10,39]. Few studies have explored the relative distribution of milk proteins in different breeds of camels, such as BC and DC. For instance, research has shown significant variations in protein composition between BC and DC milk [11,40]. Using peptidomics techniques, a study identified 622 parent proteins from 8393 peptides in camel milk. Among these proteins, 208 were from DC, and 464 were from BC. After filtering, 4464 endogenous peptides were quantified, 459 of which were common to both breeds. Notably, 170 peptides derived from 27 proteins exhibited significant differences between the two types of camel milk. For example, osteopontin and lactoperoxidase were upregulated in DC milk, whereas butyophilin subfamily member A1, perilipin, and fatty acid synthase were upregulated in BC milk [40]. Compared to other animals, camel milk exhibits distinct proportions of protein components, showing similarities primarily with human milk [41]. Camel milk proteins also demonstrate enhanced digestibility and unique peptide profiles [9], underscoring their potential suitability for infant formula applications.

6.1.1. Casein

Camel milk is known for its nutritional value and therapeutic effects, largely due to its bioactive components, including proteins and peptides. Casein, the main protein complex in BC milk, exists in a micellar form distinct from that in ruminant milk [42]. This heterogeneous protein is composed of αs_1-CN, αs_2-CN, β-CN, and κ-CN, which vary in amino acid sequence and posttranslational modifications. Although studies have reported different relative proportions of these proteins [39,43,44], they have consistently indicated that β-CN is the most abundant casein fraction. A low level of β-CN phosphorylation is associated with the formation of a softer and less cohesive curd, facilitating easier digestion. The abundance of β-CN in camel milk underscores its superior digestibility and hypoallergenic properties, making it particularly beneficial for neonatal infants [9]. Future research on camel milk casein should focus on detailed structural and functional studies to fully understand its unique properties. Exploring its potential in infant nutrition, particularly for those with allergies, is crucial given its superior digestibility and hypoallergenic nature.

6.1.2. Whey Proteins

Renowned for its exceptional nutritional profile, camel milk is abundant in whey proteins and exhibits significant variations in both the quantity and distinctive attributes of its proteins compared to those of cow's milk [10]. Among the diverse array of whey proteins detected in camel milk, the predominant ones include serum albumin (SA), α-lactalbumin (LA), lactoferrin (LF), lactoperoxidase (LPO), and glycosylation-dependent cell adhesion molecule 1 (GLYCAM1). Notably, the levels of SA, LF, and GLYCAM1 are significantly higher in camel milk than in bovine and goat milk [10]. The structural characteristics of serum albumin (SA) in camel milk include a notably high alpha-helix content, minimal beta-fold structures, and an irregular coil configuration, which are crucial for its functionality, stability, and intermolecular interactions [45]. Yang et al. [46] reported that LA, comprising 123 amino acid residues with a molecular mass of 14.6 kDa, is the most abundant whey protein in BC milk. This contrasts with bovine LA, which contains 142 amino acid residues and has a molecular mass of 16.2 kDa [47]. To elucidate the genetic and biological diversity across animal species, researchers have identified whey acidic protein and quinone oxidoreductase as distinctive traits characterizing camel milk [46]. Camel milk whey protein also contains naturally occurring heavy-chain antibodies (HCAbs), which have rehabilitative effects on infection and immunity. A study revealed that the IgG1 fraction within BC milk comprises molecules with heavy chains weighing 50 kDa and light chains weighing 36 kDa. In contrast, the IgG2 and IgG3 fractions lack light chains and consist solely of heavy chains with molecular weights of 45 kDa and 43 kDa, respectively. The abundance of heavy chains in the IgG2 and IgG3 fractions notably exceeds that in the IgG1 fraction [48].

6.1.3. Amino Acids

In terms of amino acid composition, researchers have investigated the amino acid profile of BC milk throughout the lactation period. The findings revealed the presence of 17 amino acids, excluding tryptophan. Essential amino acids (EAAs) accounted for approximately 43.52–43.87% of the total amino acids (TAAs) during lactation, surpassing the recommended standards of the FAO/WHO. The ratios of EAA/TAA and EAA/NEAA (nonessential amino acids) were above 40% and 75%, respectively. Glutamic acid and leucine were the most abundant amino acids, followed by proline and lysine, with cysteine being the least abundant. The abundance of amino acids varied with the lactation stage, peaking in early lactation and gradually decreasing as lactation progressed [19]. Amino acid analysis revealed that the total amino acids, total essential amino acids, total drug-effective amino acids, and nearly all amino acids in Inner Mongolian BC milk were significantly higher than those in bovine milk [49] This enhanced amino acid profile further highlights the nutritional superiority of camel milk over bovine milk.

6.2. Lipids

6.2.1. Lipid Composition

Camel milk has a distinctive lipid profile that differentiates it from other dairy sources and the lipid composition of BC milk in China also exhibits a diverse array of lipid classes. Zhao et al. [50] compared the lipid profiles of camel milk to those of other types of milk, identifying four main lipid classes—glycerolipids (GLs), glycerophospholipids (GPs), sphingolipids (SPs), and sterol lipids (SLs)—and 20 subclasses—triglycerides (TGs) and diglycerides (DGs). One study revealed that human milk contains 909 distinct lipids, while cow, goat, and camel milk contain 903, 918, and 907 distinct lipids, respectively. Notably, camel and human milk share the most similarities, with only 87 differential lipids, indicating a closer lipid profile. In camel milk, TGs constitute 90.08% of the lipid content, followed by phosphatidylcholine (PC, 4.28%), sphingomyelin (SM, 3.11%), phosphatidylethanolamine (PE, 1.26%), phosphatidylserine (PS, 0.38%), and other phospholipids (PLs). This study also underscores the similarities between human and camel milk lipid compositions, which could inform the development of better infant formulas. A greater proportion of functional TGs and PLs in camel milk could enhance the nutritional profile of infant formula, making it a potentially superior base [51].

6.2.2. Polar Lipids

Polar lipids, such as phospholipids and sphingolipids, are vital components of biological membranes and have potential health-promoting properties. They play significant physiological roles, including treating dyslipidemia, reducing inflammation, improving cardiovascular health, and promoting intestinal and neural development [52,53]. Recent research in China has focused on the polar lipids in BC milk, examining their unique composition, potential health benefits, and methods for quality assessment. A comprehensive study used ultra-performance liquid chromatography–tandem mass spectrometry to isolate and analyze these lipids, identifying a total of 333 polar lipids [52]. Consistent with previous research [50], these polar lipids primarily included GPs and SPs. Camel milk showed high levels of PE, SM, and PC. The lipid profile included 77 PEs, 60 PCs, 28 PIs, 20 PSs, 4 phosphatidylglycerol (PGs), 3 phosphatidic acids (PAs), 14 lysophosphatidylethanolamines (LPEs), 13 lysophosphatidylcholines (LPCs), 1 lysophosphatidylinositol (LPI), 59 SMs, and 54 ceramides (Cers) [52]. This detailed profiling underscores the nutritional and functional significance of the complex lipid composition of camel milk. Researchers have employed advanced analytical methods to study the lipid composition of BC milk. A notable study used ultra-performance supercritical fluid chromatography (UPSFC) combined with quadrupole time-of-flight mass spectrometry (Q-TOF-MS) for high-throughput analysis of polar lipids in milk samples. This technique identified a wide variety of lipid species, including PE, PC, and SM, present in all the samples. A previous study revealed that the phospholipids in camel milk are predominantly in the order of PE, followed by SM and then PC [54]. Furthermore, a study comparing the phospholipid (PL) profiles of human colostrum and colostrum from six dairy animals (mare, camel, goat, cow, yak, and buffalo) revealed that camel colostrum was most similar to human colostrum [55]. These findings highlight the unique properties and potential health benefits of camel milk lipids, suggesting further research to explore their applications in nutrition and medicine.

6.2.3. Fatty Acid Profile

The fatty acid composition of BC milk has been extensively studied and compared with that of human and other milks, revealing a unique and nutritionally significant profile (Table 2). Camel milk fat contains a wide range of fatty acids, including saturated fatty acids (SFAs), monounsaturated fatty acids (MUFAs), and polyunsaturated fatty acids (PUFAs). The composition of fatty acids in BC milk has been a subject of interest in recent studies [56–59]. A study by Zou et al. [51] revealed a distinct milk lipid composition characterized by the absence of short-chain saturated fatty acids (SC-SFAs) and a limited presence of medium-chain saturated fatty acids (MC-SFAs). The levels of long-chain saturated fatty acids (LC-SFAs) in camel milk fat are significantly greater than those in human milk fat (HMF), which is attributed to elevated levels of myristic acid (C14:0) and stearic acid (C18:0). The monounsaturated fatty acid (MUFA) content in camel milk fat ($41.00 \pm 1.20\%$) is comparable to that in HMF ($36.90 \pm 5.95\%$) due to high concentrations of palmitoleic acid (C16:1) and oleic acid (C18:1). In contrast, the PUFA content in camel milk lipids ($3.31 \pm 0.27\%$) was significantly lower than that in HMF ($10.57 \pm 4.96\%$), primarily due to the higher linoleic acid (C18:2 ω-6) content in HMF. The presence of SFAs at the sn-2 position is crucial for infant digestion, absorption, and metabolism. Despite lower SFA and PUFA levels at the sn-2 position in camel milk compared to HMF, MUFA levels are significantly greater, driven by substantial amounts of C16:1 ω-7 ($10.72 \pm 0.27\%$) and C18:1 ω-9 ($27.63 \pm 0.66\%$). The SFA and LC-SFA contents in camel milk phospholipids are considerably lower than those in HMF and other mammalian milks; however, MUFA ($32.95 \pm 1.44\%$) and PUFA ($22.42 \pm 1.14\%$) levels are significantly elevated.

In examining the effects of the lactation period on camel milk fatty acids, studies have evaluated the differences in fatty acid composition of BC milk samples collected from Inner Mongolia during lactation. They found that 23 fatty acids were detectable in BC milk, predominantly oleic (C18:1n9c), palmitic (C16:0), myristic (C14:0), and stearic (C18:0) fatty acids, which constituted 81.73% of the total fatty acids. Saturated fatty acids (SFAs) increased significantly with advanced lactation, while MUFAs and PUFAs showed the opposite trend. SFAs ranged from 50.71% to 58.84%, mainly C16:0 (29.8%), C14:0 (21.05%), and C18:0 (17.12%). MUFAs, primarily oleic acid (C18:1n-9), made up 26.81–34.79%, and PUFAs included C18:2n-6 (2.62–3.17%) and C18:3n-3 (0.72–1.03%) [19,58].

Compared to DC milk, SC-SFAs (C6:0 and C8:0) were not detected in the lipids from either camel species. In DC milk, the predominant saturated fatty acids (SFAs) were myristic (C14:0; 10.55%), palmitic (C16:0; 27.9%), and stearic (C18:0; 12.99%). These values significantly differed from those in BC milk, which had different percentage of myristic (11.54%), palmitic (31.51%), and stearic (18.67%) fatty acids. Compared with DC, BC presented a greater SFA content, corroborating previous findings by Zou et al. [51]. The MUFA content in BC milk was lower than that in DC milk, which was attributable to the higher levels of C16:1 and C18:1 in the latter. The data also revealed that the SFA content at the sn-2 position in DC was lower than that in BC. For the sn-1/3 positions, the primary fatty acids in DC milk fat were myristic (10.05%), palmitic (22.02%), and stearic acids (16.06%), whereas in BC, they were 11.90%, 23.79%, and 22.49%, respectively [60]. Interestingly, the fatty acid profile of BC milk appears closer to that of human milk than to that of cow milk. Several studies have reported that the levels of essential fatty acids, linoleic acid (C18:2n-6), α-linolenic acid (C18:3n-3), and arachidonic acid in camel milk are similar to those in human milk and higher than those in cow milk [58].

Compared with bovine and human milk, BC milk has a greater content of odd-chain saturated fatty acids (OCSFAs) and branched-chain saturated fatty acids (BCSFAs) [61], which are important for perinatal nutrition and intestinal microflora colonization in neonates [62]. Trans-monounsaturated fatty acids (trans-MUFAs), which are reported to have detrimental effects on human health, are present in greater amounts (2.46 ± 0.13 g/100 g) than in bovine milk (1.71 ± 0.02 g/100 g). Although the total amount of cis-MUFAs in camel milk is lower, the content of palmitoleic acid (C16:1 cis-9) is significantly higher than that in human and bovine milk [63]. This research also revealed the fatty acid profile of the milk, identifying that the dominant phospholipid fatty acids are palmitic acid (16:0), which ranges between 5.40% and 12.11%; stearic acid (18:0), which ranges from 10.83% to 22.60%; oleic acid (18:1), which ranges from 25.02% to 33.85%; and linoleic acid (18:2), which constitutes between 7.42% and 22.05% of the fatty acid content. This comprehensive lipidomic profile of camel milk underscores its unique nutritional characteristics and potential health benefits, as detailed by Jiang and colleagues in their 2022 study [54].

There are several differences in PUFA content, but C18:2 cis-9,12, C18:3n-3 (ALA), C20:4n-6 (AA), C20:5n-3 (EPA), and C22:5n-3 (DPA) exist in relatively high amounts [55–59].

Table 2. Fatty acid composition of camel milk from different studies.

Camel Species	SFAs (%)	MUFAs (%)	PUFAs (%)	References
Junggar BC	53.66 ± 4.85	41.00 ± 1.20	5.21 ± 0.34	[51]
Junggar BC	63.83 ± 3.11	30.27 ± 2.17	2.33 ± 1.21	[59]
Alxa BC	64.1 ± 0.87	25.7 ± 0.38	8.10 ± 0.11	[58]
Alxa BC	57.88 ± 3.88	31.35 ± 2.84	2.64 ± 0.12	[64]
Junggar BC	51.9 ± 2.20	39.6 ± 2.40	8.46 ± 0.86	[63]
Alxa BC	53.56 ± 1.86	37.99 ± 2.31	3.23 ± 0.14	[19]
Junggar BC	62.82 ± 0.05	31.72 ± 0.20	4.86 ± 0.25	[65]

Abbreviations: BC, Bactrian camel; SFAs, saturated fatty acids; MUFAs, monounsaturated fatty acids; PUFAs, polyunsaturated fatty acids.

6.3. *Milk Fat Globule Membrane Composition*

6.3.1. Lipids

Lipids are dispersed in milk in the form of spherical or globular droplets of varying sizes. These droplets are encapsulated by MFGM, which is secreted from the membrane of mammary epithelial cells [66]. The MFGM is a complex bilayer that plays a crucial role in preventing fat droplets from flocculating and coalescing in milk [67]. This membrane has a tripartite structure composed of various components, such as proteins, glycoproteins, enzymes, neutral lipids, polar lipids, and cholesterol [68]. This unique structural arrangement and compositional complexity not only have technological importance in milk processing [69–71] but also confer several biological functions. These functions are attributed to the presence of biologically important components, including polar lipids [51,72,73].

The composition of the milk fat globule membrane (MFGM) of camel milk is distinct from that of bovine milk. The total fat/MFGM and neutral fat/total fat ratios in camel milk are relatively similar to those in bovine milk. However, the polar fat/total fat ratio is significantly lower, while the phospholipid/polar fat ratio is markedly greater in camel MFGM than in bovine MFGM [74]. Phospholipids, which are natural emulsifying agents ensuring the stability of milk fat globules [75], constitute the main polar fat in camel MFGM, making up 94.31 ± 2.27% of the polar fat content [74]. These differences in MFGM composition highlight the unique nutritional and functional properties of camel milk, making it a subject of interest for further research and potential applications in the food and health industries.

6.3.2. Proteins and Amino Acids

The milk fat globule membrane (MFGM) proteome has been extensively studied across various species and lactation stages, highlighting its nutraceutical properties and potential benefits for human health. Proteomic analyses have been conducted on MFGM proteins from different mammals, including bovine, goat, human and camel milk, revealing a wide range of proteins present in these milk fractions [10,39,42,73]. Comparative proteomic analyses have been performed to examine the MFGM protein compositions of different camel breeds, specifically BC and DC. These analyses identified 911 proteins common to both breeds, along with 136 proteins unique to BC and 86 proteins unique to DC, from a total of 1047 and 997 proteins identified in the MFGM fractions of BC and DC milk, respectively. Among these proteins, lactadherin, butyrophilin subfamily 1 member A1, xanthine dehydrogenase, perilipin, and glycosylation-dependent cell adhesion molecule 1 were identified as the five most abundant proteins in the MFGM fractions of both BC and DC [11]. In another study, Han et al. [10] identified a total of 1579 MFGM proteins across bovine, goat, and camel milk samples. The findings revealed that 966, 1105, and 813 MFGM proteins were detected in bovine, goat, and camel milk, respectively. Among these proteins, 395 MFGM proteins were common across all three milk types. Specifically, 24 MFGM proteins were uniquely shared between bovine and camel milk, while 64 proteins were shared between goat and camel milk. In contrast, 120, 219, and 330 MFGM proteins were unique to bovine, goat, and camel milk, respectively. The functional analysis indicated that these MFGM proteins shared similar functional annotations, predominantly in cellular processes, intracellular anatomical structures, and binding, according to Gene Ontology (GO) annotations. A recent study assessed the protein composition of MFGM samples from camel and cow milk using a label-free quantitative proteomics approach. In total, 1427 and 1751 MFGM proteins were identified in cow and camel milk samples, respectively. This analysis revealed 1067 proteins common to both samples, while 684 proteins were unique to camel milk and 360 were unique to cow milk. The predominant MFGM proteins in both camel and cow milk included butyrophilin (BTN), xanthine dehydrogenase/xanthine oxidase (XDH/XO), fatty acid-binding protein (FABP), and fatty acid synthase (FASN) [39]. Efforts to characterize camel MFGM composition have also shown that the total dry matter and individual components, excluding moisture, are lower in camel MFGM than in bovine MFGM. Essential amino acids (EAAs), nonessential amino acids (NEAAs), flavor amino acids (FAAs), and the sum of amino acids (AAs) are present in lower amounts in camel MFGM protein than in bovine MFGM protein. In total, 17 amino acids were identified in bovine MFGM, while only 15 amino acids were detected in camel MFGM. Notably, camel MFGM contained a significantly higher amount of phenylalanine (Phe) (2.75 ± 0.29%).

Methionine (Met), tyrosine (Tyr), and proline (Pro) are absent from the camel MFGM protein, whereas only Met is absent from the bovine MFGM protein [74].

6.4. Oligosaccharides (OSs)

Oligosaccharides, often referred to as prebiotics, are long-chain carbohydrates that have gained significant attention due to their potential health benefits. BC milk has emerged as a promising source of these functional compounds, and recent years have seen extensive research into the composition [24,76] and potential applications of BC milk oligosaccharides.

One of the pioneering studies in this field was conducted by Shi et al. [77], who analyzed various OSs in milk from humans, cows, goats, sheep, and camels. They found that 30 OSs were identified in bovine milk, 42 in caprine milk, 32 in ovine milk, 34 in camel milk, and 35 in human milk. Notably, camel milk had the highest similarity to human milk in terms of the types of OSs present, sharing a total of 16 common OSs, more than any other animal milk studied. However, the concentrations of eight specific OSs in human milk were approximately six times greater than those in camel milk.

In a comprehensive study, Zhang et al. [78] investigated the variations in milk OSs across different lactation stages and species. They identified a total of 89, 97, 115, and 71 OSs in human, bovine, goat, and camel milk, respectively. The number of common OSs shared between camel and human milk was the highest, with 16 and 17 common OSs in transitional and mature milk, respectively. The distribution of different oligosaccharide types in camel milk was detailed, with camel transitional milk containing 92.6% sialylated OSs, 5.9% nonfucosylated neutral OSs, and 1.5% fucosylated neutral OSs. In mature camel milk, these proportions shifted slightly to 90.0% for sialylated OSs, 9.6% for nonfucosylated neutral OSs, and 0.7% for fucosylated neutral OSs. Among the various OSs, 3′-sialyllactose (3′-SL) stood out in mature camel milk, with the highest concentration measured at 304.5 μg/mL. This detailed analysis highlights significant species-specific differences and changes over lactation stages, providing valuable insights into the composition and potential functional roles of milk OSs.

While the OS composition of camel milk has been extensively studied, the polysaccharides present in the camel milk fat globule membrane (MFGM) are a relatively unexplored area. Current research indicates that camel MFGM contains 5.90 ± 0.17 mg/g of polysaccharide, which is lower than the 8.54 ± 0.28 mg/g of polysaccharide found in bovine MFGM [74]. However, the specific types and detailed quantities of these polysaccharides have not yet been fully characterized. Future research investigating the biological functions and nutritional properties of camel MFGM polysaccharides could reveal their unique roles in health and disease, potentially leading to new nutritional insights, therapeutic applications, and the development of novel functional foods and nutraceuticals, thereby increasing the commercial value of camel milk products.

6.5. Minerals

The mineral composition of BC milk exhibits significant variation due to factors such as water availability, lactation period, feeding practices, season, and geographical region. Studies have identified 11 essential trace minerals in BC milk, excluding vanadium, with notable differences in the levels of certain minerals compared to those in bovine milk [30]. While some researchers have reported higher levels of calcium, phosphorus, potassium, sodium, zinc, and selenium in BC milk, others have reported lower levels of magnesium and iron [79]. Recent research by Chen et al. [80] and colleagues analyzed 17 elements across different types of milk, including cow, goat, buffalo, yak, and camel milk. Their findings revealed that the levels of arsenic, calcium, cadmium, sodium, nickel, strontium, and zinc were greater in BC milk than in other milk types. Despite numerous studies on the mineral content of Bactrian camel milk, comparisons remain challenging due to varying physiological, feeding, and environmental conditions, resulting in differing mineral composition outcomes for BC reared in different regions, particularly in China.

7. Bioactivities of BC Milk

Camel milk is an excellent source of essential macromolecules and contains functional compounds with antibacterial, antiviral, anti-inflammatory, antidiabetic, and anticancer properties. These beneficial components include casein, whey, oligosaccharides, and lipids. Moreover, when camel milk proteins are hydrolyzed through enzymatic or fermentation processes [81], functional peptides embedded in the primary protein structure are released [40].

7.1. Anti-Inflammatory Effect

Inflammation, characterized by tissue injury and intestinal microbial imbalance, is a significant cause of mortality and morbidity worldwide. Considerable evidence has demonstrated the role of camel milk in regulating inflammatory factors. In a mouse model of chronic alcoholic liver disease, camel milk regulated inflammatory cytokine production, prevented colonic dysfunction, and improved alcoholic liver injury [82]. Zhu et al. [83] reported that camel milk significantly attenuated the levels of proinflammatory cytokines (tumor necrosis factor-α, IL-10, and IL-1β) in the serum of lipopolysaccharide-induced acute respiratory distress syndrome rats. Similarly, by comparing the effects of different diets (normal diet; normal diet, then ethanol; and normal diet and camel milk, then ethanol), researchers found that camel milk reduced hepatic inflammation by downregulating the expression of inflammation-related genes (IL-1β and CXCL1) in the IL-17 pathway [84]. The beneficial effects of camel milk may be due to the combined effects of its various ingredients. Dou et al. [85] investigated the effects of camel milk whey protein (CWP) on rats with streptozotocin (STZ)-induced type 2 diabetes mellitus (T2DM) and demonstrated that CWP suppressed the inflammatory response. Additionally, Du et al. [86] reported that BRL-3A hepatocytes pretreated with CWP hydrolyzed by simulated gastrointestinal fluid exhibited resistance to heat stress-induced apoptosis via activation of the Nrf2/HO-1 signaling pathway and suppression of the NF-κB/NLRP3 axis. Fermentation is one of the most commonly used methods for liberating bioactive peptides from food matrices. A metabolomics study revealed the upregulation of beneficial metabolites after fermentation of camel milk [81]. Another study investigated the bioactive peptides from fermented camel milk using network pharmacology and molecular docking, and the results indicated that fermented camel milk exerts anti-inflammatory effects via various biological mechanisms and signaling pathways [87]. Camel milk intervention also showed anti-inflammatory effects against radiation-induced intestinal injury by decreasing the levels of proinflammatory cytokine receptors (e.g., TNF-α and IL-1β) while increasing the levels of TLR4, NF-κB, HMGB1, and IL-10, indicating that camel milk exerts anti-inflammatory effects by regulating the HMGB1/TLR4/NF-κB/MyD88 inflammatory signaling pathway [88].

The anti-inflammatory properties of camel milk and its components, such as whey protein and bioactive peptides, have been demonstrated in various studies. However, further research is needed to fully understand the underlying mechanisms and potential therapeutic applications involved. Exploring the synergistic effects of different bioactive compounds present in camel milk could provide insights into developing effective anti-inflammatory strategies. Additionally, clinical studies investigating the efficacy of camel milk or its derivatives in managing inflammatory conditions in humans would be valuable. Furthermore, investigating the potential of fermented camel milk or its bioactive peptides as functional foods or nutraceuticals could open new avenues for promoting human health and well-being.

7.2. Antidiabetic Effect

Type 2 diabetes mellitus (T2DM) is primarily characterized by reduced insulin sensitivity (insulin resistance) and impaired insulin secretion. Recent studies have highlighted the potential of camel milk for managing T2DM, providing evidence from various experimental and clinical investigations. Zheng et al. [89] explored the hypoglycemic effects of camel milk powder in patients with T2DM. Participants were supplemented with 10 g of camel milk powder twice daily for four weeks. The results indicated significant reductions in fasting blood glucose and 2 h postprandial blood glucose levels. Additionally, there was a notable decrease in the serum levels of resistin and lipocalin-2, adipokines that are positively correlated with diabetes. Further investigations by Han et al. [90] examined the functional impact of camel milk on glucose homeostasis, the hepatic proteome, and the phosphoproteome in high-fat diet and streptozotocin (HFD/STZ)-induced diabetic rats. Their study demonstrated that 35 days of supplementation with camel milk improved fasting glucose levels and glucose tolerance. The positive effects on lipid metabolism were attributed to the activation of AMP-activated protein kinase (AMPK). Moreover, Manaer et al. [91] evaluated the antidiabetic effects of shubat, a traditional fermented camel milk product, on streptozotocin-induced T2DM rats. Their findings showed that a high dose of shubat improved T2DM conditions by decreasing fasting blood glucose (FBG) and glycated hemoglobin (HbA1c) levels while increasing C-peptide and glucagon-like peptide-1 (GLP-1) levels. Shubat administration also protected renal function. Researchers have hypothesized that these benefits might be due to the promotion of GLP-1 release and the improvement of β-cell function. Dou et al. [85] focused on camel milk whey protein (CWP) treatment in streptozotocin (STZ)-induced T2DM rats and insulin-resistant HepG2 cell models. Intrigastric administration of CWP mitigated body weight loss and improved liver tissue structure, including cell arrangement, congestion, edema, and vacuolization. CWP feeding also upregulated the expression of insulin receptor substrate-2 (IRS-2), phosphoinositide 3-kinase (PI3K), protein kinase B (AKT), and glycogen synthase (GS). The study suggested that the hypoglycemic effect of CWP could be due to the activation of the PI3K/AKT pathway, which inhibits gluconeogenesis and promotes glycogen synthesis. In another study, Zhang et al. [40] used peptidomics to identify endogenous peptides in camel milk with dipeptidyl peptidase-IV (DPP-IV) inhibitory activity, which could contribute to its antidiabetic effects. Yu et al. [92] further analyzed the hydrolysis of camel milk proteins using 11 different proteases and assessed their antidiabetic activities through both in vitro and in vivo methods. Camel milk protein hydrolysates (CMPHs), particularly those produced with Flavorzyme, exhibited the highest α-glucosidase and DPP-IV inhibitory activities. CMPH produced with Papain demonstrated the strongest α-amylase inhibitory activity and significant proliferative effects on STZ-induced NIT-1 cells in vitro. In vivo studies revealed that CMPH–Papain was most effective at reducing fasting blood glucose, oral glucose tolerance, proinflammatory factor (IL-1β, IL-6), and triglyceride levels while increasing insulin levels in diabetic mice. Subsequent peptide identification, molecular docking, and network pharmacology analysis identified potential antidiabetic peptides from both CMPH–Flavorzyme and CMPH–Papain. Finally, Su et al. [93] identified antidiabetic peptides from camel milk protein hydrolysates, focusing on their inhibitory activities against α-amylase and DPP-IV. Camel milk proteins were hydrolyzed using Protamex and fractionated into three categories (>10 kDa, 3–10 kDa, and <3 kDa) via ultrafiltration. The <3 kDa fraction was the most effective at inhibiting enzyme activity and was further purified through gel chromatography and identified using LC–MS/MS. Among the 20 potential bioactive peptides identified, the novel peptide QEPVPDPVRGL exhibited the highest antidiabetic activity. Molecular docking revealed that this peptide interacts with α-amylase and DPP-IV through hydrogen bonds, salt bridges, and pi-alkyl interactions, effectively occupying their active sites. In summary, camel milk (CM) and its derivatives exhibit significant potential for managing T2DM through various biochemical pathways, including enzyme inhibition, β-cell function improvement, and metabolic pathway modulation. These findings provide a strong foundation for further research and the development of functional foods and nutraceuticals aimed at diabetes management.

7.3. Lipid-Lowering Activity

Camel milk has garnered attention for its potential lipid-lowering properties, as demonstrated through various studies employing both in vitro and in vivo models. Earlier research by Manaer et al. [91] provided foundational insights into the lipid-regulating effects of fermented camel milk. In their study involving streptozotocin-induced T2DM rats, dietary supplementation with fermented camel milk led to a marked decrease in total cholesterol, triglyceride (TG), and low-density lipoprotein cholesterol (LDL-c) while also improving high-density lipoprotein cholesterol (HDL-c) levels. The regulation of lipid metabolism observed in this study underscores the benefits of fermented camel milk as a dietary intervention for managing dyslipidemia in diabetic conditions. Ming et al. [82] conducted a significant study on the effects of camel milk on lipid accumulation in a mouse model of chronic alcoholic liver disease. Their findings indicated that camel milk effectively prevented lipid build-up in the liver, suggesting its potential as a therapeutic dietary supplement for managing lipid dysregulation associated with chronic alcohol consumption. Further investigations into the lipid-lowering effects of camel milk were carried out by Dou et al. [85], who explored the impact of camel whey protein (CWP) on T2DM rats and insulin-resistant (IR) HepG2 cell models. Their research demonstrated that CWP significantly reversed dyslipidemia in T2DM rats. In the IR-treated HepG2 cells, CWP ameliorated lipid accumulation, which points to its efficacy in improving insulin sensitivity and regulating lipid metabolism. These findings are crucial, given the strong correlation between insulin resistance, T2DM, and dyslipidemia. A study by Wang et al. [87] investigated the lipid-lowering potential of camel milk by investigating the role of bioactive peptides derived from fermented camel milk. Their research revealed that these peptides are instrumental in modulating lipid and atherosclerosis signaling pathways. The peptides' ability to influence these pathways highlights the therapeutic potential of fermented camel milk in managing cardiovascular risk factors and lipid disorders.

Further research is warranted to elucidate the precise biochemical mechanisms involved and to assess the long-term benefits and safety of camel milk consumption in humans.

7.4. The Modulatory Effect of the Intestinal Microbiota

Foods can significantly influence the composition and structure of the intestinal microbiota, with changes in the abundance of certain microbial genera potentially inducing disease or providing health benefits. For example, Wang et al. [94] examined the impact of camel milk on the gut microbiota of mice, revealing a reduction in the relative abundance of *Romboutsia, Lactobacillus, Turicibacter*, and *Desulfovibrio*, while beneficial organisms such as *Allobaculum, Akkermansia*, and *Bifidobacterium* increased. Similarly, a study on mice with nonalcoholic fatty liver disease by Hao et al. [95] demonstrated that camel milk administration could enhance the structure and diversity of the intestinal flora by boosting beneficial bacteria and reducing harmful bacteria. Ming et al. [82] reported that camel milk mitigated intestinal microbial imbalance in a mouse model of chronic alcoholic liver disease. Furthermore, Ming et al. [84] explored the effects of different diets (normal diet; normal diet, then ethanol; and normal diet and camel milk, then ethanol) on mice with alcoholic liver disease (ALD). They found that camel milk (CM) modulates intestinal microbial communities by increasing the proportion of *Lactobacillus* and reducing the proportions of *Bacteroides, Alistipes*, and *Rikenellaceae* RC9. Moreover, researchers have investigated the effects of various thermal treatments of camel milk on the intestinal microbiota of mice. Shao et al. [96] reported that treatment of camel milk at low temperatures (65 °C) increased the relative abundance of probiotic genera such as *Akkermansia, Bifidobacterium*, and *Lactobacillus* while decreasing the overall diversity of the intestinal microbiota.

7.5. Anticancer Activity

The potential anticarcinogenic attributes of camel milk have garnered significant scientific interest, prompting numerous investigations to assess its effects in controlled in vitro experimental settings, as well as in vivo studies conducted on living organisms. A key focus of this research has been the TR35 fraction isolated from camel milk whey, which has shown promising results in the fight against cancer. A notable study by Yang et al. [97] investigated the impact of TR35 on Eca-109 esophageal cancer cells. Researchers have shown that TR35 significantly suppresses the proliferation of these cancer cells and induces apoptosis, which is the process of programmed cell death. In addition to in vitro experiments, the present study included in vivo tests in which mice with xenografted tumors were treated with TR35. Remarkably, TR35 treatment not only limited tumor growth but also did not cause weight loss in the mice, indicating that the treatment was effective and did not induce severe side effects commonly associated with cancer therapies. The benefits of camel milk extend beyond those of TR35 and its effects on esophageal cancer. Another study, conducted by Wang et al. [87], investigated bioactive peptides derived from fermented camel milk. These peptides were obtained from peptidoglycan recognition protein 1 (PGRP1) and were found to play significant roles in cancer-related signaling pathways. This discovery points to the intricate bioactivity of camel milk components and their potential to influence cancer cell behavior at the molecular level. In summary, the anticancer effects of camel milk, particularly the TR35 fraction, have been substantiated by various studies demonstrating its ability to suppress cancer cell proliferation, induce apoptosis, and alter gene and protein expression profiles. Additionally, the impact of camel milk on inflammatory pathways and the bioactivity of its peptides further underscore its potential as a complementary approach in cancer prevention and treatment. As research progresses, camel milk could emerge as a valuable natural supplement in the fight against cancer.

7.6. Antioxidant Activity

Several studies have highlighted the antioxidant capabilities of BC milk. For instance, a study by Chen et al. [88] showed that camel milk administration increased the survival time and rate of radiation-exposed mice, increased the levels of superoxide dismutase (SOD) and glutathione (GSH), and reduced malondialdehyde (MDA) levels, indicating its protective effects against radiation-induced damage. Further research by Ming et al. [82] revealed that camel milk could mitigate oxidative stress in mice with chronic alcoholic liver disease by enhancing the activity of antioxidant enzymes such as SOD and GSH. Similarly, Zhu et al. [83] reported that camel milk supplementation in rats with lipopolysaccharide (LPS)-induced acute respiratory distress syndrome (ARDS) alleviated the expression of oxidative stress markers in lung tissue, suggesting benefits beyond liver health. In diabetic models, camel milk has shown promising results in protecting against oxidative stress. Dou et al. [85] reported that camel milk whey protein (CWP) treatment in rats with type 2 diabetes mellitus (T2DM) induced by streptozotocin (STZ) increased the levels of superoxide dismutase (SOD) and glutathione peroxidase (GSH-Px) while decreasing malondialdehyde (MDA) levels, demonstrating its protective effects against diabetes-induced oxidative damage. A peptidomics study by Zhang et al. [40] quantified endogenous peptides from camel milk and revealed that these peptides possess antioxidative functions, further supporting the role of camel milk in mitigating oxidative stress. These findings collectively highlight the potential of camel milk and its components for regulating oxidative stress and protecting against tissue damage under various pathological conditions, including liver damage associated with T2DM and other oxidative stress-related disorders.

7.7. Antibacterial Activity

Compared to bovine milk, camel milk has slightly higher concentrations of certain bioactive ingredients, such as lactoferrin [98]. Lactoferrin in BC milk exhibits significant antibacterial activity, particularly at a pH range of 7.5–8.0 [99], which suggests potential applications in food preservation and clinical treatments targeting bacterial infections. Wang et al. [100] hydrolyzed camel and cow milk whey protein with trypsin, and both milk whey protein peptide fractions < 3 kDa showed the strongest antibacterial effects against Escherichia coli and Staphylococcus aureus. Camel milk whey peptides showed higher antibacterial activity, and in comparison with amino acid characteristics, both antibacterial peptide fractions exhibited high amounts of alkaline amino acids and hydrophobic amino acids.

8. Technological Properties

Camel milk, a newcomer to the dairy market, holds significant promise as an alternative to traditional bovine milk for meeting the growing global demand for functional food ingredients. Despite its potential for the development of new consumer products, the unique composition of camel milk presents challenges in terms of producing high-quality yogurt and cheese. To fully harness the benefits of this valuable food resource, it is essential to gain a thorough understanding of how different processing parameters—such as pH, heat treatment, high-pressure processing, microwave treatment, and homogenization—affect its physical and functional properties. These parameters play a crucial role in the production of camel milk products. Therefore, advanced processing technologies are necessary to enhance the functional properties of camel milk, unlocking its full potential for a variety of applications.

8.1. Buffering Capacity

The ability of milk to resist changes in pH when acids or alkalis are added or removed is due to its proteins, minerals, and organic acids [101]. This property, known as buffering capacity, varies between BC milk and bovine milk. Studies have shown distinct buffering behaviors in these two types of milk. When acid is added to camel milk, followed by a base, a loop in the pH range of 6.6 to 4.4 is observed, with the maximum buffering index occurring at approximately pH 4.4. Conversely, bovine milk exhibits a similar loop in the pH range of 5.0 to 6.6, with the highest buffering capacity occurring at approximately pH 5.1 [26]. However, different trends appear when the titration order is reversed—base followed by acid—with no loops observed and various buffering pH levels [102]. These findings suggest that camel milk has a stronger buffering capacity than bovine milk, which can be significant in the development of fermented products and human nutrition [103].

8.2. Effect of Heat Treatment

Camel milk, which contains essential proteins, lipids, sugars, vitamins, and minerals, is highly sought after for its rich nutrient profile. However, this nutrient richness also increases susceptibility to microbial spoilage, posing potential health risks. To address this concern, heat treatment is commonly employed to ensure microbiological safety and extend the shelf life of camel milk and its products. Nevertheless, this process can alter the sensory, physical, and chemical properties of milk, as well as degrade some of its functional components [96].

Recent research has revealed valuable insights into the effects of heat treatment on camel milk. While heat treatment may diminish the milk's alcohol stability, it has no appreciable effect on its density, viscosity, refractive index, or pH. Pasteurization at temperatures between 60 and 75 °C did not significantly alter the protein content. However, it does lead to a notable reduction in lactose content and affects the stability of whey proteins at temperatures above 65 °C. Treating milk at 75 °C for 8–15 min can reduce the microbial load by 81.9%, with moderate impacts on protein structure, such as decreased levels of lactoferrin and albumin [98]. A study by Zhang et al. [38] examined the effects of microwave heating on camel and bovine milk and revealed that the heat resistance of whey proteins in camel milk was greater than that of proteins in bovine milk. Microwave heating between 300 and 700 W caused a reduction in κ-casein and the absence of β-lactoglobulin in camel milk, suggesting that microwave heating at 300 W for 2 min is optimal for retaining nutrients and maintaining protein stability. Proteomic analysis revealed significant changes in proteins involved in various biological processes and molecular functions in high-temperature processed liquid camel milk and camel milk powder, highlighting the impact of thermal treatment on protein functions [104]. Heat treatment also affects the formation of furosine (FRS) and 5-hydroxymethylfurfural (5-HMF), which increase significantly with increasing temperature, peaking at 135 °C. A greater degree of heat treatment also increases the aldehyde and ketone contents, affecting the flavor of camel milk [105]. Different heat treatments impact the physicochemical properties of camel milk, which in turn affects the gut microbiota in mice. Ultrahigh-temperature (UHT) treatment significantly reduces nutrient levels and alters gut microbiota diversity, whereas low-temperature, long-time (LTLT) treatment preserves more nutrients without affecting microbiota diversity [106]. Metabolomic analysis revealed significant changes in metabolites, including saccharides, glycosylamines, adenosines, and phospholipids, in raw, heated, and powdered camel milk. Thermal treatment slightly increases D-lactose and significantly increases dipeptides such as His-Pro and Lys-Trp, improving the flavor profile of camel milk [107]. Furthermore, the OS profiles of milk from dairy cows, camels, yaks, sheep, buffaloes, and horses were similar, characterized by a richness of sialylated OSs. Treatment at 65 °C had no significant effect on the concentration or distribution of OSs, whereas 135 °C heating was associated with their decrease, suggesting that more attention to temperature control is needed in milk product processing [108]. In a study by Han et al. [108], changes in the proteins of MFGM in Holstein, buffalo, yak, goat, and camel milk samples following heat treatment were investigated using an LC–MS/MS approach. The results showed that the Holstein, yak, and buffalo milk samples had similar MFGM protein components, followed by the goat and camel milk samples. Changes in lipoprotein lipase and α-lactalbumin in MFGM were dependent on the intensity of the heat treatment and were similar among the studied species, whereas changes in κ-casein, lactoferrin, and apolipoprotein A-I differed among different types of milk [109].

Overall, these studies highlight the critical balance between ensuring microbiological safety through heat treatment and preserving the nutritional and sensory qualities of camel milk. Proper heat treatment is essential to maximize the health benefits of camel milk while maintaining its safety and quality for consumption.

8.3. Coagulation

Milk coagulation, achievable through enzymatic methods [110] or acidic gelation [111], is essential in the production of most dairy products. However, achieving the desired gel consistency in DC milk is challenging due to its unique casein composition, micelle size, and levels of total solids, calcium, and phosphorus [112]. There are few studies on the coagulation of BC in milk. Recent studies have shown promising results with camel milk yogurt, which exhibits high pH, antioxidant activity, and crude protein content, as well as low bacterial counts and titratable acidity. However, the texture remains fragile [21]. Innovations such as the synergistic use of trisodium citrate and microbial transglutaminase are expected to enhance the gelation process by breaking down casein micelles and promoting their crosslinking into a stable gel [113]. Our recent study (unpublished) showed that camel milk yogurt could be produced with good coagulation by the addition of pectin, starch, and sodium caseinate. In the future, milk coagulation, achieved through enzymatic methods or acidic gelation, will continue to be a pivotal step in dairy product development. Advances in understanding the differences in casein composition, micelle size, total solids, calcium, and phosphorus content in dromedary camel milk will help overcome current challenges in achieving desirable gelation.

9. Camel Milk Products

BC milk and its processed products are gaining increasing popularity in China because of their perceived health benefits and unique nutritional composition. While traditionally consumed fresh or fermented, modern processing techniques are enabling a wider range of BC milk products to enter the Chinese market. One of the most important camel milk products is fermented camel milk, known as shubat, which has been a staple in the diets of nomadic communities in northern China for centuries. The traditional stages of shubat processing are shown in Figure 2.

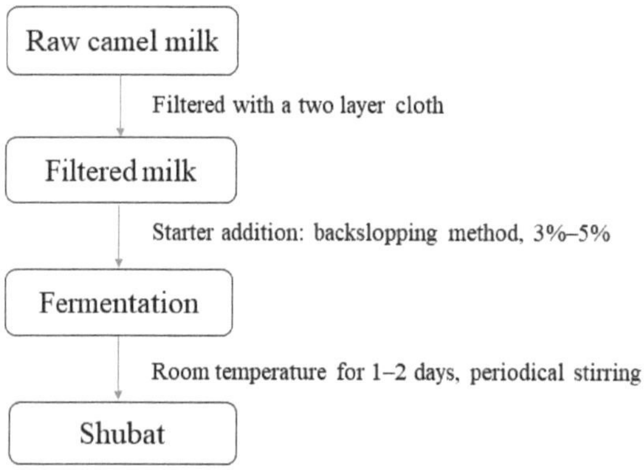

Figure 2. Traditional shubat processing procedure.

This traditional beverage is created through spontaneous fermentation by dominant lactic acid bacteria present in fresh milk [114], a technique passed through generations with minimal changes. The spontaneous fermentation of Bactrian camel milk is characterized by a dynamic shift in microbial community composition over a period of several days at ambient temperature. This process results in a significant transformation of both bacterial and fungal populations. Initially, the bacterial community is dominated by the genus *Lactococcus*, which subsequently gives way to *Lactobacillus* as the predominant genus. Concurrently, the fungal community undergoes a transition from a diverse assemblage comprising *Apiotrichum*, *Cutaneotrichosporon*, and *Candida* to one primarily dominated by *Kazachstania* and *Kluyveromyces*. The culmination of this microbial succession is the production of a distinctive fermented beverage. This end product is notable for its snow-white appearance, sour taste profile, and pleasant flavor characteristics. Moreover, the fermentation process imparts a relatively high viscosity to the final product, further distinguishing it from the original unfermented milk [115]. Recently, this traditionally prepared fermented beverage has gained popularity among all age groups in both rural and urban areas because of its reported nutritional benefits and medicinal properties, including antidiabetic, anticancer, and antituberculosis effects. However, the traditional method of spontaneous fermentation of unheated milk raises concerns about the product's safety and potential quality issues, such as excessive sourness, creaming, and precipitation.

As awareness of these potential benefits has spread, so has the demand for BC milk and its associated products. In China, a growing number of consumers are seeking camel milk as a healthier alternative to traditional dairy products, fueling the expansion of the market and driving innovation within the industry. One notable development has been the emergence of camel milk powder, a product that has facilitated the distribution and consumption of camel milk beyond traditional production areas. By dehydrating the milk and transforming it into a shelf-stable powder, producers have been able to tap into new markets and reach consumers in urban centers across China and beyond. As the market continues to evolve, there is a growing emphasis on value-added products, such as fermented camel milk beverages, dried camel milk products, functional camel milk products, milk-based skincare and cosmetic products, and camel milk infant formula. These products not only cater to the diverse preferences of consumers but also provide opportunities for innovation and product diversification, ensuring the long-term sustainability of the industry.

10. Economic Potential

The economic potential of Bactrian camel (BC) milk in China is set for substantial growth, with conservative estimates forecasting a market value of approximately 2.8 billion USD by 2026. This expanding industry is driven by several critical factors, including increasing consumer awareness of BC milk's unique properties and perceived health benefits, rising disposable incomes, evolving dietary preferences in China, and a focused interest in health and wellness due to BC milk's rich nutritional profile and potential therapeutic properties. Additionally, as consumers seek alternatives to traditional dairy products, BC milk is positioning itself as a premium option.

Despite its current modest market share compared to conventional dairy products, BC milk's niche status presents significant opportunities for growth and product diversification. The industry, however, faces challenges such as limited production capacity, higher price points, and the need for consumer education [116]. These challenges also signal areas for potential improvement and investment.

The development of the BC milk industry in China carries broader economic implications. It could become a significant economic driver for China's northwestern regions, where most Bactrian camels are raised, and foster increased job creation and enhanced rural livelihoods. The industry's growth may assist in preserving traditional herding practices and align with China's goals for sustainable agriculture and food security. Furthermore, ongoing studies into BC milk's nutritional and therapeutic properties are advancing scientific knowledge and potentially opening new market segments. Producers are investing in modernizing production techniques and improving supply chain efficiency to meet growing demand, while government agencies are implementing supportive policies and regulations to foster industry growth, recognizing its potential economic impact [117].

As the market matures, it has the potential not only to meet domestic demand but also to position China as a significant player in the global specialty dairy market. The unique convergence of tradition, culture, and modern innovation within the BC milk industry presents a promising economic opportunity for China, with potential positive impacts across various sectors of the economy.

11. Future Perspectives

BC milk possesses unique nutritional and functional properties, yet a comprehensive understanding of its composition and potential health benefits remains elusive. Extensive characterization of major and minor proteins/peptides through proteomics and peptidomics is crucial for revealing their nutritional and functional attributes. Further research on the fat composition, fatty acid profiles, distribution of different lipid classes, and role of minor lipids such as polar lipids and MFGM components in conferring health benefits is warranted, as is elucidating their cellular and molecular mechanisms. Investigating the effects of various processing techniques, such as thermal, nonthermal, fermentation, and renneting, on the composition, structure, and bioactivity of BC milk has attracted increased amounts of attention. Comprehensive studies elucidating the nutritional makeup and potential health effects of BC milk are pivotal to realizing its full potential in the dairy industry. Innovations in product development, including the use of milk powder, fermented beverages, dried products, and infant formula and the exploration of new value-added functional products, are key areas of focus. The growing Chinese market for BC milk, driven by perceived health benefits and demand for alternative dairy sources, presents opportunities for economic growth through sustainable production and commercialization of this unique dairy stream. Extensive multidimensional research coupled with product innovations tailored to consumer demands can unlock the tremendous untapped potential of BC milk as a nutritious and therapeutic food resource for the expanding Chinese market.

12. Conclusions

This comprehensive review highlights the unique nutritional profile and bioactive properties of Bactrian camel (BC) milk, positioning it as a promising functional food with significant potential in the Chinese market. The composition of camel milk, characterized by high protein content, distinctive fatty acid profile, and bioactive compounds, contributes to its various health benefits, including anti-inflammatory, antidiabetic, lipid-lowering, and anticancer properties. The modulatory effects on intestinal microbiota and antioxidant activities further underscore its potential as a therapeutic food.

However, the review also reveals several challenges and limitations in the current state of research. While numerous studies have explored the composition and bioactivities of BC milk, there remains a need for more comprehensive, standardized analyses across different breeds and geographical locations. The technological properties of BC milk, particularly in relation to product development, require further investigation to overcome challenges in coagulation and texture formation.

In conclusion, while BC milk shows great promise as a functional food, continued research and innovation are necessary to fully harness its potential and address current limitations in understanding and application.

Author Contributions: Conceptualization, review collection, writing and funding acquisition, S.S.; correction, editing, G.Y.; conceptualization, review collection and writing, A.K. All authors have read and agreed to the published version of the manuscript.

Funding: This work was funded by Xinjiang University Cultivation Program (Research on the High-quality Development Mode and Countermeasures of Featured Animal Husbandry in Northern Xinjiang under the Background of Rural Revitalization, No. 22CPY053).

Institutional Review Board Statement: Not applicable.

Informed Consent Statement: Not applicable.

Data Availability Statement: Data are contained within the article.

Acknowledgments: The authors are grateful for the financial support and encouragement from Xinjiang University.

Conflicts of Interest: The authors declare no conflicts of interest. The funders had no role in the design of the study; in the collection, analyses, or interpretation of the data; in the writing of the manuscript; or in the decision to publish the results.

References

1. Ming, L.; Li, Y.; Lu, H.; Hosblig, L.Y. Chemical Composition and Proteomics of Camel Milk at Different Lactation Stages. *Food Sci.* **2024**, *45*, 205–211.
2. Farah, Z. *Camel Milk: Properties and Products*; Swiss Centre for Development Cooperation in Technology and Management: St. Gallen, Switzerland, 1996; ISBN 3908001528.
3. Mohandesan, E.; Speller, C.F.; Peters, J.; Uerpmann, H.P.; Uerpmann, M.; De Cupere, B.; Hofreiter, M.; Burger, P.A. Combined Hybridization Capture and Shotgun Sequencing for Ancient DNA Analysis of Extinct Wild and Domestic Dromedary Camel. *Mol. Ecol. Resour.* **2017**, *17*, 300–313. [CrossRef] [PubMed]
4. Wu, H.; Guang, X.; Al-Fageeh, M.B.; Cao, J.; Pan, S.; Zhou, H.; Zhang, L.; Abutarboush, M.H.; Xing, Y.; Xie, Z.; et al. Camelid Genomes Reveal Evolution and Adaptation to Desert Environments. *Nat. Commun.* **2014**, *5*, 5188. [CrossRef] [PubMed]
5. Faye, B. The Camel, New Challenges for a Sustainable Development. *Trop. Anim. Health Prod.* **2016**, *48*, 689–692. [CrossRef]
6. Farah, Z.; Mollet, M.; Younan, M.; Dahir, R. Camel Dairy in Somalia: Limiting Factors and Development Potential. *Livest. Sci.* **2007**, *110*, 187–191. [CrossRef]
7. FAOSTAT Live Animals (Data). Available online: https://www.fao.org/faostat/en/#data/QCL (accessed on 6 June 2024).
8. Li, Y.; Liu, L.; Fan, Y.; Liu, W.; Yang, Q.; Wen, W.; Li, W.; Hao, L.; Zou, H.; Xieermaola, Y.; et al. Quality Characteristics and Nutrient Contents of Bactrian Camel Milk as Determined by Mid-Infrared Spectroscopy. *Int. J. Dairy Technol.* **2024**, *77*, 304–312. [CrossRef]
9. Jiang, H.; Xu, Y.; Chen, G.; Liu, T.; Yang, Y.; Mao, X. Digestive Properties and Peptide Profiles Exhibited Significant Differences between Skim Camel Milk and Bovine Milk Powder after Static in Vitro Simulated Infant Gastrointestinal Digestion. *Food Res. Int.* **2023**, *178*, 113860. [CrossRef]
10. Han, B.; Zhang, L.; Zhou, P. Comparison of Milk Fat Globule Membrane Protein Profile among Bovine, Goat and Camel Milk Based on Label Free Proteomic Techniques. *Food Res. Int.* **2022**, *162*, 112097. [CrossRef]
11. Han, B.; Zhang, L.; Luo, B.; Ni, Y.; Bansal, N.; Zhou, P. Comparison of Milk Fat Globule Membrane and Whey Proteome between Dromedary and Bactrian Camel. *Food Chem.* **2022**, *367*, 130658. [CrossRef]
12. NBS of China Camel Population. Available online: https://www.stats.gov.cn/sj/ndsj/2023/indexch.htm (accessed on 7 June 2024).
13. Ministry of Agriculture and Rural Affairs of China; National List of Animal Genetic Resources and Varieties. Available online: http://www.moa.gov.cn/govpublic/nybzzj1/202101/t20210114_6359937.htm (accessed on 12 March 2024).
14. Miao, J.; Xiao, S.; Wang, J. Comparative Study of Camel Milk from Different Areas of Xinjiang Province in China. *Food Sci. Anim. Resour.* **2023**, *43*, 674–684. [CrossRef] [PubMed]

15. Faye, B.; Konuspayeva, G.; Messad, S.; Loiseau, G. Discriminant Milk Components of Bactrian Camel (*Camelus bactrianus*), Dromedary (*Camelus dromedarius*) and Hybrids. *Dairy Sci. Technol.* **2008**, *88*, 607–617. [CrossRef]
16. Nurseitova, M.; Konuspayeva, G.; Jurjanz, S. Comparison of Dairy Performances between Dromedaries, Bactrian and Crossbred Camels in the Conditions of South Kazakhstan. *Emirates J. Food Agric.* **2014**, *26*, 366–370. [CrossRef]
17. Yao, H.; Dou, Z.; Zhao, Z.; Liang, X.; Yue, H.; Ma, W.; Su, Z.; Wang, Y.; Hao, Z.; Yan, H.; et al. Transcriptome Analysis of the Bactrian Camel (*Camelus bactrianus*) Reveals Candidate Genes Affecting Milk Production Traits. *BMC Genom.* **2023**, *24*, 660. [CrossRef] [PubMed]
18. Luo, X.; Ma, W.; Lu, D.; Wang, W.; Li, X.; Xu, M. Determination and Analysis of the Chemical Components and Physicochemical Indexes of Xinjiang Bactrian Camel Milk. *Grass-Feed. Livest.* **2014**, *5*, 79–82. [CrossRef]
19. Xiao, Y.; Yi, L.; Ming, L.; He, J.; Ji, R. Changes in Milk Components, Amino Acids, and Fatty Acids of Bactrian Camels in Different Lactation Periods. *Int. Dairy J.* **2022**, *131*, 105363. [CrossRef]
20. Zhang, M.; Lu, D.; Dong, J.; Chen, G.; Huang, X. The Analysis of Chemical Composition and Physico-Chemical Indexes of the Junggar Bactrian Camel Milk. *China Dairy* **2016**, *8*, 52–55. [CrossRef]
21. Wang, L.; Wu, T.; Zhang, Y.; Yang, K.; He, Y.; Deng, K.; Liang, C.; Gu, Y. Comparative Studies on the Nutritional and Physicochemical Properties of Yoghurts from Cows', Goats', and Camels' Milk Powder. *Int. Dairy J.* **2023**, *138*, 105542. [CrossRef]
22. Ryskaliyeva, A.; Henry, C.; Miranda, G.; Faye, B.; Konuspayeva, G.; Martin, P. Combining Different Proteomic Approaches to Resolve Complexity of the Milk Protein Fraction of Dromedary, Bactrian Camels and Hybrids, from Different Regions of Kazakhstan. *PLoS ONE* **2018**, *13*, e0197026. [CrossRef] [PubMed]
23. Zhang, H.; Yao, J.; Zhao, D.; Liu, H.; Li, J.; Guo, M. Changes in Chemical Composition of Alxa Bactrian Camel Milk during Lactation. *J. Dairy Sci.* **2005**, *88*, 3402–3410. [CrossRef]
24. Wang, H.; Zhang, M.; Huo, Y.; Cui, X.; He, R.; Han, B.; Wang, Z.; Song, Y.; Lv, X.; Zhang, J.; et al. Comprehensive Investigation of Milk Oligosaccharides in Different Mammalian Species and the Effect of Breed and Lactation Period on Sheep Milk Oligosaccharides. *Food Res. Int.* **2023**, *172*, 113132. [CrossRef]
25. Chen, B.; Zhu, H.; Zhang, Y.; Wang, X.; Zhang, W.; Wang, Y.; Pang, X.; Zhang, S.; Lv, J. Comparison of Species and Lactation of Different Mammalian Milk: The Unique Composition and Stereospecificity of Fatty Acids of Mare Milk. *Int. Dairy J.* **2024**, *150*, 105822. [CrossRef]
26. Zhao, D.; Mutu, J.; Liu, H.; Shao, Y.; Zhang, H. Study on the Physicochemical Property of Alxa Bactrian Camel Milk. *J. Dairy Sci. Technol.* **2005**, *3*, 112–117. [CrossRef]
27. Zhao, D.B.; Bai, Y.H.; Niu, Y.W. Composition and Characteristics of Chinese Bactrian Camel Milk. *Small Rumin. Res.* **2015**, *127*, 58–67. [CrossRef]
28. Chen, Z.; Feng, D.; Wang, X.; Wu, H.; Hua, E.; Xu, X.; Ye, F.; Li, L. Study on the Changes of Milk Composition of Alashan Bactrian Camel during Different Season. *Mod. Anim. Husb. Vet. Med.* **2023**, *5*, 10–14.
29. Jiangong, G. Studies on Nutritional Ingredient and Function of Active Substances In Sonid Camel Milk. Ph.D. Thesis, Inner Mongolia Agricultural University, Hohhot, China, 2009.
30. Zhang, M.; Chen, G.; Zang, C.; Dong, J.; Huang, X.; Lu, D. Analysis of Physicochemical Indexes of Junggar Bactrian Camel Milk. *China Anim. Husb. Vet. Med.* **2016**, *43*, 2628–2633.
31. Song, W.; Chen, Z.; Li, X.; Chen, L.; Shu, Z.; Li, Z.; Guo, X.; Ye, Q.; Ma, X.; Su, Z.; et al. A Comparative Analysis of Milk Nutritional and Functional Components between Junggar Bactrian Camel and Holstein Cow. *Xinjiang Agric. Sci.* **2023**, *60*, 234–241.
32. Jiang, X.; Lazat, A.; Chen, G.; Zhang, Y. Comparative Analysis of Chemical Composition and Nutritional Value of Xinjiang Junggar Bactrian Camel Milk. *Herbiv. Livest.* **2016**, *5*, 35–40.
33. Ming, L.; Zheng, Z.; Guo, C.E. The Study of Milk Nutrients Dynamic Changes of Zhungeer Camel during the Lacation Period. *J. Chin. Inst. Food Sci. Technol.* **2015**, *15*, 240–245.
34. Amr, M.; Mohie-Eldinn, M.; Farid, A. Evaluation of Buffalo, Cow, Goat and Camel Milk Consumption on Multiple Health Outcomes in Male and Female Sprague Dawley Rats. *Int. Dairy J.* **2023**, *146*, 105760. [CrossRef]
35. Amalfitano, N.; Patel, N.; Haddi, M.-L.; Benabid, H.; Pazzola, M.; Vacca, G.M.; Tagliapietra, F.; Schiavon, S.; Bittante, G. Detailed Mineral Profile of Milk, Whey, and Cheese from Cows, Buffaloes, Goats, Ewes and Dromedary Camels, and Efficiency of Recovery of Minerals in Their Cheese. *J. Dairy Sci.* 2024, *in press*. [CrossRef] [PubMed]
36. Haddadin, M.S.Y.; Gammoh, S.I.; Robinson, R.K. Seasonal Variations in the Chemical Composition of Camel Milk in Jordan. *J. Dairy Res.* **2008**, *75*, 8–12. [CrossRef]
37. Anusha Siddiqui, S.; Mahmood Salman, S.H.; Ali Redha, A.; Zannou, O.; Chabi, I.B.; Oussou, K.F.; Bhowmik, S.; Nirmal, N.P.; Maqsood, S. Physicochemical and Nutritional Properties of Different Non-Bovine Milk and Dairy Products: A Review. *Int. Dairy J.* **2024**, *148*, 105790. [CrossRef]
38. Zhang, J.; Wang, S.; Lu, Q.; Kong, L.; Ge, W. Effect of Microwave Heating on Physicochemical Properties, Protein Composition and Structure, and Micromorphology of Camel and Bovine Milk Samples. *J. Food Compos. Anal.* **2023**, *122*, 105468. [CrossRef]
39. He, J.; Si, R.; Wang, Y.; Ji, R.; Ming, L. Lipidomic and Proteomic Profiling Identifies the Milk Fat Globule Membrane Composition of Milk from Cows and Camels. *Food Res. Int.* **2024**, *179*, 113816. [CrossRef] [PubMed]
40. Zhang, L.; Han, B.; Luo, B.; Ni, Y.; Bansal, N.; Zhou, P. Characterization of Endogenous Peptides from Dromedary and Bactrian Camel Milk. *Eur. Food Res. Technol.* **2022**, *248*, 1149–1160. [CrossRef]

41. Xiao, T.; Zeng, J.; Zhao, C.; Hou, Y.; Wu, T.; Deng, Z.; Zheng, L. Comparative Analysis of Protein Digestion Characteristics in Human, Cow, Goat, Sheep, Mare, and Camel Milk under Simulated Infant Condition. *J. Agric. Food Chem.* **2023**, *71*, 15035–15047. [CrossRef] [PubMed]
42. Wang, X.; Zhao, X.; Huang, D.; Pan, X.; Qi, Y.; Yang, Y.; Zhao, H.; Cheng, G. Proteomic Analysis and Cross Species Comparison of Casein Fractions from the Milk of Dairy Animals. *Sci. Rep.* **2017**, *7*, 43020. [CrossRef]
43. Kappeler, S.; Farah, Z.; Puhan, Z. Sequence Analysis of *Camelus dromedarius* Milk Caseins. *J. Dairy Res.* **1998**, *65*, 209–222. [CrossRef]
44. Pauciullo, A.; Giambra, I.J.; Iannuzzi, L.; Erhardt, G. The β-Casein in Camels: Molecular Characterization of the CSN2 Gene, Promoter Analysis and Genetic Variability. *Gene* **2014**, *547*, 159–168. [CrossRef]
45. Guo, Y.; Ming, L.; Yi, L.; Yi, R.; Gao, W. Purification and Thermal Denaturation Kinetics of Serum Albumin in Bactrian Camel Milk. *J. Camel Pract. Res.* **2018**, *25*, 99–108. [CrossRef]
46. Yang, Y.; Bu, D.; Zhao, X.; Sun, P.; Wang, J.; Zhou, L. Proteomic Analysis of Cow, Yak, Buffalo, Goat and Camel Milk Whey Proteins: Quantitative Differential Expression Patterns. *J. Proteome Res.* **2013**, *12*, 1660–1667. [CrossRef] [PubMed]
47. Hinz, K.; O'Connor, P.M.; Huppertz, T.; Ross, R.P.; Kelly, A.L. Comparison of the Principal Proteins in Bovine, Caprine, Buffalo, Equine and Camel Milk. *J. Dairy Res.* **2012**, *79*, 185–191. [CrossRef] [PubMed]
48. Yao, H.; Zhang, M.; Li, Y.; Yao, J.; Meng, H.; Yu, S. Purification and Quantification of Heavy-Chain Antibodies from the Milk of Bactrian Camels. *Anim. Sci. J.* **2017**, *88*, 1446–1450. [CrossRef] [PubMed]
49. Gao, D.; Mu, Q.; Liu, L.; Guo, J. Determination and Comparison on Protein and Amino Acids Profile of Four Kinds of Livestock Milk in Inner Mongolian. *Food Sci.* **2017**, *42*, 267–272. [CrossRef]
50. Zhao, L.; Zhang, J.; Ge, W.; Wang, J. Comparative Lipidomics Analysis of Human and Ruminant Milk Reveals Variation in Composition and Structural Characteristics. *J. Agric. Food Chem.* **2022**, *70*, 8994–9006. [CrossRef] [PubMed]
51. Zou, X.; Huang, J.; Jin, Q.; Guo, Z.; Liu, Y.; Cheong, L.; Xu, X.; Wang, X. Lipid Composition Analysis of Milk Fats from Different Mammalian Species: Potential for Use as Human Milk Fat Substitutes. *J. Agric. Food Chem.* **2013**, *61*, 7070–7080. [CrossRef]
52. He, J.; Wang, D.; Guo, K.; Ji, R. Camel Milk Polar Lipids Ameliorate Dextran Sulfate Sodium (DSS)-Induced Colitis in Mice by Modulating the Gut Microbiota. *J. Dairy Sci.* **2024**, *107*, 6413–6424. [CrossRef]
53. Pawar, A.; Zabetakis, I.; Gavankar, T.; Lordan, R. Milk Polar Lipids: Untapped Potential for Pharmaceuticals and Nutraceuticals. *PharmaNutrition* **2023**, *24*, 100335. [CrossRef]
54. Jiang, C.; Zhang, X.; Yu, J.; Yuan, T.; Zhao, P.; Tao, G.; Wei, W.; Wang, X. Comprehensive Lipidomic Analysis of Milk Polar Lipids Using Ultraperformance Supercritical Fluid Chromatography-Mass Spectrometry. *Food Chem.* **2022**, *393*, 133336. [CrossRef]
55. Wang, M.; Zhang, F.; Fan, J.; Yu, W.; Yuan, Q.; Hou, H.; Du, Z. Quantitative Phospholipidomics and Screening for Significantly Different Phospholipids in Human Colostrum and Milk, and Dairy Animal Colostrum. *Int. Dairy J.* **2023**, *146*, 105741. [CrossRef]
56. Xu, M.; Lu, D.; Luo, X.; Li, X.; Wang, W.; Ma, W. Analysis of the Composition and Property of Xinjiang Bactrian Camel Milk. *China Dairy Cattle* **2014**, *15*, 49–52.
57. Lu, D.; Ye, D.; Li, J.; Xu, M. Comparative on the Concentration and the Chemical Composition of the Fatty Acid in Milk Powder of 4 Kinds of Livestock in Xinjiang. *Grass-Feed. Livest.* **2017**, *4*, 7–14.
58. Teng, F.; Wang, P.; Yang, L.; Ma, Y.; Day, L. Quantification of Fatty Acids in Human, Cow, Buffalo, Goat, Yak, and Camel Milk Using an Improved One-Step GC-FID Method. *Food Anal. Methods* **2017**, *10*, 2881–2891. [CrossRef]
59. Li, Y.I.; Zhi-qiang, Z.; Liang, M. Study on Seasonal Variation of Fatty Acid Composition of Bactrain Camel's Milk in Zhunger. *China Dairy Ind.* **2014**, *42*, 18–21.
60. Bakry, I.A.; Ali, A.H.; Abdeen, E.-S.M.; Ghazal, A.F.; Wei, W.; Wang, X. Comparative Characterisation of Fat Fractions Extracted from Egyptian and Chinese Camel Milk. *Int. Dairy J.* **2020**, *105*, 104691. [CrossRef]
61. Wang, F.; Chen, M.; Luo, R.; Huang, G.; Wu, X.; Zheng, N.; Zhang, Y.; Wang, J. Fatty Acid Profiles of Milk from Holstein Cows, Jersey Cows, Buffalos, Yaks, Humans, Goats, Camels, and Donkeys Based on Gas Chromatography–Mass Spectrometry. *J. Dairy Sci.* **2022**, *105*, 1687–1700. [CrossRef]
62. Ran-Ressler, R.R.R.; Khailova, L.; Arganbright, K.K.M.; Adkins-Rieck, C.K.; Jouni, Z.E.; Koren, O.; Ley, R.E.; Brenna, J.T.; Dvorak, B.; Egge, H.; et al. Branched Chain Fatty Acids Reduce the Incidence of Necrotizing Enterocolitis and Alter Gastrointestinal Microbial Ecology in a Neonatal Rat Model. *PLoS ONE* **2011**, *6*, e29032. [CrossRef] [PubMed]
63. Yang, J.; Zheng, N.; Wang, J.; Yang, Y. Comparative Milk Fatty Acid Analysis of Different Dairy Species. *Int. J. Dairy Technol.* **2017**, *71*, 12443. [CrossRef]
64. Gu, X.; Guo, J.; Sha-Sha, L.I. Comparison on Milk Fat Fatty Acids Profile of Inner Mongolian Cow, Horse and Bactrian Camel. *China Dairy Ind.* **2016**, *44*, 16–19.
65. Liu, B.; Liang, Y.-H.; He, Y.-Z.; Ye, W.; Deng, Z.-Y.; Li, J.; Guo, S. Differences in Fat Digestion from Milk of Different Species: In Vitro Gastrointestinal Digestion Model for Infants. *Food Res. Int.* **2023**, *174*, 113571. [CrossRef] [PubMed]
66. Wooding, F.B.P.; Schwab, A.J.; Werner, S.; Rajagopalan, K.V.; Franke, W.W.; Zulak, I.M.; Bushway, A.A. The Structure of the Milk Fat Globule Membrane. *J. Ultrastruct. Res.* **1971**, *37*, 388–400. [CrossRef] [PubMed]
67. Keenan, T.W. Milk Lipid Globules and Their Surrounding Membrane: A Brief History and Perspectives for Future Research. *J. Mammary Gland. Biol. Neoplasia* **2001**, *6*, 365–371. [CrossRef] [PubMed]
68. Argov, N.; Lemay, D.G.; German, J.B. Milk Fat Globule Structure and Function: Nanoscience Comes to Milk Production. *Trends Food Sci. Technol.* **2008**, *19*, 617–623. [CrossRef] [PubMed]

69. Ye, A.; Singh, H.; Oldfield, D.J.; Anema, S. Kinetics of Heat-Induced Association of β-Lactoglobulin and α-Lactalbumin with Milk Fat Globule Membrane in Whole Milk. *Int. Dairy J.* **2004**, *14*, 389–398. [CrossRef]
70. Sakkas, L.; Moutafi, A.; Moschopoulou, E.; Moatsou, G. Assessment of Heat Treatment of Various Types of Milk. *Food Chem.* **2014**, *159*, 293–301. [CrossRef] [PubMed]
71. Sharma, P.; Oey, I.; Everett, D.W. Interfacial Properties and Transmission Electron Microscopy Revealing Damage to the Milk Fat Globule System after Pulsed Electric Field Treatment. *Food Hydrocoll.* **2015**, *47*, 99–107. [CrossRef]
72. Li, T.; Gao, J.; Du, M.; Mao, X. Milk Fat Globule Membrane Supplementation Modulates the Gut Microbiota and Attenuates Metabolic Endotoxemia in High-Fat Diet-Fed Mice. *J. Funct. Foods* **2018**, *47*, 56–65. [CrossRef]
73. Yang, Y.; Zheng, N.; Zhao, X.; Zhang, Y.; Han, R.; Ma, L.; Zhao, S.; Li, S.; Guo, T.; Wang, J. Proteomic Characterization and Comparison of Mammalian Milk Fat Globule Proteomes by ITRAQ Analysis. *J. Proteom.* **2015**, *116*, 34–43. [CrossRef]
74. Ji, X.; Ma, Y. Compositions and Properties of Milk Fat Globule Membrane from Five Different Species Milk. *China Dairy Ind.* **2017**, *45*, 19–23.
75. Lopez, C.; Ménard, O. Human Milk Fat Globules: Polar Lipid Composition and in Situ Structural Investigations Revealing the Heterogeneous Distribution of Proteins and the Lateral Segregation of Sphingomyelin in the Biological Membrane. *Colloids Surf. B Biointerfaces* **2011**, *83*, 29–41. [CrossRef]
76. Yan, J.; Ding, J.; Jin, G.; Yu, D.; Yu, L.; Long, Z.; Guo, Z.; Chai, W.; Liang, X. Profiling of Sialylated Oligosaccharides in Mammalian Milk Using Online Solid Phase Extraction-Hydrophilic Interaction Chromatography Coupled with Negative-Ion Electrospray Mass Spectrometry. *Anal. Chem.* **2018**, *90*, 3174–3182. [CrossRef] [PubMed]
77. Shi, Y.; Han, B.; Zhang, L.; Zhou, P. Comprehensive Identification and Absolute Quantification of Milk Oligosaccharides in Different Species. *J. Agric. Food Chem.* **2021**, *69*, 15585–15597. [CrossRef] [PubMed]
78. Zhang, L.; Lin, Q.; Zhang, J.; Shi, Y.; Pan, L.; Hou, Y.; Peng, X.; Li, W.; Wang, J.; Zhou, P. Qualitative and Quantitative Changes of Oligosaccharides in Human and Animal Milk over Lactation. *J. Agric. Food Chem.* **2023**, *71*, 15553–15568. [CrossRef]
79. Xu, M.; Lu, D.; Ma, W.; Wang, W.; Qian, H. Measure on the Mass Concentration of Minerals and Vitamins in Xinjiang Bactrian Camel Milk. *Grass-Feed. Livest.* **2014**, *4*, 68–71. [CrossRef]
80. Chen, L.; Li, X.; Li, Z.; Deng, L. Analysis of 17 Elements in Cow, Goat, Buffalo, Yak, and Camel Milk by Inductively Coupled Plasma Mass Spectrometry (ICP-MS). *RSC Adv.* **2020**, *10*, 6736–6742. [CrossRef]
81. Xu, W.; Dong, Q.; Zhao, G.; Han, B. Analysis of Metabolites of Bactrain Camel Milk in Alxa of China before and after Fermentation with Fermenting Agent TR1 Based on Untargeted LC-MS/MS Based Metabolomics. *Heliyon* **2023**, *9*, e18522. [CrossRef]
82. Ming, L.; Qi, B.; Hao, S.; Ji, R. Camel Milk Ameliorates Inflammatory Mechanisms in an Alcohol-Induced Liver Injury Mouse Model. *Sci. Rep.* **2021**, *11*, 22811. [CrossRef]
83. Zhu, W.W.; Kong, G.Q.; Ma, M.M.; Li, Y.; Huang, X.; Wang, L.P.; Peng, Z.Y.; Zhang, X.H.; Liu, X.Y.; Wang, X.Z. Short Communication: Camel Milk Ameliorates Inflammatory Responses and Oxidative Stress and Downregulates Mitogen-Activated Protein Kinase Signaling Pathways in Lipopolysaccharide-Induced Acute Respiratory Distress Syndrome in Rats. *J. Dairy Sci.* **2016**, *99*, 53–56. [CrossRef]
84. Ming, L.; Qiao, X.Y.; Yi, L.; Siren, D.; He, J.; Hai, L.; Guo, F.; Xiao, Y.; Ji, R. Camel Milk Modulates Ethanol-Induced Changes in the Gut Microbiome and Transcriptome in a Mouse Model of Acute Alcoholic Liver Disease. *J. Dairy Sci.* **2020**, *103*, 3937–3949. [CrossRef]
85. Dou, Z.; Liu, C.; Feng, X.; Xie, Y.; Yue, H.; Dong, J.; Zhao, Z.; Chen, G.; Yang, J. Camel Whey Protein (CWP) Ameliorates Liver Injury in Type 2 Diabetes Mellitus Rats and Insulin Resistance (IR) in HepG2 Cells: Via Activation of the PI3K/Akt Signaling Pathway. *Food Funct.* **2022**, *13*, 255–269. [CrossRef]
86. Du, D.; Lv, W.; Su, R.; Yu, C.; Jing, X.; Bai, N.; Hasi, S. Hydrolyzed Camel Whey Protein Alleviated Heat Stress-Induced Hepatocyte Damage by Activated Nrf2/HO-1 Signaling Pathway and Inhibited NF-KB/NLRP3 Axis. *Cell Stress Chaperones* **2021**, *26*, 387–401. [CrossRef] [PubMed]
87. Wang, Y.; Liang, Z.; Shen, F.; Zhou, W.; Manaer, T.; Jiaerken, D.; Nabi, X. Exploring the Immunomodulatory Effects and Mechanisms of Xinjiang Fermented Camel Milk-Derived Bioactive Peptides Based on Network Pharmacology and Molecular Docking. *Front. Pharmacol.* **2023**, *13*, 1038812. [CrossRef] [PubMed]
88. Chen, Y.Z.; Li, C.; Gu, J.; Lv, S.C.; Song, J.Y.; Tang, Z.B.; Duan, G.X.; Qin, L.Q.; Zhao, L.; Xu, J.Y. Anti-Oxidative and Immuno-Protective Effect of Camel Milk on Radiation-Induced Intestinal Injury in C57BL/6 J Mice. *Dose-Response* **2021**, *19*, 1–8. [CrossRef] [PubMed]
89. Zheng, Y.; Wu, F.; Zhang, M.; Fang, B.; Zhao, L.; Dong, L.; Zhou, X.; Ge, S. Hypoglycemic Effect of Camel Milk Powder in Type 2 Diabetic Patients: A Randomized, Double-Blind, Placebo-Controlled Trial. *Food Sci. Nutr.* **2021**, *9*, 4461–4472. [CrossRef] [PubMed]
90. Han, B.; Zhang, L.; Hou, Y.; Zhong, J.; Hettinga, K.; Zhou, P. Phosphoproteomics Reveals That Camel and Goat Milk Improve Glucose Homeostasis in HDF/STZ-Induced Diabetic Rats through Activation of Hepatic AMPK and GSK3-GYS Axis. *Food Res. Int.* **2022**, *157*, 111254. [CrossRef] [PubMed]
91. Manaer, T.; Yu, L.; Zhang, Y.; Xiao, X.J.; Nabi, X.H. Anti-Diabetic Effects of Shubat in Type 2 Diabetic Rats Induced by Combination of High-Glucose-Fat Diet and Low-Dose Streptozotocin. *J. Ethnopharmacol.* **2015**, *169*, 269–274. [CrossRef] [PubMed]

92. Yu, Y.; Sun, P.; Liu, Y.; Zhao, W.-L.; Wang, T.-J.; Yu, S.-X.; Tian, L.-K.; Zhao, L.; Zhang, M.-M.; Zhang, Q.-Y.; et al. Characterization and Evaluation of the in Vitro and in Vivo Anti-Diabetic Activities of Camel Milk Protein Hydrolysates Derived with Different Protease Digestions. *J. Funct. Foods* **2024**, *117*, 106227. [CrossRef]
93. Su, N.; Yi, L.; He, J.; Ming, L.; Jambal, T.; Mijiddorj, B.; Maizul, B.; Enkhtuul, T.; Ji, R. Identification and Molecular Docking of a Novel Antidiabetic Peptide from Protamex-Camel Milk Protein Hydrolysates against α-Amylase and DPP-IV. *Int. Dairy J.* **2024**, *152*, 105884. [CrossRef]
94. Wang, Z.; Zhang, W.; Wang, B.; Zhang, F.; Shao, Y. Influence of Bactrian Camel Milk on the Gut Microbiota. *J. Dairy Sci.* **2018**, *101*, 5758–5769. [CrossRef] [PubMed]
95. Hao, S.; Ming, L.; Li, Y.; Lv, H.; Li, L.; Jambal, T.; Ji, R. Modulatory Effect of Camel Milk on Intestinal Microbiota of Mice with Non-Alcoholic Fatty Liver Disease. *Front. Nutr.* **2022**, *9*, 1072133. [CrossRef]
96. Shao, Y.; Wang, Z. Changes in the Nutrients of Camels' Milk Alter the Functional Features of the Intestine Microbiota. *Food Funct.* **2018**, *9*, 6484–6494. [CrossRef] [PubMed]
97. Yang, J.; Dou, Z.; Peng, X.; Wang, H.; Shen, T.; Liu, J.; Li, G.; Gao, Y. Transcriptomics and Proteomics Analyses of Anti-Cancer Mechanisms of TR35–An Active Fraction from Xinjiang Bactrian Camel Milk in Esophageal Carcinoma Cell. *Clin. Nutr.* **2019**, *38*, 2349–2359. [CrossRef] [PubMed]
98. Yang, Y.; Ye, L. Selection of Sterilization Conditions for Xinjiang Bactrian Camel Milk and Its Thermal Stability. *Food Sci.* **2013**, *34*, 36–41. [CrossRef]
99. Zhao, R.G.T.; Bao, H.J.; Ha, S.S.R. Study on Antibacterial Activity and Its Influential Factor of Lactoferrin in Bactrian Camel Milk in Vitro. Progress of Veterinary. *Med. China* **2007**, *28*, 23–26.
100. Wang, R.; Han, Z.; Ji, R.; Xiao, Y.; Si, R.; Guo, F.; He, J.; Hai, L.; Ming, L.; Yi, L. Antibacterial Activity of Trypsin-Hydrolyzed Camel and Cow Whey and Their Fractions. *Animals* **2020**, *10*, 337. [CrossRef]
101. Ismail, A.A.; Deeb, S.A.; Difrawi, E.A. The Buffering Properties of Cow and Buffalo Milks. *Z. Leb. Unters.-Forsch.* **1973**, *152*, 25–31. [CrossRef]
102. Bai, Y.-H.; Zhao, D.-B. The Acid-Base Buffering Properties of Alxa Bactrian Camel Milk. *Small Rumin. Res.* **2015**, *123*, 287–292. [CrossRef]
103. Sukhov, S.V.; I Kalamkarova, L.; A Il'Chenko, L.; Zhangabylov, A.K. Microfloral Changes in the Small and Large Intestines of Chronic Enteritis Patients on Diet Therapy Including Sour Milk Products. *Vopr. Pitan.* **1986**, *4*, 14–17.
104. Li, R.R.; Yue, H.T.; Shi, Z.Y.; Shen, T.; Yao, H.B.; Zhang, J.W.; Gao, Y.; Yang, J. Protein Profile of Whole Camel Milk Resulting from Commercial Thermal Treatment. *LWT* **2020**, *134*, 110256. [CrossRef]
105. Zhao, X.; Guo, Y.; Zhang, Y.; Pang, X.; Wang, Y.; Lv, J.; Zhang, S. Effects of Different Heat Treatments on Maillard Reaction Products and Volatile Substances of Camel Milk. *Front. Nutr.* **2023**, *10*, 1072261. [CrossRef]
106. He, J.; Sun, R.; Hao, X.; Battulga, A.; Juramt, N.; Yi, L.; Ming, L.; Rimutu, J. The Gut Microbiota and Its Metabolites in Mice Are Affected by High Heat Treatment of Bactrian Camel Milk. *J. Dairy Sci.* **2020**, *103*, 11178–11189. [CrossRef] [PubMed]
107. Li, R.; Wang, S.; Zhang, J.; Miao, J.; Chen, G.; Dong, J.; Wu, L.; Yue, H.; Yang, J. Untargeted Metabolomics Allows to Discriminate Raw Camel Milk, Heated Camel Milk, and Camel Milk Powder. *Int. Dairy J.* **2022**, *124*, 105140. [CrossRef]
108. Yao, Q.; Gao, Y.; Wang, F.; Delcenserie, V.; Wang, J.; Zheng, N. Label-Free Quantitation of Milk Oligosaccharides from Different Mammal Species and Heat Treatment Influence. *Food Chem.* **2024**, *430*, 136977. [CrossRef]
109. Han, R.; Shi, R.; Yu, Z.; Ho, H.; Du, Q.; Sun, X.; Wang, J.; Jiang, H.; Fan, R.; Yang, Y. Distribution and Variation in Proteins of Casein Micellar Fractions Response to Heat-Treatment from Five Dairy Species. *Food Chem.* **2021**, *365*, 130640. [CrossRef]
110. Shieh, C.J.; Phan Thi, L.A.; Shih, I.L. Milk-Clotting Enzymes Produced by Culture of Bacillus Subtilis Natto. *Biochem. Eng. J.* **2009**, *43*, 85–91. [CrossRef]
111. Zouari, A.; Marchesseau, S.; Chevalier-Lucia, D.; Raffard, G.; Ayadi, M.A.; Picart-Palmade, L. Acid Gelation of Raw and Reconstituted Spray-Dried Dromedary Milk: A Dynamic Approach of Gel Structuring. *Int. Dairy J.* **2018**, *81*, 95–103. [CrossRef]
112. Hailu, Y.; Hansen, E.B.; Seifu, E.; Eshetu, M.; Ipsen, R. Factors Influencing the Gelation and Rennetability of Camel Milk Using Camel Chymosin. *Int. Dairy J.* **2016**, *60*, 62–69. [CrossRef]
113. Chen, C.; Wang, P.; Zhang, N.; Zhang, W.; Ren, F. Improving the Textural Properties of Camel Milk Acid Gel by Treatment with Trisodium Citrate and Transglutaminase. *LWT* **2019**, *103*, 53–59. [CrossRef]
114. Zhao, J.; Fan, H.; Kwok, L.Y.; Guo, F.; Ji, R.; Ya, M.; Chen, Y. Analyses of Physicochemical Properties, Bacterial Microbiota, and Lactic Acid Bacteria of Fresh Camel Milk Collected in Inner Mongolia. *J. Dairy Sci.* **2020**, *103*, 106–116. [CrossRef] [PubMed]
115. Bao, W.; He, Y.; Yu, J.; Yang, X.; Liu, M.; Ji, R. Diversity Analysis and Gene Function Prediction of Bacteria and Fungi of Bactrian Camel Milk and Naturally Fermented Camel Milk from Alxa in Inner Mongolia. *LWT* **2022**, *169*, 114001. [CrossRef]
116. Wang, Y.; Wu, X.; Wang, L.; Feng, H.; Wan, Z. Research Progress on Functional Properties and Product Development of Camel Milk in Xinjiang. *Herbiv. Livest.* **2024**, *224*, 1–8.
117. Luo, J.; Li, W.; Pan, Y. The Current Situation, Ideas of Development, and Countermeasures of Industrialization of Dairy Bactrian Camels. *Heilongjiang Anim. Breed.* **2024**, *32*, 29–32.

molecules

MDPI

Article

Polyphenols vs. Caffeine in Coffee from Franchise Coffee Shops: Which Serving of Coffee Provides the Optimal Amount of This Compounds to the Body

Regina Ewa Wierzejska [1,*], Iwona Gielecińska [2], Ewelina Hallmann [3,4] and Barbara Wojda [1]

[1] Department of Nutrition and Nutritional Value of Food, National Institute of Public Health NIH—National Research Institute, Chocimska St. 24, 00-791 Warsaw, Poland; bwojda@pzh.gov.pl
[2] Department of Food Safety, National Institute of Public Health NIH—National Research Institute, Chocimska St. 24, 00-791 Warsaw, Poland; igielecinska@pzh.gov.pl
[3] Institute of Human Nutrition Sciences, Department of Functional and Organic Food, Warsaw University of Life Sciences, Nowoursynowska St. 159c, 02-776 Warsaw, Poland; ewelina_hallmann@sggw.edu.pl
[4] Bioeconomy Research Institute, Agriculture Academy, Vytautas Magnus University, K. Donelaičio Str. 58, 44248 Kanuas, Lithuania
* Correspondence: rwierzejska@pzh.gov.pl

Abstract: The scientific literature indicates that there is a limited number of data on the content of bioactive components in coffees consumed "on the go". Therefore, this study examined the polyphenol and caffeine content of different types of coffee from franchise coffee shops, and the caffeine/total polyphenol ratio. The five most popular types of coffee purchased in six franchise coffee shops in Warsaw were analysed. A total of 120 coffee samples were tested. A significant positive ($r = 0.7407$, $p < 0.001$) correlation was found between the total polyphenol and caffeine content in all coffee types tested. Per unit volume, espresso coffee had the highest significant ($p < 0.005$) average total polyphenol and caffeine contents (232.9 ± 63.9 mg/100 mL and 198.6 ± 68.3 mg/100 mL, respectively). After taking into account the coffee's serving size, a serving of Americano provided significantly ($p < 0.05$) the most total polyphenol (average 223.5 ± 81.5 mg), while the highest caffeine content was provided by a serving of ice latte/latte frappe (average 136 ± 57.0 mg). The most favourable ratio of caffeine to total polyphenols (0.56) was found in a serving of Americano coffee; therefore, it seems that this coffee can be considered optimal in terms of the content of both compounds. These findings demonstrate that the polyphenol and caffeine contents of coffees offered in franchise coffee shops are closely related to the serving size.

Keywords: polyphenols; caffeine; coffee; franchise coffee shops; coffees serving size

1. Introduction

Coffee is currently considered one of the food items that support overall health. Drinking moderate amounts of coffee on a regular basis is believed to reduce the risk of type 2 diabetes, cardiovascular diseases, some cancers, etc. [1–4]. The coffee infusion is a complex mixture of many biologically active substances, of which the components with physiological effects—polyphenols and caffeine—have attracted the most attention from researchers. Although caffeine is supposed to be responsible for improving general well-being, after drinking coffee [5,6] the caffeine intake should be kept in moderation, and in light of current knowledge, the safe intake of this alkaloid for adults is up to 400 mg per day and up to 200 mg for pregnant women [7,8]. Single doses of caffeine, up to 200 mg (about 3 mg/kg bw for a 70-kg adult), do not give rise to safety concerns [7]. Polyphenols, on the other hand, are highly beneficial dietary components with anti-inflammatory and antioxidant properties, and it is currently presumed that those found in coffee are as valuable to the body as those in vegetables and fruits [9,10]. A number of factors affect the caffeine and polyphenol content of coffee infusions, including the botanical species of coffee, growing conditions, roasting method, grind size, brewing method and the strength of the infusion [11–13]. All of this makes it difficult to predict the amount of these ingredients in a coffee serving, not only the one we order at a coffee shop but also the one we make every day in the same way at home. Although it is believed that there is no "perfect" coffee, the proportion of caffeine and polyphenols affects its health properties [14,15]. It is worth mentioning that the European Food Safety Authority (EFSA) considers it scientifically proven that a single intake of at least 75 mg of caffeine increases alertness, which has not been confirmed for lower doses of caffeine [16]. Therefore, from the point of view of mental performance, an overly light brew of coffee may not be the optimal choice.

Experts recommend drinking 3–5 cups of coffee per day [1,17], but scientific publications do not specify the size of a cup or a single serving, yet some people drink coffee in mugs. Only US recommendations specify the cup volume (8 oz, i.e., ~240 mL) [18]. Thus, the literature emphasises that a cup of coffee is not a sufficiently characterised measure of either the volume of coffee consumption or its components [19–21], and data on the caffeine content of the servings of coffee that consumers buy are scarce and need to be supplemented [6,11,22]. What is more, the content of polyphenols and caffeine in a coffee serving reported in the literature generally does not take into account either the method of brewing coffee or the serving size, which is highly relevant. Therefore, the more information there is about the composition of differently prepared coffees, the more accurately it will be possible to estimate the dietary intake of these compounds in scientific research. This is particularly crucial for coffee consumed by vulnerable populations, such as pregnant women, with careful consideration given to the specific type of coffee being consumed. Moreover, taking into account the impact of coffee components on the body, the ratio of caffeine to total polyphenols may be a useful parameter for the health properties of coffee [15].

Therefore, the aim of this study was to determine the polyphenols and caffeine content of coffee infusions available at franchise coffee shops, and then calculate the caffeine/total polyphenol ratio in order to select the coffee that is optimal in terms of both compounds. The optimal serving of coffee was considered to be coffee with a low caffeine/total polyphenol ratio, and, taking into account the position of EFSA, also containing, on average, not less than 75 mg but not more than 200 mg of caffeine [7,16].

2. Results

2.1. Polyphenol Content per 100 mL of Coffee

The total polyphenol content in 100 mL of the coffees tested averaged 101.3 mg (min–max: 21.3–343.1 mg/100 mL). Espresso coffee had the highest ($p < 0.005$) polyphenol level (Table 1). Each type of coffee showed a large variation in the content of polyphenols within the group. Cappuccino exhibited an approximately 2-fold difference, while Americano displayed a variance of more than 4-fold.

Table 1. Content of polyphenols and caffeine in different types of coffee collected from franchise coffee shops (in mg/100 mL) *.

Components	Coffee Brew					*p*-Value
	Espresso	Caffè latte/Latte Macchiato	Cappuccino	Americano	Ice Latte/Latte Frappe	
total polyphenols	232.9 ± 63.9 [c] (114.3–343.1)	62.3 ± 10.8 [ab] (42.1–77.5)	66.9 ± 10.8 [ab] (51.9–85.8)	101.3 ± 45.2 [b] (46.0–194.1)	43.3 ± 16.2 [a] (21.3–72.9)	<0.005
gallic acid	56.3 ± 27.6 [b] (16.0–108.7)	33.8 ± 20.0 [bc] (2.3–74.9)	3.3 ± 1.4 [a] (1.8–5.2)	47.0 ± 47.6 [b] (1.8–128.6)	5.0 ± 4.3 [ac] (1.4–13.2)	<0.05
chlorogenic acid	45.6 ± 19.2 [b] (13.3–72.2)	4.9 ± 2.2 [a] (2.2–8.7)	14.4 ± 11.0 [b] (1.3–33.4)	43.8 ± 25.7 [b] (13.5–92.1)	5.6 ± 2.0 [a] (2.5–8.6)	<0.0001

Table 1. *Cont.*

Components	Coffee Brew					p-Value
	Espresso	Caffè latte/Latte Macchiato	Cappuccino	Americano	Ice Latte/Latte Frappe	
caffeic acid	3.10 ± 1.79 [b] (0.34–5.46)	0.24 ± 0.17 [a] (0.08–0.65)	0.21 ± 0.10 [a] (0.10–0.40)	0.64 ± 0.54 [a] (0.09–1.60)	0.09 ± 0.09 [a] (0.02–0.26)	<0.0001
salicylic acid	1.23 ± 1.42 (0.08–4.99)	0.31 ± 0.32 (0.12–1.23)	0.68 ± 0.92 (0.08–2.80)	1.40 ± 1.50 (0.09–5.70)	0.24 ± 0.16 (0.08–0.60)	n.s.
epigallocatechin	0.67 ± 0.89 [b] (0.01–2.15)	0.07 ± 0.04 [a] (0.02–0.14)	0.09 ± 0.07 [a] (0.01–0.28)	0.75 ± 0.36 [b] (0.36–1.57)	0.06 ± 0.03 [a] (0.01–0.10)	<0.05
quercetin-3-O-rutinoside	0.15 ± 0.06 [b] (0.04–0.24)	0.05 ± 0.02 [a] (0.03–0.09)	0.08 ± 0.06 [ab] (0.02–0.23)	0.23 ± 0.11 [c] (0.12–0.48)	0.05 ± 0.02 [a] (0.02–0.09)	<0.05
kaempferol-3-O-glucoside	0.28 ± 0.17 (0.12–0.73)	0.33 ± 0.20 (0.09–0.58)	0.28 ± 0.77 (0.10–0.22)	0.27 ± 0.20 (0.12–0.82)	0.43 ± 0.21 (0.10–0.78)	n.s.
quercetin	0.05 ± 0.01 (0.03–0.06)	0.05 ± 0.01 (0.03–0.07)	0.05 ± 0.02 (0.03–0.10)	0.05 ± 0.01 (0.03–0.07)	0.05 ± 0.01 (0.03–0.07)	n.s.
quercetin-3-O-glucoside	5.95 ± 2.92 [b] (1.03–11.20)	4.69 ± 3.06 [ab] (1.39–9.06)	2.75 ± 0.88 [a] (1.52–4.01)	3.05 ± 1.85 [ab] (1.33–7.63)	4.46 ± 3.92 [ab] (0.85–11.03)	<0.05
kaempferol	1.64 ± 1.02 (0.63–3.76)	1.26 ± 1.03 (0.60–4.31)	1.38 ± 1.06 (0.53–4.31)	1.57 ± 0.86 (0.62–2.87)	1.14 ± 0.73 (0.53–2.62)	n.s.
caffeine	198.6 ± 68.3 [b] (109.9–370.4)	48.0 ± 16.7 [a] (20.5–88.9)	52.3 ± 17.5 [a] (27.1–96.7)	55.5 ± 26.9 [a] (33.2–133.6)	43.9 ± 14.2 [a] (21.8–70.6)	<0.0001

* Data are presented as the mean ± SD (min–max) with Tukey test *p*-value; means in rows followed by the same letter are not significantly different ($p < 0.05$); n.s. not significant statistically.

Gallic acid and chlorogenic acid were found to be the predominant polyphenols in all the coffee samples analysed. When it comes to gallic acid, significantly ($p < 0.05$) the lowest content was found in cappuccino coffee, while the highest ($p < 0.05$) was in espresso and Americano. With regards to chlorogenic acid, caffè latte/latte macchiato and ice latte/latte frappe coffee had significantly ($p < 0.0001$) the lowest content of this compound, compared with the other coffees tested (Table 1).

2.2. Polyphenol Content per Coffee Serving

The total polyphenol content in a serving of the coffees tested averaged 137.0 mg (min–max: 35.4–321.1 mg). The highest statistically significant ($p < 0.005$) total polyphenol content was found in a serving of Americano coffee (Figure 1). Differences in total polyphenol level, depending on the coffee type, were 3 fold (espresso, caffè latte/latte macchiato, cappuccino and Americano) or more than 4-fold (ice latte/latte frappe).

As shown in Figure 1, a serving of Americano coffee also had the highest statistically significant ($p < 0.05$) content of gallic acid and chlorogenic acid.

2.3. Caffeine Content per 100 mL of Coffee

The average caffeine amount in 100 mL of coffee was found to be 79.7 mg (min–max: 20.5–370.4 mg/100 mL). Espresso coffee had the highest statistically significant ($p < 0.0001$) mean caffeine content (198.6 mg/100 mL), while ice latte/latte frappe had the lowest (43.9 mg/100 mL) (Table 1). It should be noted that the mean caffeine level of espresso coffee was about 4-times higher ($p < 0.0001$) compared to the other coffees, while there were no statistically significant differences in the caffeine content among the other types of coffees tested. Similar to polyphenols, there was considerable variability observed in the caffeine content among the different coffee groups tested. The difference between the smallest and largest amount of caffeine in the samples of most coffees was more than 3-fold, while for caffè latte/latte macchiato and Americano—it was more than 4-fold.

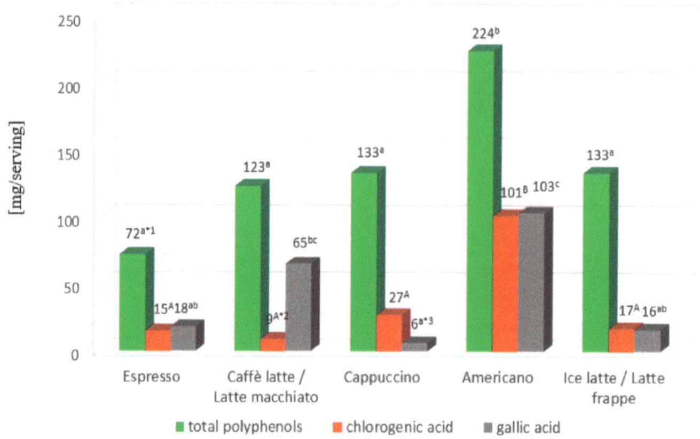

Figure 1. The content of total polyphenols and selected phenolic acids in coffee serving. *1 Statistically significant difference ($p < 0.005$) in the total polyphenols level between various coffee brew types. Values marked with different letters were significantly different. *2 Statistically significant differences ($p < 0.0001$) in chlorogenic acid levels between various coffee brew types. Values marked with different letters were significantly different. *3 Statistically significant differences ($p < 0.05$) in gallic acid levels between various coffee brew types. Values for a given compound marked with different letters were significantly different.

2.4. *Caffeine Content per Coffee Serving*

The average caffeine content per serving of coffee was found to be 103.7 mg (min–max: 32.5–309.4 mg). The order of coffees in terms of caffeine content per serving was opposite compared to caffeine level per 100 mL of coffee. As shown in Figure 2, a serving of espresso provided statistically significantly ($p < 0.05$) the least caffeine, while a serving of ice latte/latte frappe was the most ($p < 0.05$). In all types of coffees tested, there was a large variation in caffeine content per serving (3- to 4-fold for espresso, caffè latte/latte macchiato and Americano, and almost 7-fold for cappuccino and ice latte/latte frappe).

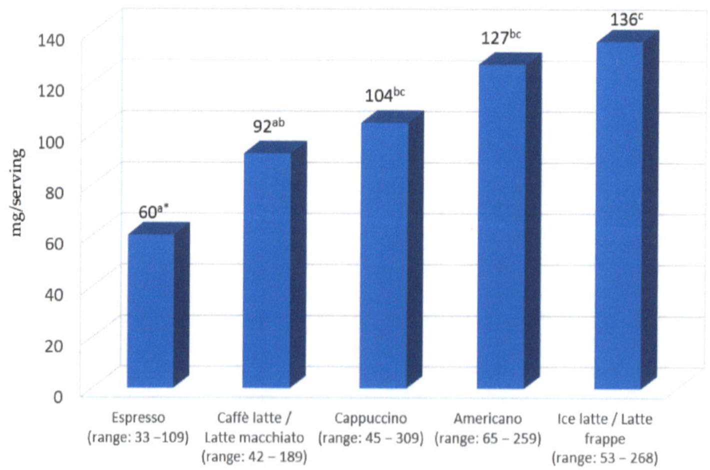

Figure 2. Caffeine content in coffee servings. * Statistically significant difference ($p < 0.05$) between various coffee brew types. Values marked with different letters were significantly different.

2.5. Relationship between Polyphenol and Caffeine Content

Statistical analysis showed a strong positive correlation (r = 0.7407, $p < 0.001$) between the content of total polyphenols and caffeine in all coffee types (Figure 3). Furthermore, a significant correlation was found between the caffeine level and the content of two out of the ten polyphenols examined, specifically gallic acid (r = 0.3902, $p < 0.05$) and caffeic acid (r = 0.6774, $p < 0.001$). For the other polyphenolic compounds tested, the correlations were much weaker and were not statistically significant.

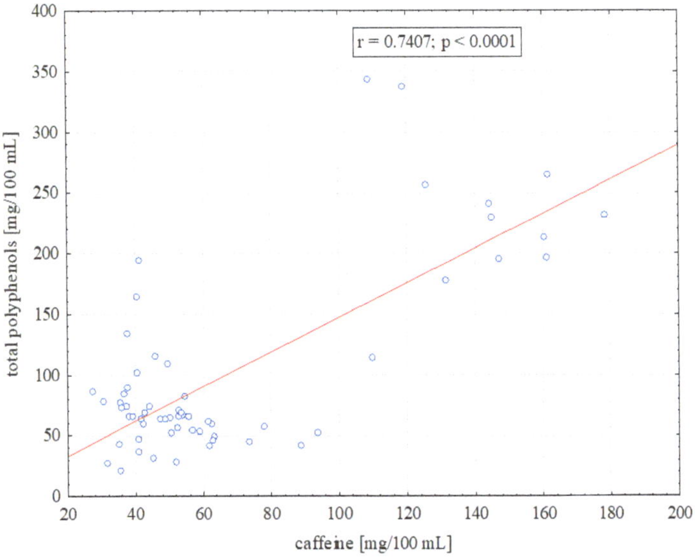

Figure 3. Correlation between total polyphenol and caffeine levels in coffee.

The lowest significant ($p < 0.05$) caffeine/total polyphenol ratio was found for Americano and espresso coffee, while the highest ($p < 0.05$) was for ice latte/latte frappe (Table 2). Therefore, also taking into account the average caffeine content in a serving of Americano (127 mg), which provides the alertness-increasing dose of caffeine (minimum 75 mg), a serving of this coffee can be considered optimal in terms of both analysed components.

Table 2. Ratio of the caffeine content to selected polyphenols' level *.

Components	Type of Coffee					p-Value
	Espresso	Caffè Latte/Latte Macchiato	Cappuccino	Americano	Ice Latte/Latte Frappe	
caffeine/total polyphenols	0.65 ± 0.19 [a] (0.32–0.96)	0.99 ± 0.48 [ab] (0.40–2.11)	0.78 ± 0.39 [ab] (0.32–1.81)	0.56 ± 0.34 [a] (0.21–1.37)	1.17 ± 0.42 [b] (0.49–1.85)	<0.05
caffeine/ chlorogenic acid	4.3 ± 3.5 [ab] (1.8–13.4)	14.8 ± 10.1 [c] (3.6–39.5)	8.4 ± 9.4 [ac] (1.1–32.0)	1.5 ± 1.0 [a] (0.4–3.6)	9.2 ± 3.7 [bc] (3.7–14.6)	<0.05
caffeine/gallic acid	3.4 ± 2.1 [a] (1.5–6.9)	2.9 ± 3.4 [a] (1.0–13.6)	18.2 ± 11.3 [b] (7.5–46.0)	5.0 ± 9.5 [a] (0.3–34.3)	15.7 ± 10.4 [b] (3.2–29.9)	<0.05

* Data are presented as the mean ± SD (min–max) with Tukey test p-value; means in rows followed by the same letter are not significantly different ($p < 0.05$).

In terms of chlorogenic acid, Americano coffee demonstrated the lowest ratio, whereas caffè latte/latte macchiato was the highest. When it comes to gallic acid, the ratio of caffeine to this compound was lower for caffè latte/latte macchiato, espresso and Americano compared with cappuccino and ice latte/latte frappe. The aforementioned values were statistically significant ($p < 0.05$).

3. Discussion

Polyphenols and caffeine are natural components of coffee, and as with all substances present in foods of plant origin, their content is subject to significant fluctuations. When it comes to polyphenols, it is sometimes assumed that the total polyphenols amount to 200 mg/100 mL of coffee, and there are 300 mg in a serving [23,24], but there are also data indicating that the polyphenol content in a serving of coffee is much higher (500–1200 mg) [11,14]. This shows a high variability in the polyphenol amount, which depends on numerous factors. In our study, the total polyphenol content was lower and in 100 mL of coffee it was on average 101 mg, while in a serving of coffee, which for most coffees was higher than 100 mL, the average was 137 mg. When relating our results to the literature data, it should also be borne in mind that published studies differ in terms of the analytical methods used (including the limit of quantification), and the manner in which results are presented (such as average/mean or median).

Comparison of the content of individual polyphenolic compounds is difficult, as published studies deal with different groups of polyphenols. The main coffee polyphenols are chlorogenic acids (CGAs), including caffeoylquinic acids (CQAs) as a subclass of CGAs [25–27]. According to the literature data, the content of CGAs in a cup of coffee ranges from 20 mg to 675 mg [26] or from 70 mg to 350 mg [23,28]. The content of CGAs in a serving of espresso coffee determined in our study (mean 15 mg) is relatively small compared to the results obtained by other authors. In a study by Crozier et al. [25], Scottish espresso from Costa Coffee had an average of 227 mg of total CQA, but from Starbucks only 24 mg. In a study by Ludwig et al. [21], Scottish, Spanish and Italian espresso from coffee shops had a median total CQA in a serving of coffee of 59 mg, 142 mg and 46 mg, respectively. In Americano coffee, the CGA content we found (mean 101 mg) is comparable to the results of a Korean study [15], in which the total CGAs in a serving of coffee purchased from coffee shops averaged 99.4 mg. In the case of cappuccino coffee, the CGA content (27 mg per serving) fell within the range of CQA's content for coffee purchased in Scottish coffee shops, where the mean content of these compounds ranged from 23 to 135 mg per serving, depending on the type of coffee shop [21].

With regard to caffeine, many authors assume that a standard cup of coffee (~240 mL) contains 100 mg [29–31], 107 mg [32] or 135 mg [33] of caffeine. Still, other authors believe that an average cup of coffee is 150 mL in size and provides 80–90 mg [7,34,35], 60 mg [36] or only 50 mg [37] of caffeine. In our study, without taking into account the coffee type, an average serving of coffee from franchise coffee shops (192 mL) contained 104 mg of caffeine. As far as coffee type is concerned, the caffeine content of a serving of coffee (22–420 mL) averaged between 60 mg and 136 mg and increased as the serving size increased. According to a US study [34], a coffee serving, depending on its type, provides from 58 mg to 259 mg of caffeine, and similarly, the largest differences stem from the serving size (30–473 mL). These data support the view that assuming a constant caffeine content for a cup of different coffees may be either overestimated or underestimated, making it difficult to assess caffeine intake, as well as compare it to a safe level.

One of the most widely consumed coffees is espresso. In Poland, it is consumed by more than 90% of coffee drinkers [23], so data on the caffeine content in this coffee type seems particularly useful. Espresso, compared to other coffees, is considered a strong coffee, i.e., a coffee with a high caffeine content per unit volume, which is due to the use of a large amount of ground coffee relative to the amount of water [6,11,14]. This was also confirmed by our study. However, it should be noted that the study did not cover coffees brewed by other methods, such as drip coffee or Turkish coffee. From the consumer's point of view, of much greater importance is the amount of caffeine supplied to the body with a coffee serving, and with this approach, the strength of the coffees studied took the reverse order compared with 100 mL. Espresso coffee, served in portions several times smaller than other coffees (an average of 31 mL in our study), contained the least caffeine (mean 60 mg). There have been no studies in Poland to date on the caffeine content of "takeaway" coffees, so the authors cannot compare their results at the local level. In light of the world literature, the mean caffeine content we found is highly consistent with many, but not all, of the data. O'Keefe et al. [38] cite that espresso usually has about 60 mg per shot. The same result was obtained by McCusker et al. [34] in a serving of American espresso from coffee shops (58 mg of caffeine in a 30 mL serving), which also coincides with the USDA Food Composition Databases [39] and, according to which, a cup of espresso (1 oz; ~30 mL) contains an average of 63 mg of caffeine. A considerably higher content of this ingredient, compared with our result, was found in Scottish espresso (median 140 mg/serving, median serving 43 mL) [25]. However, it is worth noting that among 20 different outlet coffee shops in Glasgow where the authors of the aforementioned study purchased espresso, only two—Starbucks and Costa Coffee—overlapped with our study. Ludwig et al. [21], analysed Scottish, Italian and Spanish espresso from coffee shops and found comparable amounts of caffeine in these coffees (median 100 mg, 102 mg and 116 mg, respectively), but their results are higher than in our study. Other authors report a lower caffeine content. According to Crozier et al. [25], a cup of classic espresso (30 mL) supplies 30–50 mg of caffeine, while according to Heckman et al. [29]—40 mg. All studies reveal significant differences in the caffeine content of an espresso serving, depending on the coffee shops. The values vary more than 6-fold [25], more than 5-fold [21], and—in our study—more than 3-fold. This may not be surprising, since more than 2-fold differences were found even in the same type of coffee, purchased from the same coffee shop for six consecutive days [34].

When it comes to Americano coffee, it is worthwhile to compare the caffeine content found (an average of 127 mg/serving) with the results of a study by Jeon et al. [15], who bought Americano coffee at eleven coffee shops, four fast-food restaurants and three bakery stores. The caffeine content of coffee from coffee shops averaged 166 mg/serving, from fast-food restaurants—108 mg/serving, and from bakery stores—94 mg/serving. Considering only coffees from coffee shops and fast-food restaurants, i.e., places included in our study as well, the mean caffeine content in both studies is highly consistent (137 mg/serving vs. 127 mg/serving). According to van Dam et al. [40], a serving of Americano from coffee shops (12 oz; ~340 mL) contains 150 mg of caffeine, while according to the USDA Food Composition Databases—94–150 mg [39]. Slightly smaller amounts in a serving of Americano (8 oz; ~240 mL) are reported by Mitchell et al.—63–126 mg of caffeine [41]. It is worth mentioning that Americano coffee offered in coffee shops (referred to as American coffee imitation) is espresso-based, often including double espresso. In our survey, the use of double espresso was declared by baristas from three out of six coffee shops, which, also considering the much larger serving size of such coffee, explains the higher caffeine content compared with espresso.

In the case of cappuccino, the mean caffeine content in our study (104 mg) is congruent with data reported by Mattioli and Farinetti—110 mg [42]. Cappuccino coffee purchased in coffee shops in Scotland had a higher caffeine content, with a median of 180 mg [21].

With regard to other types of coffee offered in coffee shops, there is little data in the literature. Caffè latte, according to the USDA Food Composition Databases [39], contains as much caffeine as cappuccino (mean 86 mg/serving), which coincides with the results of our study (92 mg). In Norway, caffè latte and cappuccino are also assumed to contain the same amount of caffeine but at a much lower level—21 mg/100 mL [43], which would be 50 mg for a 240 mL serving. A serving of ice latte/latte frappe coffee had the highest caffeine content in our study (mean 136 mg), but note that the serving size was expressed as the volume of the coffee alone after separating the ice immediately after purchasing the coffee (coffee with ice had up to 40% more volume than coffee without ice). According to US charts, a serving (8 oz; ~240 mL) of iced coffee provides less caffeine (74 mg), but there is no information on the preparation method, which constitutes a factor that determines the content of this ingredient [39]. It should be noted that in our study, in three out of the six coffee shops, ice latte/latte frappe was prepared based on double espresso, which may explain the higher caffeine content.

When considering the caffeine content of coffees from franchise coffee shops, it is worth relating it to the dose that benefits mental performance. According to EFSA [16], in most people, an intake of 75 mg of caffeine in a single dose is necessary for increased alertness, so such an effect cannot be produced by either light coffee infusions or too small a coffee serving. Given the average caffeine content in the coffees tested, this would apply to a serving of espresso, which may come as a surprise to many, since such coffee is generally considered strong. However, taking into account the caffeine content of individual coffee samples, a dose lower than 75 mg/serving was found in 36% of all coffee samples tested (in 75% of espresso coffee samples, 50% of caffè latte/latte macchiato, 33% of cappuccino, 17% of Americano and 8% of ice latte/latte frappe samples).

Our study showed a strong correlation between the level of total polyphenol and caffeine in all coffee types tested. This means that coffee supplying a large amount of caffeine also contains a high level of polyphenols. Polyphenols are compounds with beneficial effects on the body and it is polyphenols that are mainly responsible for the health-enhancing effects of coffee [9,44]. Thus, it would be fair to say that for people who like coffee and tolerate it well, coffee is a much better choice than caffeinated sodas. For the average consumer, and especially for people who should limit their caffeine intake, such as pregnant women, the most suitable option would be coffee containing more polyphenols than caffeine. The low caffeine/total polyphenols ratio means that the level of polyphenols is relatively high compared to the caffeine level. The most favourable ratio of caffeine to total polyphenols in our study was found for Americano coffee (mean 0.56), and the least favourable was for ice latte/latte frappe coffee (mean 1.17). Therefore, coming back to the original question in the title of the article about the optimal serving of coffee, in terms of polyphenols and caffeine, it should be stated, that for an average consumer, it would be a serving of Americano. It seems that espresso which also had a low caffeine-to-polyphenol ratio (0.65) cannot be considered optimal coffee because the average caffeine content in a serving of this coffee (60 mg) was lower than alertness-increasing dose of caffeine (minimum 75 mg). Moreover, the minimal caffeine content found in the Americano coffee samples (65 mg) is the closest to the dose of 75 mg, compared to the minimal caffeine content of espresso (33 mg). This means that the risk of buying Americano containing very little caffeine is low. Since the total polyphenol content consists of a whole range of compounds, the calculated ratio of caffeine to individual polyphenols takes a value above one. The ratio of caffeine to CGA (chlorogenic acid) for espresso averaged 4.3, and for Americano 1.5, indicating that the amount of caffeine far exceeds the amount of CGA. Also, in Scottish espresso from coffee shops, purchased four times, the ratio of caffeine to CGAs (chlorogenic acids) exceeded the value of 1 in each case, ranging from 1.5 to 10.4 depending on the coffee shop [21]. In Americano coffee from franchise coffee shops in the Republic of Korea, this ratio averaged 2.1 [15].

A limitation of our study is the lack of detailed information on how the coffee is prepared (the amount of coffee beans used, brewing time, and, for coffees with milk, the amount of milk used). The differences between the smallest and the largest content of polyphenols and caffeine confirm the high variability of these components, but the authors are unable to identify the most significant factor. Data obtained from coffee shops show that four of them used only Arabica coffee, and two used a mix of Arabica and Robusta, which could theoretically have contributed to the resulting variability. However, the purpose of the study was not to analyse individual factors affecting coffee composition but to estimate polyphenol and caffeine content from a consumer point of view. One of the advantages of our study was that we bought coffee several times in the same coffee shops located in different parts of the city, which, in addition to the fact that the coffee was prepared by different staff, also increased the chances of including different batches of coffee beans. Furthermore, the study included five types of coffee most frequently chosen by Poles from six popular franchise coffee shops located throughout Warsaw. In Poland, there have been no studies so far on the polyphenol and caffeine content of coffees bought in coffee shops; therefore, with the growing popularity of such establishments, the data obtained may serve to estimate the dietary intake of the components in question more accurately.

4. Materials and Methods

4.1. Study Material

Five types of the most popular coffees in Poland [45] were selected for the study. These types included espresso, cappuccino, caffè latte/latte macchiato, Americano and ice latte/latte frappe. No sugar, cream, etc. was added to the coffee. The samples of coffee were purchased at six different franchise coffee shops in Warsaw, including four popular chain coffee shops (Costa Coffee, Starbucks, Tchibo, Green Caffè Nero), one fast-food restaurant—McCaffé (McDonald's) and one outlet located at BP gas stations—Wild Bean Cafe. In order to obtain a comprehensive and reliable assessment of caffeine and polyphenol levels, which depend on the type of coffee bean, among other factors, samples were collected four times (July, September, October and November 2021) from the same establishments. The caffeine content was determined in all samples purchased, while polyphenol levels were determined only in samples taken in the months of July and September. A total of 120 coffee samples were tested.

The purchased coffee samples were consistently delivered to the laboratory of the Department of Food Safety at the National Institute of Public Health NIH—National Research Institute. Here, average laboratory samples were meticulously prepared. One portion of these samples underwent caffeine content analysis on the same day, while the remainder was subsequently transported to the Department of Functional and Organic Food at Warsaw University of Life Sciences. In the laboratory, the total polyphenol content, as well as the levels of specific phenolic acids and flavonols, were meticulously determined.

4.2. Total Polyphenol Content Determination

The total polyphenol content, expressed as gallic acid equivalent, was performed using the Folin colorimetric method. The sample preparation for testing followed the procedure described by Singleton et al. [46]. Briefly, 5 mL of coffee brew sample was measured into a 250 mL beaker and water was added. Extraction was carried out in an ultrasonic bath (30 °C, 6000 Hz, 20 min). Subsequently, the sample was subjected to vacuum filtration and further diluted at a ratio of 1:15 with water. From this prepared solution, 1 mL was transferred into a 50 mL volumetric flask. Then, to this flask, 2.5 mL of Folin-Ciocalteu reagent (F-C) and 5 mL of 20% sodium carbonate were added, followed by topping off with water. The sample was left to incubate under room temperature, shielded from light, for a period of 45 min. Afterwards, the absorbance at a wavelength of $\lambda = 750$ nm was measured using a spectrophotometer (Helios γ, Thermo Scientific, Waltham, MA, USA).

4.3. Selected Phenolic Acid and Flavonol Content Determination

The quantitative and qualitative analysis of polyphenolic compounds was performed using the high-performance liquid chromatography (HPLC) method described earlier by Król et al. [47]. Briefly, a 3 mL sample of coffee brew was combined with 5 mL of 80% methanol. The mixture was then mechanically shaken and subjected to extraction in an ultrasonic bath (30 °C, 10 min). Following extraction, the sample was centrifuged (5 °C, 3780× g, 10 min) and was subsequently subjected to chromatographic analysis (HPLC-DAD using two LC-20AD pumps, a CBM-20A controller, an SIL-20AC column oven, and UV/Vis SPD-20 AV, and SPD-M20A spectrometers; Shimadzu, Kyoto, Japan).

Chromatographic separation was carried out on a Phenomenex Fusion-RP 80A column (Torrance, CA, USA, 250 mm × 4.60 mm). The mobile phase consisted of two components: (A) 90% water and 10% acetonitrile, and (B) 45% water and 55% acetonitrile, both with a pH of 3.0. The applied gradient program is detailed in Table 3. The injection volume was set at 100 µL, with the sample temperature maintained at 30 °C, and the column temperature also held at 30 °C. UV detection was performed at wavelengths: $\lambda = 250$ nm for flavonoids (quercetin-3-*O*-rutinoside, kaempferol-3-*O*-glucoside, quercetin, quercetin-3-O-glucoside, kaempferol, epigallocatechin) and λ = 370 nm for phenolic acids (gallic, chlorogenic, caffeic, salicylic).

Table 3. Liquid chromatography gradient program.

Time (min)	Flow (mg/mL)	% A	% B
Initial	1.00	95	5
23	1.00	50	50
28	1.00	80	20
29	1.00	95	5
Total runtime: 38 min			

4.4. *Caffeine Content Determination*

The caffeine content in coffee samples was determined using high-performance liquid chromatography with a diode array detector (HPLC-DAD) according to an in-house test procedure. The study sample was prepared as previously described [48]. Briefly, a coffee brew ranging from 2.5 to 7.5 mL (depending on the expected caffeine content) was measured into a 25 mL volumetric flask. To effectively remove high-molecular compounds, we added 1 mL each of Carrez I (15 g of potassium hexacyanoferrate (II) dissolved in 100 mL of distilled water) and Carrez II (30 g of zinc sulphate dissolved in 100 mL of distilled water) to the sample. The sample was stirred and gently slightly shaken and then allowed to stand for 10 min. After replenishing the flask with water, the sample was thoroughly stirred before being transferred in its entirety to a polypropylene centrifuge tube. The centrifuge operated at 10,000 rpm for 10 min at approximately 10 °C. The clear solution was swiftly and quantitatively transferred to an Erlenmeyer flask. Subsequently, 1 mL of the supernatant was filtered using a PVDF syringe filter (0.45 µm) into a chromatography vial.

Caffeine quantification in coffee was performed using an Alliance 2695 high-performance liquid chromatograph with a diode array detector (HPLC-DAD; Waters, Milford, MA, USA). This analysis was conducted on a Lichrospher RP-18 column (125 mm × 4 mm; 5 µm; Agilent Technologies, Santa Clara, CA, USA), including a pre-column (4 mm × 4 mm; 5 µm; Agilent Technologies), with the same packing material. Chromatographic analysis conditions: mobile phase—water and methanol HPLC (70:30; *v*/*v*), isocratic flow 0.9 mL/min, sample injection volume 20 µL, sample temperature 20 °C, column temperature 40 °C, UV detection at a wavelength of λ = 273 nm, total analysis duration 8 min. The identification of the test compound was carried out based on its retention time and a comparison between the acquired UV spectrum and the spectrum of the caffeine standard. The result was taken as an average of three parallel determinations, corrected for recovery for each analysis series.

The test method described above had undergone prior validation [48], and has received accreditation for compliance with requirements of EN ISO 17025 [49] by the Polish Centre for Accreditation. Our laboratory regularly participates in proficiency testing (PT), achieving satisfactory results.

4.5. *Calculation of Polyphenol and Caffeine Content in a Coffee Serving*

To calculate the polyphenol and caffeine content in a coffee serving, an individual serving size were employed for each type of coffee. The volume of purchased coffee servings was meticulously measured in the laboratory and, depending on the franchise coffee shops, was as follows range: 22–49 mL for espresso, 142–305 mL for caffè latte/latte macchiato, 140–320 mL for cappuccino, 155–360 mL for Americano and 245–420 mL for ice latte/latte frappe.

4.6. *Statistical Analysis*

The obtained analytical results underwent evaluation, including Dixon's and Grubbs's tests to eliminate outliers (with coarse errors). Subsequently, the data were subjected to comprehensive statistical analysis using the following software tools: Statistica version 6.0 (Statsoft, Inc., Tulsa, OK, USA) and Stata/SE 17.0. The results for polyphenol and caffeine content in coffee were expressed in two ways: as mg/100 mL of coffee brew and as mg per coffee serving. These values were presented as means, along with minimum and maximum values.

To evaluate the significance of differences between various coffee types, a one-way analysis of variance (ANOVA) was conducted. Subsequently, multiple comparisons were performed using Tukey's test to determine significant differences between coffee types. Pearson's linear correlation coefficient was use to evaluate the association between polyphenol and caffeine content. In all conducted statistical tests, a significance level of $p \leq 0.05$ was applied to determine the presence of meaningful differences and relationships among the variables under examination.

5. Conclusions

This study showed a strong positive correlation between total polyphenol and caffeine content in all studied types of coffee from franchise coffee shops. It was concluded that the most important measure of the above-mentioned components supplied to the body with coffee is the serving size. Although espresso coffee had the highest total polyphenol and caffeine content per unit volume, given the serving size, up to ten times smaller than other coffees; the Americano serving provided the most polyphenols, while the ice latte/latte frappe serving had the highest caffeine content. In terms of the proportion of both bioactive compounds, it seems that the optimal coffee would be a serving of Americano, which contained the most polyphenols and the average caffeine content covers the minimum dose increasing alertness. While cups are a common measure of coffee intake in both scientific publications and everyday life, such a measurement method is not accurate for coffee bought in coffee shops, as most coffees are served in drinking vessels larger than a typical cup.

Author Contributions: R.E.W.: conceptualization, data curation, formal analysis, supervision, writing—original draft, writing—review and editing, and funding acquisition; I.G.: methodology, formal analysis, writing—original draft, writing—review and editing, validation, and visualization; E.H.: analytical research; B.W.: data collection. All authors have read and agreed to the published version of the manuscript.

Funding: This study was financed by the NIPH NIH—NRI research fund (Research No. 4FŻBW/2020).

Institutional Review Board Statement: Not applicable.

Informed Consent Statement: Not applicable.

Data Availability Statement: Data are contained within the article.

Conflicts of Interest: The authors declare no conflicts of interest.

References

1. Carlström, M.; Larsson, S.C. Coffee consumption and reduced risk of developing type 2 diabetes: A systematic review with meta analysis. *Nutr. Rev.* **2018**, *76*, 395–417. [CrossRef] [PubMed]
2. Wierzejska, R. Coffee consumption and cardiovascular diseases—Has the time come to change dietary advice? A mini review. *Pol. J. Food Nutr. Sci.* **2016**, *66*, 5–10. [CrossRef]
3. Pauwels, E.K.J.; Volterrani, D. Coffee consumption and cancer risk: An assessment of the health implications based on recent knowledge. *Med. Princ. Pract.* **2021**, *30*, 401–411. [CrossRef]
4. Socała, K.; Szopa, A.; Serefko, A.; Poleszak, E.; Wlaź, P. Neuroprotective effects of coffee bioactive compounds: A review. *Int. J. Mol. Sci.* **2021**, *22*, 107. [CrossRef] [PubMed]
5. Wachamo, H.L. Review on health benefit and risk of coffee consumption. *Med. Aromat. Plants* **2017**, *6*, 4. [CrossRef]
6. Olechno, E.; Puścion-Jakubik, A.; Zujko, M.E.; Socha, K. Influence of various factors on caffeine content in coffee brews. *Foods* **2021**, *10*, 1208. [CrossRef] [PubMed]
7. European Food Safety Authority—EFSA. Panel on Dietetic Products, Nutrition and Allergies. Scientific opinion on the safety of caffeine. *EFSA J.* **2015**, *13*, 4102. [CrossRef]

8. Reyes, C.M.; Cornelis, M.C. Caffeine in the diet: Country-level consumption and guidelines. *Nutrients* **2018**, *10*, 1772. [CrossRef] [PubMed]
9. Kolb, H.; Kempf, K.; Martin, S. Health effects of coffee: Mechanism unraveled? *Nutrients* **2020**, *12*, 1842. [CrossRef]
10. Rojas-González, A.; Figueroa-Hernández, C.Y.; González-Rios, O.; Suárez-Quiroz, M.L.; González-Amaro, R.M.; Hernández-Estrada, Z.J.; Rayas-Duarte, P. Coffee chlorogenic acids incorporation for bioactivity enhancement of foods: A review. *Molecules* **2022**, *27*, 3400. [CrossRef]
11. Derossi, A.; Ricci, I.; Caporizzi, R.; Fiore, A.; Severini, C. How grinding level and brewing method (Espresso, American, Turkish) could affect the antioxidant activity and bioactive compounds in a coffee cup. *J. Sci. Food Agric.* **2018**, *98*, 3198–3207. [CrossRef]
12. Górecki, M.; Hallmann, E. The antioxidant content of coffee and its in vitro activity as an effect of its production method and roasting and brewing time. *Antioxidants* **2020**, *9*, 308. [CrossRef]
13. Bastian, F.; Hutabarat, O.S.; Dirpan, A.; Nainu, F.; Harapan, H.; Emran, T.B.; Simal-Gandara, J. From plantation to cup: Changes in bioactive compounds during coffee processing. *Foods* **2021**, *10*, 2827. [CrossRef] [PubMed]
14. Severini, C.; Derossi, A.; Ricci, I.; Caporizzi, R.; Fiore, A. Roasting conditions, grinding level and brewing method highly affect the healthy benefits of a coffee cup. *Int. J. Clin. Nutr. Diet.* **2018**, *4*, 127. [CrossRef] [PubMed]
15. Jeon, J.S.; Kim, H.T.; Jeong, I.H.; Hong, S.R.; Oh, M.S.; Yoon, M.H.; Shim, J.H.; Jeong, J.H.; Abd El-Aty, A.M. Contents of chlorogenic acids and caffeine in various coffee-related products. *J. Adv. Res.* **2019**, *17*, 85–94. [CrossRef] [PubMed]
16. European Food Safety Authority—EFSA. Scientific Opinion on the substantiation of a health claim related to caffeine and increased alertness pursuant to Article *13*(5) of Regulation (EC) No 1924/2006. *EFSA J.* **2014**, *12*, 3574. [CrossRef]
17. Ding, M.; Bhupathiraju, S.N.; Satija, A.; van Dam, R.M.; Hu, F.B. Long-term coffee consumption and risk of cardiovascular disease: A systematic review and dose-response meta-analysis of prospective cohort studies. *Circulation* **2014**, *129*, 643–659. [CrossRef] [PubMed]
18. Rehm, C.D.; Ratliff, J.C.; Riedt, C.S.; Drewnowski, A. Coffee consumption among adults in the United States by demographic variables and purchase location: Analyses of NHANES 2011–2016 Data. *Nutrients* **2020**, *12*, 2463. [CrossRef] [PubMed]
19. Surma, S.; Sahebkar, A.; Banach, M. Coffee or tea: Anti-inflammatory properties in the context of atherosclerotic cardiovascular disease prevention. *Pharmacol. Res.* **2023**, *187*, 106596. [CrossRef]
20. Poole, R.; Ewings, S.; Parkes, J.; Fallowfield, J.A.; Roderick, P. Misclassification of coffee consumption data and the development of a standardised coffee unit measure. *BMJ Nutr. Prev. Health* **2019**, *2*, 11–19. [CrossRef]
21. Ludwig, I.A.; Mena, P.; Calani, L.; Cid, C.; del Rio, D.; Lean, M.E.J.; Crozier, A. Variations in caffeine and chlorogenic acid contents of coffees: What are we drinking? *Food Funct.* **2014**, *5*, 1718–1726. [CrossRef]
22. Lean, M.E.J.; Crozier, A. Coffee, caffeine and health: What's in your cup? *Maturitas* **2012**, *72*, 171–172. [CrossRef] [PubMed]
23. Yamagata, K. Do coffee polyphenols have a preventive action on metabolic syndrome associated endothelial dysfunctions? An assessment of the current evidence. *Antioxidants* **2018**, *7*, 26. [CrossRef]
24. Fukushima, Y.; Tashiro, T.; Kumagai, A.; Ohyanagi, H.; Horiuchi, T.; Takizawa, K.; Sugihara, N.; Kishimoto, Y.; Taguchi, C.; Tani, M.; et al. Coffee and beverages are the major contributors to polyphenol consumption from food and beverages in Japanese middle-aged women. *J. Nutr. Sci.* **2014**, *3*, e48. [CrossRef] [PubMed]
25. Crozier, T.W.M.; Stalmach, A.; Lean, M.E.J.; Croizer, A. Espresso coffees, caffeine and chlorogenic acid intake: Potential health implications. *Food Funct.* **2012**, *3*, 30–33. [CrossRef]
26. Liang, N.; Kitts, D.D. Antioxidant property of coffee components: Assessment of methods that define mechanisms of action. *Molecules* **2014**, *19*, 19180–19208. [CrossRef]
27. Liczbiński, P.; Bukowska, B. Tea and coffee polyphenols and their biological properties based on the latest in vitro investigations. *Ind. Crops Prod.* **2022**, *175*, 114265. [CrossRef] [PubMed]
28. Tajik, N.; Tajik, M.; Mack, I.; Enck, P. The potential effects of chlorogenic acid, the main phenolic components in coffee, on health: A comprehensive review of the literature. *Eur. J. Nutr.* **2017**, *56*, 2215–2244. [CrossRef]
29. Heckman, M.A.; Weil, J.; de Mejia, E.G. Caffeine (1, 3, 7-trimethylxanthine) in foods: A comprehensive review on consumption, functionality, safety, and regulatory matters. *J. Food Sci.* **2010**, *75*, 77–87. [CrossRef]
30. Jee, H.J.; Lee, S.G.; Bormate, K.J.; Jung, Y. Effect of caffeine consumption on the risk for neurological and psychiatric disorders: Sex differences in human. *Nutrients* **2020**, *12*, 3080. [CrossRef]
31. Higdon, J.V.; Frei, B. Coffee and health: A review of recent human research. *Crit. Rev. Food Sci. Nutr.* **2006**, *46*, 101–123. [CrossRef] [PubMed]
32. Balat, O.; Balat, A.; Ugur, M.G.; Pençe, S. The effect of smoking and caffeine on the fetus and placenta in pregnancy. *Clin. Exp. Obstet. Gynecol.* **2003**, *30*, 57–59. [PubMed]
33. Addicott, M.A.; Yang, L.L.; Peiffer, A.M.; Burnett, L.R.; Burdette, J.H.; Chen, M.Y.; Hayasaka, S.; Kraft, R.A.; Maldjian, J.A.; Laurienti, P.J. The effect of daily caffeine use on cerebral blood flow: How much caffeine can we tolerate? *Hum. Brain Mapp.* **2009**, *30*, 3102–3114. [CrossRef] [PubMed]
34. McCusker, R.R.; Goldberger, B.A.; Cone, E.J. Caffeine content of specialty coffees. *J. Anal. Toxicol.* **2003**, *27*, 520–522. [CrossRef] [PubMed]
35. Nehlig, A. Effects of coffee/caffeine on brain health and disease: What should I tell my patients? *Pract. Neurol.* **2016**, *16*, 89–95. [CrossRef] [PubMed]

36. Błaszczyk-Bębenek, E.; Piórecka, B.; Kopytko, M.; Chadzińska, Z.; Jagielski, P.; Schlegel-Zawadzka, M. Evaluation of caffeine consumption among pregnant women from southern Poland. *Int. J. Environ. Res. Public Health* **2018**, *15*, 2373. [CrossRef] [PubMed]
37. Lamy, S.; Houivet, E.; Benichou, J.; Marret, S.; Thibaut, F. Caffeine use during pregnancy: Prevalence of use and newborn consequences in a cohort of French pregnant women. *Eur. Arch. Psychiatry Clin. Neurosci.* **2021**, *271*, 941–950. [CrossRef] [PubMed]
38. O'Keefe, J.H.; DiNicolantonio, J.J.; Lavie, C.J. Coffee for cardioprotection and longevity. *Prog. Cardiovasc. Dis.* **2018**, *61*, 38–42. [CrossRef] [PubMed]
39. USDA Food Composition Databases. Available online: https://fdc.nal.usda.gov/fdc-app.html#/?query=ndbNumber:14209 (accessed on 4 July 2019).
40. van Dam, R.M.; Hu, F.B.; Willett, W.C. Coffee, caffeine, and health. *N. Engl. J. Med.* **2020**, *383*, 369–378. [CrossRef]
41. Mitchell, D.C.; Knight, C.A.; Hockenberry, J.; Teplansky, R.; Hartman, T.J. Beverage caffeine intakes in the U.S. *Food Chem. Toxicol.* **2014**, *63*, 136–142. [CrossRef]
42. Mattioli, A.V.; Farinetti, A. Espresso coffee, caffeine and colon cancer. *World J. Gastrointest. Oncol.* **2020**, *12*, 601–603. [CrossRef] [PubMed]
43. Sengpiel, V.; Elind, E.; Bacelis, J.; Nilsson, S.; Grove, J.; Myhre, R.; Haugen, M.; Meltzer, H.M.; Alexander, J.; Jacobsson, B.; et al. Maternal caffeine intake during pregnancy is associated with birth weight but not with gestational length: Results from a large prospective observational cohort study. *BMC Med.* **2013**, *11*, 42. [CrossRef] [PubMed]
44. Wierzejska, R. Can coffee consumption lower the risk of Alzheimer's disease and Parkinson's disease? A literature review. *Arch. Med. Sci.* **2017**, *13*, 507–514. [CrossRef] [PubMed]
45. Czarniecka-Skubina, E.; Pielak, M.; Sałek, P.; Korzeniowska-Ginter, R.; Owczarek, T. Consumer choices and habits related to coffee consumption by Poles. *Int. J. Environ. Res. Public Health* **2021**, *18*, 3948. [CrossRef]
46. Singleton, V.L.; Orthofer, R.; Lamuela-Raventos, R.M. Analysis of total phenols and other oxidation substrates and antioxidants by means of Folin-Ciocalteu reagent. *Methods Enzymol.* **1999**, *299*, 152–178. [CrossRef]
47. Król, K.; Gantner, M.; Tatarak, A.; Hallmann, E. The effect of roasting, storage, origin on the bioactive compounds in organic and conventional coffee (Caffea arabica). *Eur. Food Res. Technol.* **2019**, *246*, 33–39. [CrossRef]
48. Gielecińska, I.; Cendrowski, A.; Mojska, H. Opracowanie i walidacja metody oznaczania zawartości kofeiny w napojach bezalkoholowych techniką HPLC-DAD. *Żyw. Człow. Metabol.* **2017**, *44*, 61–76.
49. *European Standard EN ISO/IEC 17025*; General Requirements for the Competence of Testing and Calibration Laboratories. CEN-CENELEC: Brussels, Belgium, 2017.

molecules

MDPI

Article

Protective Effects and Mechanisms of Esculetin against H_2O_2-Induced Oxidative Stress, Apoptosis, and Pyroptosis in Human Hepatoma HepG2 Cells

Ying Luo, Tenglong Chang, Shiting Huang, Jing Xiang, Shuangyang Tang and Haiyan Shen *

The Institute of Biochemistry and Molecular Biology, Hunan Provincial Key Laboratory for Special Pathogens Prevention and Control, Hengyang Medical College, University of South China, Hengyang 421001, China; luoyingps@163.com (Y.L.); changtl168@163.com (T.C.); shea1119@163.com (S.H.); xxj19980705@163.com (J.X.); tsy7909@126.com (S.T.)

* Correspondence: 2015002076@usc.edu.cn

Abstract: Oxidative stress plays a crucial role in the pathogenesis of many diseases. Esculetin is a natural coumarin compound with good antioxidant and anti-inflammatory properties. However, whether esculetin can protect HepG2 cells through inhibiting H_2O_2-induced apoptosis and pyroptosis is still ambiguous. Therefore, this study aimed to investigate the protective effects and mechanisms of esculetin against oxidative stress-induced cell damage in HepG2 cells. The results of this study demonstrate that pretreatment with esculetin could significantly improve the decrease in cell viability induced by H_2O_2 and reduce intracellular ROS levels. Esculetin not only apparently reduced the apoptotic rates and prevented MMP loss, but also markedly decreased cleaved-Caspase-3, cleaved-PARP, pro-apoptotic protein (Bax), and MMP-related protein (Cyt-c) expression, and increased anti-apoptotic protein (Bcl-2) expression in H_2O_2-induced HepG2 cells. Meanwhile, esculetin also remarkably reduced the level of LDH and decreased the expression of the pyroptosis-related proteins NLRP3, cleaved-Caspase-1, Il-1β, and GSDMD-N. Furthermore, esculetin pretreatment evidently downregulated the protein expression of p-JNK, p-c-Fos, and p-c-Jun. Additionally, anisomycin, a specific activator of JNK, blocked the protection of esculetin against H_2O_2-induced HepG2 cells apoptosis and pyroptosis. In conclusion, esculetin can protect HepG2 cells against H_2O_2-induced oxidative stress, apoptosis, and pyroptosis via inhibiting the JNK signaling pathway. These findings indicate that esculetin has the potential to be used as an antioxidant that improves oxidative stress-related diseases.

Keywords: esculetin; oxidative stress; apoptosis; pyroptosis; JNK; liver diseases

1. Introduction

The liver is an important organ, which can transform and clean up endogenous metabolites and exogenous toxins, maintaining the homeostasis of the body. It plays a vital role in metabolism, detoxification, bile secretion, and immune defense. However, the liver is also susceptible to exogenous drugs, viruses, alcohol, metabolic disorders, and other factors, leading to various liver diseases [1].

Oxidative stress is caused by an imbalance between the production and clearance of reactive oxygen species (ROS) in the body. The abundant mitochondria that exist in the liver and its vigorous metabolism make it easy for liver cells to accumulate a large amount of ROS. Excessive ROS can result in oxidative damage to biofilm systems, proteins, and DNA, leading to cell death in serious cases, which plays an important role in the pathogenesis of many chronic liver diseases, such as liver injury, hepatitis, cirrhosis, and liver cancer [2]. Numerous studies demonstrated that natural antioxidants, such as quercetin [3] and curcumin [4], are widely used in daily life, and can effectively inhibit

ROS-induced oxidative stress, which may help to improve the severity of these liver diseases.

Esculetin (Esc), a natural coumarin (Figure 1), is a major active ingredient of the traditional Chinese medicine *Cortex Fraxini* [5]. Esc is a natural antioxidant with good antioxidant properties, which possesses a strong free radical-scavenging capacity [6]. Esc can protect ARPE-19 cells and HEK293 cells from t-BHP-induced oxidative stress and apoptosis [7,8]. Esc can also inhibit amyloid protein-induced oxidative stress and neuronal death in SH-SY5Y cells through the activation of Nrf2 and increase in GSH [9]. Esc improves H_2O_2, t-BHP or ethanol-induced oxidative damage of liver cells by reducing ROS and malondialdehyde (MDA) production and regulating the redox system through the Nrf2/NQO1 pathway [10–12]. Besides, Esc also protects against t-BHP, CCl4, or ethanol-induced liver injury in animal models by reducing oxidative stress and lowering the levels of alanine transaminase (ALT) and aspartate transaminase (AST) in serum [11–13]. However, to date, there has been no report regarding whether Esc improves H_2O_2-induced oxidative stress in HepG2 cells and thus inhibits mitochondrial apoptosis and pyroptosis.

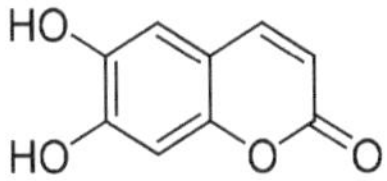

Figure 1. Molecular structure of esculetin.

Therefore, this study aimed to investigate the hepatoprotective effect and mechanism of Esc against H_2O_2-induced oxidative damage, apoptosis, and pyroptosis in HepG2 cells by MTT assays, DCFH-DA assays, LDH assays, flow cytometry, and Western blot assays. Our findings will provide a theoretical basis for the application of Esc as an antioxidant to improve oxidative stress.

2. Results

2.1. Esc Protected HepG2 Cells against H_2O_2-Induced Oxidative Stress

Firstly, HepG2 cells were treated with different concentrations (0~200 μM) of Esc for 12 h to test the cytotoxicity via MTT assay. As shown in Figure 2A, the viability of HepG2 cells was not affected when the concentration of Esc was at 0~50 μM. Therefore, pretreatment with Esc (12.5, 25, and 50 μM) for 12 h was used for the following experiments. Then, HepG2 cells were exposed to the different concentrations of H_2O_2 (0~1100 μM) for 6 h to choose the optimal dose of H_2O_2 treatment (Figure 2B). The results revealed that when the concentration of H_2O_2 was at 700 μM, the viability of HepG2 cells was 51.54 ± 3.83%. So, 700 μM H_2O_2 was used to establish the HepG2 cell oxidative damage model. Next, the protective effect of Esc on H_2O_2-induced oxidative damage in HepG2 cells was shown in Figure 2C. The cell viability of the H_2O_2 model group decreased to 54.27 ± 2.36% compared with that of the control group. On the contrary, compared with the H_2O_2 model group, the cell viability of the Esc (12.5, 25, and 50 μM) group increased to 58.98 ± 1.29%, 68.52 ± 0.45%, and 70.42 ± 3.26% respectively. This indicated that Esc can protect HepG2 cells from H_2O_2-induced repression of cell viability in a dose-dependent manner. Finally, the levels of intracellular ROS were tested by the 2′, 7′-dichlorofluorescin diacetate (DCFH-DA) probe. As presented in Figure 2D, compared with the control group, the intracellular ROS level significantly increased after HepG2 cells exposure to 700 μM H_2O_2 alone for 6 h, while the production of intracellular ROS was notably reduced by pretreatment with 25 μM or 50 μM Esc. These data suggested that Esc could ameliorate H_2O_2-induced oxidative stress in HepG2 cells.

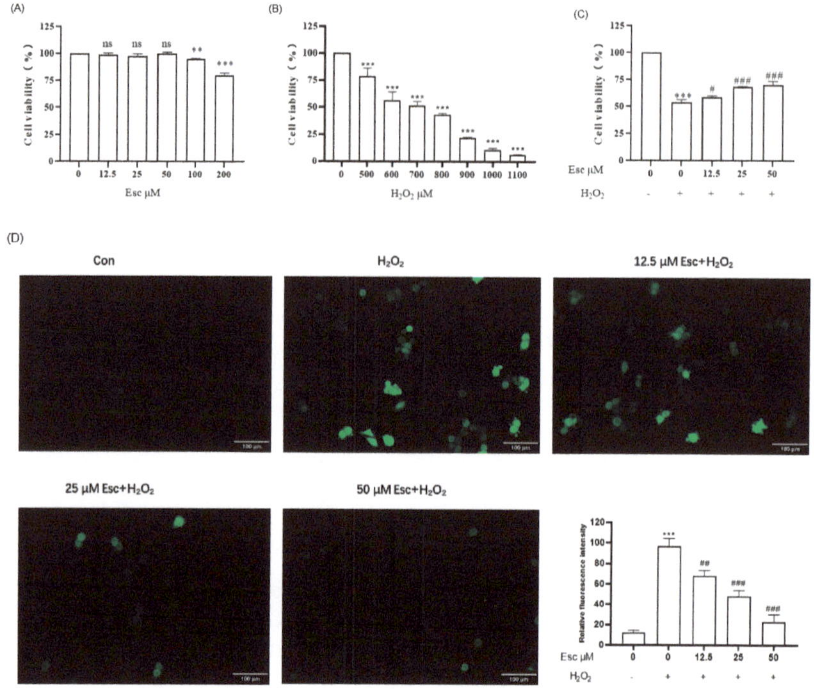

Figure 2. Esc protected HepG2 cells against H_2O_2−induced oxidative stress. (**A**) Effect of Esc on the viability of HepG2 cells. (**B**) Effect of H_2O_2 on the viability of HepG2 cells. (**C**) Effect of Esc on the viability of H_2O_2−induced HepG2 cells. (**D**) Effect of Esc on intracellular ROS levels of H_2O_2−induced HepG2 cells (200×, bar scale = 100 μm). All data are presented as mean ± SD (n = 3). ** $p < 0.01$, *** $p < 0.001$ vs. Control. # $p < 0.05$, ## $p < 0.01$, ### $p < 0.001$ vs. H_2O_2 group. Con: control; Esc: esculetin.

2.2. Esc Protected HepG2 Cells from H_2O_2-Induced Apoptosis

We investigated whether Esc plays a cytoprotective effect in HepG2 cells through suppressing cell apoptosis. Firstly, the apoptosis rate of HepG2 cells was tested by flow cytometry. As shown in Figure 3A, with exposure to H_2O_2 alone for 6 h, the apoptosis rate was increased to 32.88 ± 1.35%, whereas after pretreatment with 50 μM Esc, the apoptotic rates apparently reduced to 12.14 ± 0.90%. Secondly, mitochondrial membrane potential (MMP) was detected by JC-1 staining. Compared with the control group, the red fluorescence intensity of the H_2O_2 mode group was dramatically attenuated, suggesting HepG2 cells exhibited remarkable mitochondrial dysfunction and damage. In contrast, pretreatment with Esc effectively prevented H_2O_2-induced MMP loss (Figure 3B). Finally, to further confirm the above results, the expression of apoptosis-related proteins was measured by Western blot analysis. Compared with those in the H_2O_2 mode group, Esc pretreatment significantly decreased cleaved-Caspase-3, cleaved-PARP, pro-apoptotic protein (Bax) and MMP-related protein (Cyt-c) expression, and increased anti-apoptotic protein (Bcl-2) expression in HepG2 cells (Figure 3C). Accordingly, the ratio of Bax/Bcl-2 and cleaved-PARP/RAPR was markedly decreased in the presence of Esc (Figure 3C). Taken together, these data confirmed that Esc protected HepG2 cells from H_2O_2-induced mitochondrial apoptosis.

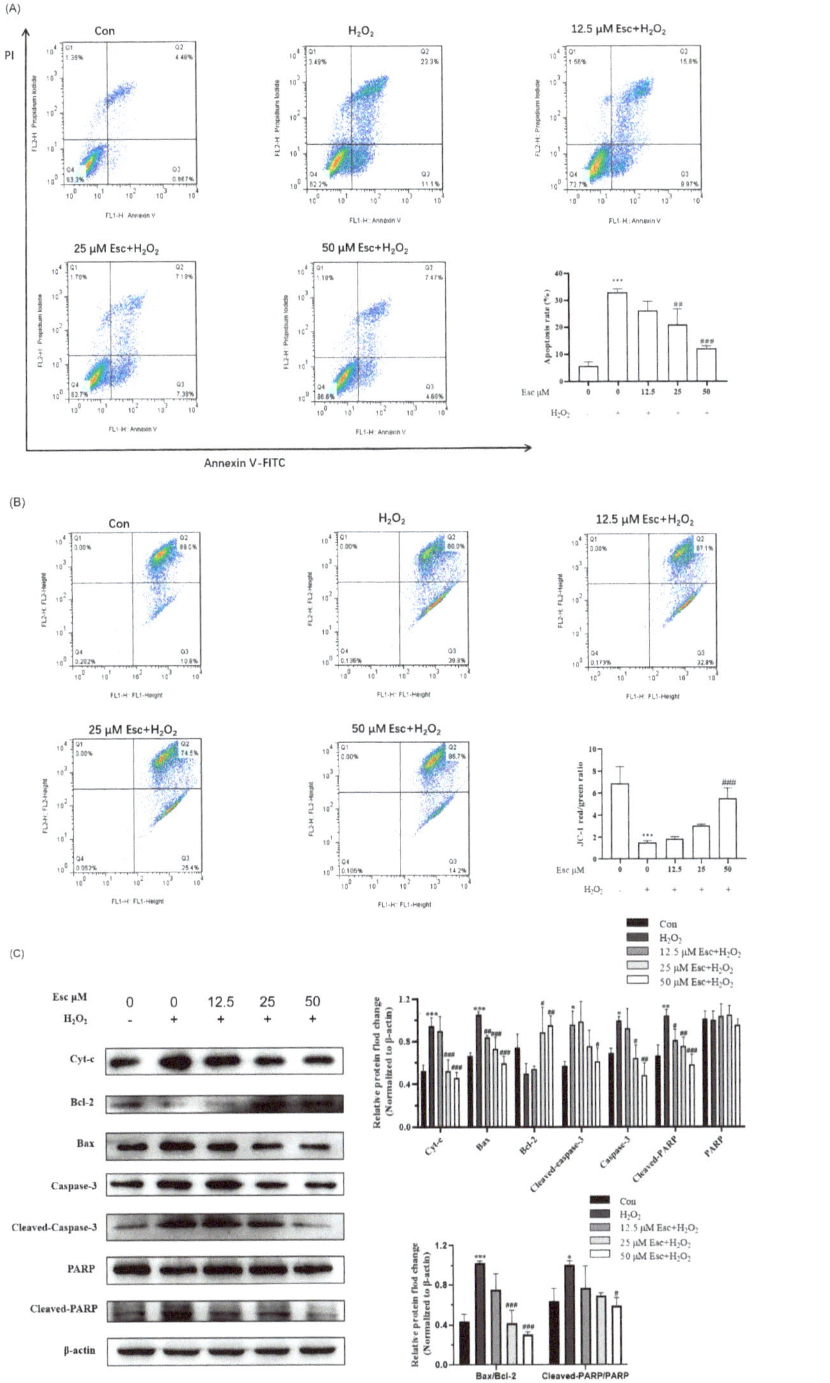

Figure 3. Esc protected HepG2 cells from H_2O_2−induced apoptosis. HepG2 cells were exposed to H_2O_2 for 6 h with Esc (0, 12.5, 25, and 50 μM) pretreatment for 12 h. (**A**) The apoptosis rate of HepG2

cells was determined by flow cytometry. (**B**) The mitochondrial membrane potential of HepG2 cells was detected by flow cytometry. (**C**) The apoptosis proteins levels in each group were measured by Western blot analysis. All data are presented as mean ± SD (n = 3). * $p < 0.05$, ** $p < 0.01$, *** $p < 0.001$ vs. Control. # $p < 0.05$, ## $p < 0.01$, ### $p < 0.001$ vs. H_2O_2 group. Con: control; Esc: esculetin.

2.3. Esc Protected HepG2 Cells from H_2O_2-Induced Pyroptosis

Furthermore, we investigated whether Esc plays a cytoprotective effect in HepG2 cells through inhibiting cell pyroptosis. LDH assay showed the LDH release of the H_2O_2 model group significantly increased to 56.13 ± 1.59%, while pretreatment with 50 μM Esc remarkably reduced the level of LDH to 16.50 ± 0.83% (Figure 4A). Meanwhile, the expression of pyroptosis-related proteins was measured by Western blot analysis. Compared with the H_2O_2 mode group, Esc pretreatment significantly decreased the expression of the pyroptosis-related proteins NLRP3, cleaved-Caspase-1, Il-1β, and GSDMD-N (Figure 4B). Accordingly, the ratio of cleaved-Caspase-1/Caspase-1, Il-1β/Pro-Il-1β, and GSDMD-N/GSDMD-FL was markedly decreased in the presence of Esc (Figure 4B). In short, these results suggested that Esc protected HepG2 cells from H_2O_2-induced pyroptosis.

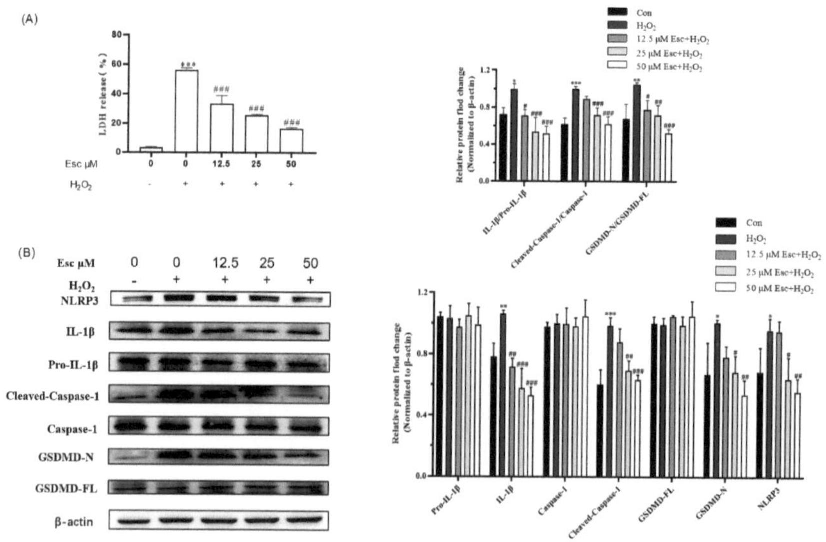

Figure 4. Esc protected HepG2 cells from H_2O_2−induced pyroptosis. HepG2 cells were exposed to H_2O_2 for 6 h with Esc (0, 12.5, 25, and 50 μM) pretreatment for 12 h. (**A**) The LDH release of HepG2 cells was tested by LDH assay. (**B**) The pyroptosis−related proteins of HepG2 cells were measured by Western blot analysis. All data are presented as mean ± SD (n = 3). * $p < 0.05$, ** $p < 0.01$, *** $p < 0.001$ vs. control. # $p < 0.05$, ## $p < 0.01$, ### $p < 0.001$ vs. H_2O_2 group. Con: control; Esc: esculetin.

2.4. Esc Protected HepG2 Cells against H_2O_2-Induced Oxidative Stress via the JNK Signaling Pathway

To determine whether JNK is involved in Esc protected HepG2 cells against H_2O_2-induced oxidative stress, the JNK signaling pathway related proteins were measured by Western blot analysis. As shown in Figure 5, compared with those in the H_2O_2 mode group, the protein expression of p-JNK, p-c-Fos, and p-c-Jun was significantly decreased by Esc pretreatment. Accordingly, the ratio of p-JNK/JNK, p-c-Fos/c-Fos, and p-c-Jun/c-Jun was markedly decreased in the presence of Esc (Figure 5). These illustrated that Esc protected HepG2 cells against H_2O_2-induced oxidative stress was closely related to the JNK signaling pathway.

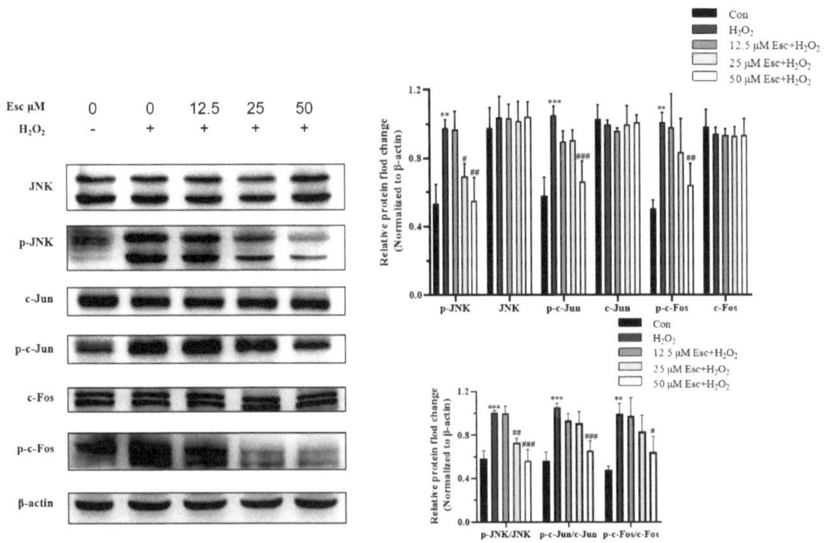

Figure 5. Esc protected HepG2 cells against H_2O_2−induced oxidative damage via the JNK signaling pathway. HepG2 cells were exposed to H_2O_2 for 6 h with Esc (0, 12.5, 25, and 50 μM) pretreatment for 12 h. The JNK signaling pathway−related proteins of HepG2 cells were measured by Western blot analysis. All data are presented as mean ± SD (n = 3). ** $p < 0.01$, *** $p < 0.001$ vs. Control. $^{\#}$ $p < 0.05$ or $^{\#\#}$ $p < 0.01$ or $^{\#\#\#}$ $p < 0.001$ vs. H_2O_2 group. Con: control; Esc: esculetin.

2.5. Ani Reversed the Protection of Esc against H_2O_2-Induced HepG2 Cells' Oxidative Stress, Apoptosis, and Pyroptosis

To further verify that the protection of Esc against H_2O_2-induced HepG2 cells oxidative stress was correlated with the JNK signaling pathway, the JNK-specific activator anisomycin (Ani) was used for revalidation (Figure 6). As shown in Figure 6A, when the concentration of Ani was at 12.5 nM or below, it had no cytotoxic effect on HepG2 cells, so 12.5 nM Ani was selected for subsequent experiments. Firstly, our results demonstrated that the combined treatment (Ani and Esc) obviously decreased the cell viability of HepG2 compared with the Esc treatment (Figure 6B). Then, the percentage of apoptotic cells (Figure 6C), and the apoptotic protein (cleaved-Caspase-3) expression (Figure 6D) in the combined treatment group (Ani and Esc) was significantly increased compared with that of the Esc treatment group. Likewise, the generation of LDH release (Figure 6E), and the pyroptosis related-protein (GSDMD-N) expression (Figure 6F) in the combined treatment group (Ani and Esc) was also increased. These results indicated that Ani can reverse the protection of Esc against H_2O_2-induced HepG2 cells apoptosis and pyroptosis. Moreover, Western blot analysis of the JNK signaling pathway related-proteins showed that the combined treatment group (Ani and Esc) markedly inhibited the downregulation of p-JNK, p-c-Fos, and p-c-Jun expression in H_2O_2-induced HepG2 cells compared with the Esc treatment group (Figure 6G). Collectively, Esc can protect HepG2 cells against H_2O_2-induced oxidative stress, apoptosis, and pyroptosis by inhibiting the JNK signaling pathway.

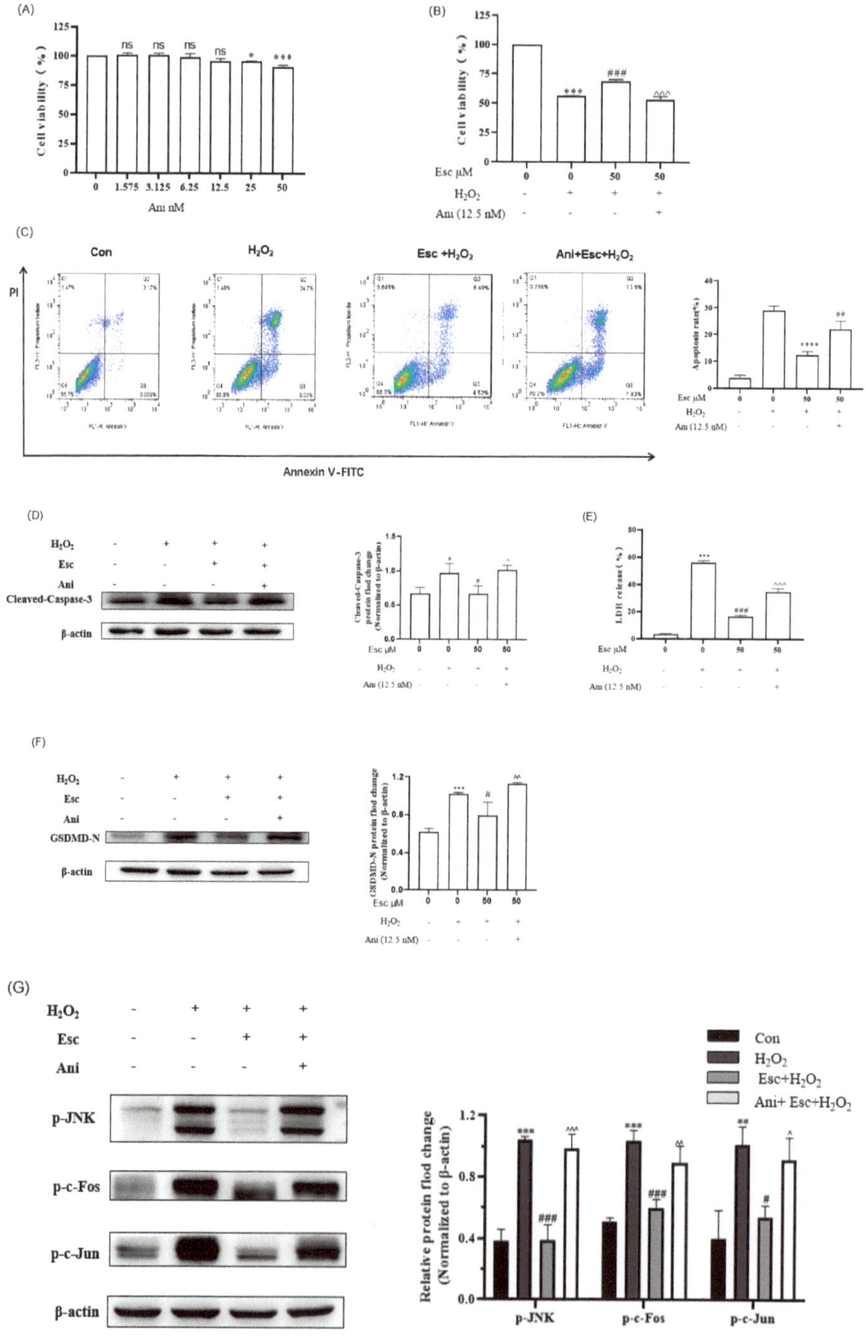

Figure 6. Ani reversed the protection of Esc against H_2O_2−induced HepG2 cells' oxidative stress, apoptosis, and pyroptosis. (**A**) The cytotoxicity of Ani on HepG2 cells. (**B**) The effect of Ani and Esc on the viability of H_2O_2−induced HepG2 cells. (**C**) The apoptosis rate of HepG2 cells was determined by flow cytometry. (**D**) The protein expression level of cleaved−Caspase-3 was measured by Western blot analysis. (**E**) LDH release from HepG2 cells was tested by LDH assay. (**F**) The protein expression

level of GSDMD-N was measured by Western blot analysis. (**G**) The protein expression levels of p−JNK, p−c−Fos, and p−c−Jun were measured by Western blot analysis. Data are presented as mean ± SD (n = 3). * $p < 0.05$, ** $p < 0.01$, *** $p < 0.001$, **** $p < 0.0001$ vs. Control. # $p < 0.05$, ## $p < 0.01$, ### $p < 0.001$ vs. H_2O_2 group. ^ $p < 0.05$, ^^ $p < 0.01$, ^^^ $p < 0.001$ vs. Esc experimental group. Con: control; Esc: esculetin; Ani: anisomycin.

3. Discussion

Oxidative stress is a negative factor caused by excessive free radicals in the body, which is crucial in the pathogenesis of many chronic liver diseases, such as liver injury, hepatitis, cirrhosis, nonalcoholic fatty liver disease, and liver cancer. Accumulating evidence has shown that natural antioxidants are commonly used in daily life, playing vital roles in antioxidative stress and thus improving these chronic liver diseases. Esculetin (Esc) is a major active coumarin isolated from the traditional Chinese medicine *Cortex Fraxini* [5], which exhibits notable antioxidant, anti-inflammatory, anti-bacterial, and anti-cancer pharmacological effects [14]. Previous studies have demonstrated that Esc not only improves H_2O_2-, t-BHP- or ethanol-induced oxidative damage of liver cells, but also protects t-BHP-, CCl4-, or ethanol-induced liver injury in animal models [10–13]. However, to date, there has been no report of whether Esc improving H_2O_2-induced oxidative stress in HepG2 cells is related to the inhibition of mitochondrial apoptosis and pyroptosis. Therefore, we aimed to investigate the protective function and the mechanism of Esc against H_2O_2-induced oxidative stress, apoptosis and pyroptosis in HepG2 cells for the purpose of exploring the potential use of Esc in attenuating oxidative stress.

H_2O_2 is the main component of intracellular ROS that are produced during many physiological and pathological processes, which can easily cross the cell membrane, directly destroy the structural stability of DNA and protein, induce lipid peroxidation, and cause oxidative stress [15]. As an ideal inducer, H_2O_2 is often used to establish cell oxidative damage models and analyze the mechanism of oxidative stress. In our experiments, when HepG2 cells were exposed to 700 μM H_2O_2 for 6 h, the cell viability of HepG2 cells was notably decreased, and the ROS production was significantly accumulated. It was determined that Esc pretreatment can effectively decrease the ROS level and improve the cell viability in a dose-dependent manner. These results confirmed that Esc can alleviate H_2O_2-induced oxidative stress in HepG2 cells.

Excessive ROS will cause lipid peroxidation of the mitochondrial membrane, which will lead to opening of the mitochondrial permeability transition pore and will depolarize the mitochondrial membrane. Cyt-c is released from the mitochondria and binds to apoptotic protease activating factor 1 (Apaf-1), triggering the caspase family cascade reaction that initiates the mitochondrial apoptosis pathway [16–18]. It was reported that H_2O_2 can reduce the mitochondrial membrane potential and increase the apoptotic protein expression of cleaved Caspase-3, thus inducing apoptosis in HepG2 cells. Previous studies indicated that Esc could improve oxidative stress in HEK293 cells and H9c2 cells by suppressing mitochondrial apoptosis pathways [7,19]. These findings implied that Esc improves H_2O_2-induced oxidative stress in HepG2 cells, which might be related to inhibition of the mitochondrial apoptosis pathway. Our results revealed that pretreatment with Esc not only apparently increased anti-apoptotic protein (Bcl-2) expression and prevented MMP loss, but also markedly reduced the apoptotic rates and the expression of cleaved-Caspase-3, cleaved-PARP, pro-apoptotic protein (Bax), and MMP-related protein (Cyt-c) in H_2O_2-induced HepG2 cells. In short, it can be concluded that Esc can alleviate H_2O_2-induced oxidative stress in HepG2 cells by inhibiting mitochondrial apoptosis pathway.

Pyroptosis is initiated by the activation of inflammasomes. The NLRP3 inflammasome triggers the activation of Caspase-1, and cleavage of pro-IL-1β and Gasdermin D (GSDMD), thereby resulting in the formation of membrane openings and the release of inflammatory cytokine (IL-18 and IL-1β), ultimately leads to cell pyroptosis. A large amount of studies confirmed that ROS can cause mitochondrial damage in various liver cells (HL7702, AML12,

LO2, LM3, and Huh7) though activating the NLRP3 mediated-pyroptosis pathway [20–22]. Recent evidence demonstrated that Esc has good anti-inflammation effects and can treat intestinal inflammatory diseases (IBD) by inhibiting the NLRP3 mediated-pyroptosis pathway [23,24]. Based on this, we speculate that Esc improves H_2O_2-induced oxidative stress in HepG2 cells, which might be associated with the NLRP3 mediated-pyroptosis pathway. Our results suggested that pretreatment with Esc remarkably decreased the level of LDH, but also significantly reduced the expression of the pyroptotic proteins NLRP3, cleaved-Caspase-1, Il-1β, and GSDMD-N in H_2O_2-induced HepG2 cells. Therefore, these results indicated that Esc alleviates H_2O_2-induced oxidative stress in HepG2 by inhibiting the NLRP3 mediated-pyroptosis pathway.

Previous studies have shown that H_2O_2 can induce cell apoptosis through the JNK/AP-1 signaling pathway [25,26]. Meanwhile, H_2O_2 can also result in cell pyroptosis by activating Caspase-1 and GSDMD proteins [27]. In recent years, it has been reported that Esc can improve oxidative stress and apoptosis in H9c2 cells by reducing the phosphorylation level of JNK in order to attenuate myocardial ischemia/reperfusion (I/R) injury [28]. In addition, Esc can protect oxidative stress-induced aging in human HaCaT keratinocytes by decreasing the phosphorylation level of c-Jun and c-Fos and inhibiting the JNK signaling pathway [29]. Thus, we hypothesized that the protective effect of Esc in H_2O_2-induced HepG2 cells might be closely associated with the JNK signaling pathway. Our Western blot results signified that Esc pretreatment evidently downregulated the proteins expression of p-JNK, p-c-Fos, and p-c-Jun. Then, using the JNK specific activator (anisomycin) to reverse verification, it was revealed that anisomycin upregulated the expression of the apoptotic protein (cleaved-Caspase-3), the pyroptosis related-protein (GSDMD-N), and the JNK signaling pathway related-proteins of p-JNK, p-c-Fos, and p-c-Jun. This indicated that anisomycin reversed the protection of Esc in H_2O_2-induced HepG2 cells. Taken together, Esc was able to attenuate oxidative stress, apoptosis, and pyroptosis in H_2O_2-induced HepG2 cells by inhibiting the JNK signaling pathway.

4. Materials and Methods

4.1. Reagents and Antibodies

Esculetin was purchased from Yuanyebio (Shanghai, China). Anisomycin (Ani), and methylthiazolyldiphenyl-tetrazolium bromide (MTT) was purchased from Solarbio (Beijing, China). A LDH Detection Kit and a JC-1 Staining Kit were purchased from Beyotime (Shanghai, China). An Annexin V-FITC/PI Apoptosis Detection Kit was purchased from Key gen Biotech (Nanjing, China). DMEM was purchased from Sigma Aldrich (St. Louis, IL, USA). Fetal bovine serum was purchased from ExCell Bio (Taicang, China). Antibodies against Bax, Bcl-2, Caspase-1, and p-c-Jun were purchased from Proteintech (Chicago, MO, USA). Antibodies against cleaved-Caspase-3, and p-c-Fos were purchased from Cell Signaling Technology (Danvers, MA, USA). Antibodies against c-Fos, c-Jun, Caspase-3, JNK, and PARP were purchased from Beyotime (Shanghai, China). Anti-β-actin antibody was purchased from Solarbio (Beijing, China). Antibodies against cleaved-PARP, Cytochrome c (Cyt-c), GSDMD, IL-1β, NLRP3, and p-JNK were purchased from Abcam (Cambridge, UK).

4.2. Cell Culture

HepG2 cells were purchased from the Cell Bank of the Chinese Academy of Sciences (Shanghai, China) and cultured at 37 °C in high-glucose DMEM, supplemented with 10% FBS, 1% 100 unit/mL penicillin and 100 μg/mL streptomycin with 5% CO_2.

4.3. Cell Viability Assay

Cell viability was evaluated by MTT assay. HepG2 cells treated with neither Esc nor H_2O_2 were used as the control group. Briefly, HepG2 cells were seeded in a 96-well plate at a density of 1×10^4 cells/100 μL in DMEM medium and cultured for 12 h. Next, the cells were treated with 700 μM H_2O_2 for 6 h with or without 12 h pretreatment with different concentrations of Esc or Ani. Then, 10 μL 5 mg/mL MTT solution was added to each well and incubated at 37 °C in the dark for 4 h. The liquid was discarded and 150 μL DMSO was added to each well. The absorbance at 490 nm was determined using a microplate reader (Bio-Rad, Hercules, CA, USA). The cell viability was calculated using the following formula:

$$\text{cell viability (\%)} = (A_{\text{experimental group}} - A_{\text{blank group}})/(A_{\text{control group}} - A_{\text{blank group}}) \times 100\%$$

4.4. ROS Production Assay

The ROS productions in HepG2 cells were tested using the dichloro dihydrofluorescein diacetate (DCFH-DA) Detection Kit method. HepG2 cells were seeded in six-well plates at a density of 2.5×10^5 cells/well and incubated with different concentrations of Esc for 12 h. Next, the cells were exposed to 700 μM H_2O_2 for 6 h. Then, the cells were washed twice and incubated with DCFH-DA for 30 min at 37 °C in the dark. The ROS relative intensity of cells was observed under a fluorescence microscope (Olympus, Tokyo, Japan). Green fluorescence intensity was measured by imageJ 1.8.0 software.

4.5. Apoptosis Rate Assay

Reh Cell apoptosis was tested by using an Annexin-V-FITC Apoptosis Detection Kit. HepG2 cells were seeded in 6-well plates at 2.5×10^5 cells/well and exposed to 700 μM H_2O_2 for 6 h, with or without 12 h of pretreatment with different concentrations of Esc. Cells were harvested and double-stained with Annexin V-FITC and PI for 10 min in the dark. The apoptotic rate (%) of HepG2 cells was detected via flow cytometry (Piscataway, NJ, USA).

4.6. Mitochondrial Membrane Potential Assay

Mitochondrial membrane potential (MMP) was tested by JC-1 assay. HepG2 cells were seeded in six-well plates at 2.5×10^5 cells/well and exposed to 700 μM H_2O_2 for 6 h, with or without 12 h of pretreatment with different concentrations of Esc. After being washed with PBS, the cells were harvested and stained with JC-1 for 40 min in the dark. Then, the cells were washed twice with staining buffer. Changes in the mitochondria of each group were detected via flow cytometry (Piscataway, NJ, USA).

4.7. LDH Release Assay

The LDH content in the culture medium was tested with the LDH Release Assay kit. HepG2 cells were seeded in 96-well plates at 1×10^4 cells/well and exposed to 700 μM H_2O_2 for 6 h, with or without 12 h of pretreatment with different concentrations of Esc. Then, the LDH release reagents were added to the plate 1 h before the test. After centrifuging, the supernatant was collected and incubated with LDH working solution in the dark for 30 min. The absorbance at 490 nm was determined using a microplate reader.

4.8. Western Blot Analysis

The total protein of HepG2 cells was extracted with RIPA buffer (Solarbio) and the protein concentration was quantified using a BCA protein assay kit (ComWin Biotech, Beijing, China). Protein (20 μg) in different groups was separated on 10–15% SDS-PAGE (Beyotime) and transferred to PVDF membrane (Millipore, USA). After being blocked with 5% BSA (Soiarbio) at room temperature for 2 h, membranes were incubated with primary antibodies (Bax, Bcl-2, Caspase-3, cleaved-Caspase-3, PARP, cleaved-PARP, Cyt-c, Caspase-1, GSDMD, IL-1β, Pro-Il-1β, NLRP3, JNK, p-JNK, c-Fos, p-c-Fos, c-Jun, p-c-Jun, and β-actin) overnight at 4 °C. Subsequently, the membranes were washed with Tris-buffered saline and Tween 20 (TBST) six times and incubated with the corresponding secondary antibodies at room temperature for 1 h. Finally, protein bands were visualized using an ECL developer (New Cell & Molecular Biotech, Suzhou, China) and scanned in a chemiluminescence imager (Tanon, Shanghai, China). The protein levels were calculated relative to that of β-actin.

4.9. Statistical Analysis

Data are presented as the mean ± SD. Each experiment was repeated at least three times. GraphPad Prism 8 software was used for data analysis. Statistical analysis was performed by *t*-test or one-way analysis of variance (ANOVA), and a $p < 0.05$ and below was regarded as statistically significant.

5. Conclusions

In conclusion, we verified that esculetin could protect HepG2 cells against H_2O_2-induced oxidative stress by attenuating apoptosis and pyroptosis by inhibiting the JNK signaling pathway. These results suggest that esculetin has the potential to be used as an antioxidant to improve oxidative stress. In the future, the protective effects and mechanism of esculetin against oxidative stress in animal models will be investigated to provide a theoretical basis for applying Esc to prevent oxidative stress-related diseases.

Author Contributions: Conceptualization, Y.L.; methodology, H.S.; software, J.X.; validation, Y.L.; formal analysis, S.H.; investigation, S.T.; data curation, S.T.; writing—original draft preparation, Y.L. and T.C.; writing—review and editing, Y.L. and H.S.; visualization, S.T.; supervision, J.X.; project administration, T.C. and S.H.; funding acquisition, H.S. All authors have read and agreed to the published version of the manuscript.

Funding: This work was financially supported by the Science Fund of Hunan Provincial Education Department (Nos. 23B0395, 23A0328, and 20B504), the Natural Science Foundation of Hunan Province (No. 2018JJ3433), the Health Research Project of Hunan Provincial Health Commission (No. D202302047808), the Undergraduate Training Programs for Innovation and Entrepreneurship of Hunan Province (No. D202305162258543382), and the Undergraduate Training Programs for Innovation and Entrepreneurship of University of South China (No. 2022x10555110).

Institutional Review Board Statement: Not applicable.

Informed Consent Statement: Not applicable.

Data Availability Statement: The data that support the findings of this study are available from the corresponding author upon reasonable request.

Conflicts of Interest: The authors declare no conflicts of interest.

References

1. Lister, I.N.E.; Ginting, C.N.; Girsang, E.; Nataya, E.D.; Azizah, A.M.; Widowati, W. Hepatoprotective properties of red betel (*Piper crocatum Ruiz and Pav*) leaves extract towards H_2O_2-induced HepG2 cells via anti-inflammatory, antinecrotic, antioxidant potency. *Saudi Pharm. J.* **2020**, *28*, 1182–1189. [CrossRef]
2. Jian, L.Y.; Xue, Y.; Gao, Y.F.; Wang, B.; Qu, Y.H.; Li, S.H.; Li, H.Q.; Li, Z.; Wang, B.; Luo, H.L. Vitamin E can ameliorate oxidative damage of ovine hepatocytes in vitro by regulating genes expression associated with apoptosis and pyroptosis, but not ferroptosis. *Molecules* **2021**, *26*, 4520. [CrossRef]
3. Doustimotlagh, A.H.; Taheri, S.; Mansourian, M.; Eftekhari, M. Extraction and identification of two flavonoids in *Phlomoides hyoscyamoides* as an endemic plant of Iran: The role of quercetin in the activation of the glutathione peroxidase, the improvement of the hydroxyproline and protein oxidation in bile duct-ligated rats. *Curr. Comput.-Aided Drug Des.* **2020**, *16*, 629–640.
4. Wang, Y.J.; Liu, F.J.; Liu, M.R.; Zhou, X.; Wang, M.; Cao, K.X.; Jin, S.J.; Shan, A.S.; Feng, X.J. Curcumin mitigates aflatoxin B1-induced liver injury via regulating the NLRP3 inflammasome and Nrf2 signaling pathway. *Food Chem. Toxicol.* **2022**, *161*, 112823. [CrossRef]
5. Wu, C.R.; Huang, M.Y.; Lin, Y.T.; Ju, H.Y.; Ching, H. Antioxidant properties of Cortex Fraxini and its simple coumarins. *Food Chem.* **2007**, *104*, 1464–1471. [CrossRef]
6. Kim, S.H.; Kang, K.A.; Zhang, R.; Piao, M.J.; Ko, D.O.; Wang, Z.H.; Chae, S.W.; Kang, S.S.; Lee, K.H.; Kang, H.K.; et al. Protective effect of esculetin against oxidative stress-induced cell damage via scavenging reactive oxygen species. *Acta Pharmacol. Sin.* **2008**, *29*, 1319–1326. [CrossRef] [PubMed]
7. Jung, W.K.; Park, S.B.; Yu, H.Y.; Kim, Y.H.; Kim, J. Antioxidant efficacy of esculetin against tert-butyl hydroperoxide-induced oxidative stress in HEK293 cells. *Curr. Issues Mol. Biol.* **2022**, *44*, 5986–5994. [CrossRef]
8. Jung, W.K.; Park, S.B.; Yu, H.Y.; Kim, Y.H.; Kim, J. Effect of esculetin on tert-butyl hydroperoxide-induced oxidative injury in retinal pigment epithelial cells in vitro. *Molecules* **2022**, *27*, 8970. [CrossRef] [PubMed]
9. Pruccoli, L.; Morroni, F.; Sita, G.; Hrelia, P.; Tarozzi, A. Esculetin as a bifunctional antioxidant prevents and counteracts the oxidative stress and neuronal death induced by amyloid protein in SH-SY5Y cells. *Antioxidants* **2020**, *9*, 551. [CrossRef] [PubMed]

10. Subramaniam, S.R.; Ellis, E.M. Esculetin-induced protection of human hepatoma HepG2 cells against hydrogen peroxide is associated with the Nrf2-dependent induction of the NAD(P)H: Quinone oxidoreductase 1 gene. *Toxicol. Appl. Pharmacol.* **2011**, *250*, 130–136. [CrossRef]
11. Lin, W.L.; Wang, C.J.; Tsai, Y.Y.; Liu, C.L.; Hwang, J.M.; Tseng, T.H. Inhibitory effect of esculetin on oxidative damage induced by t-butyl hydroperoxide in rat liver. *Arch. Toxicol.* **2000**, *74*, 467–472. [CrossRef]
12. Lee, J.; Yang, J.; Yang, J.; Jeon, J.; Jeong, Y.; Jeong, H.S.; Lee, J.; Sun, J. Hepatoprotective effect of esculetin on ethanol-induced liver injury in human HepG2 cells and C57BL/6J mice. *J. Funct. Foods* **2018**, *40*, 536–543. [CrossRef]
13. Murat Bilgin, H.; Atmaca, M.; Deniz Obay, B.; Ozekinci, S.; Taşdemir, E.; Ketani, A. Protective effects of coumarin and coumarin derivatives against carbon tetrachloride-induced acute hepatotoxicity in rats. *Exp. Toxicol. Pathol.* **2011**, *63*, 325–330. [CrossRef]
14. Garg, S.S.; Gupta, J.; Sahu, D.; Liu, C.J. Pharmacological and therapeutic applications of esculetin. *Int. J. Mol. Sci.* **2022**, *23*, 12643. [CrossRef] [PubMed]
15. Xu, J.J.; Zhang, Y.W.; Ren, G.F.; Yang, R.G.; Chen, J.F.; Xiang, X.J.; Qin, H.; Chen, J.H. Inhibitory effect of delphinidin on oxidative stress induced by H_2O_2 in HepG2 Cells. *Oxid. Med. Cell. Longev.* **2020**, *2020*, 4694760. [CrossRef] [PubMed]
16. Basu, C.; Sur, R. S-Allyl cysteine alleviates hydrogen peroxide induced oxidative injury and apoptosis through upregulation of Akt/Nrf-2/HO-1 signaling pathway in HepG2 cells. *BioMed Res. Int.* **2018**, *2018*, 3169431. [CrossRef]
17. Jiang, J.Y.; Yu, S.N.; Jiang, Z.C.; Liang, C.H.; Yu, W.B.; Li, J.; Du, X.D.; Wang, H.L.; Gao, X.H.; Wang, X. N-acetyl-serotonin protects HepG2 cells from oxidative stress injury induced by hydrogen peroxide. *Oxid. Med. Cell. Longev.* **2014**, *2014*, 310504. [CrossRef] [PubMed]
18. Su, M.; Yu, T.F.; Zhang, H.; Wu, Y.; Wang, X.Q.; Li, G. The antiapoptosis effect of glycyrrhizate on HepG2 cells induced by hydrogen peroxide. *Oxid. Med. Cell. Longev.* **2016**, *2016*, 6849758. [CrossRef] [PubMed]
19. Xu, F.; Li, X.; Liu, L.F.; Xiao, X.; Zhang, L.; Zhang, S.L.; Lin, P.P.; Wang, X.J.; Wang, Y.W.; Li, Q.S. Attenuation of doxorubicin-induced cardiotoxicity by esculetin through modulation of Bmi-1 expression. *Exp. Ther. Med.* **2017**, *14*, 2216–2220. [CrossRef]
20. Mai, W.J.; Xu, Y.Z.; Xu, J.H.; Zhao, D.; Ye, L.Y.; Yu, G.X.; Wang, Z.L.; Lu, Q.T.; Lin, J.E.; Yang, T.; et al. Berberine inhibits Nod-like receptor family pyrin domain containing 3 inflammasome activation and pyroptosis in nonalcoholic steatohepatitis via the ROS/TXNIP axis. *Front. Pharmacol.* **2020**, *11*, 185. [CrossRef]
21. Xie, W.H.; Ding, J.; Xie, X.X.; Yang, X.H.; Wu, X.F.; Chen, Z.X.; Guo, Q.L.; Gao, W.Y.; Wang, X.Z.; Li, D. Hepatitis B virus X protein promotes liver cell pyroptosis under oxidative stress through NLRP3 inflammasome activation. *Inflamm. Res.* **2020**, *69*, 683–696. [CrossRef]
22. Yao, T.; Chen, J.M.; Shen, L.E.; Yu, Y.S.; Tang, Z.H.; Zang, G.Q.; Zhang, Y.; Chen, X.H. Astragalus polysaccharide alleviated hepatocyte senescence via autophagy pathway. *Kaohsiung J. Med. Sci.* **2022**, *38*, 457–468. [CrossRef]
23. Wang, S.K.; Chen, T.X.; Wang, W.; Xu, L.L.; Zhang, Y.Q.; Jin, Z.; Liu, Y.B.; Tang, Y.Z. Aesculetin exhibited anti-inflammatory activities through inhibiting NF-κB and MAPKs pathway in vitro and in vivo. *J. Ethnopharmacol.* **2022**, *296*, 115489. [CrossRef]
24. Ren, W.; Zhou, Q.Y.; Yu, R.Y.; Liu, Z.J.; Hu, Y.S. Esculetin inhibits the pyroptosis of microvascular endothelial cells through NF-κB /NLRP3 signaling pathway. *Arch. Biochem. Biophys.* **2022**, *720*, 109173. [CrossRef]
25. Ishikawa, Y.; Kitamura, M. Anti-apoptotic effect of quercetin: Intervention in the JNK- and ERK-mediated apoptotic pathways. *Kidney Int.* **2000**, *58*, 1078–1087. [CrossRef]
26. Kitamura, M.; Ishikawa, Y.; Moreno-Manzano, V.; Xu, Q.; Konta, T.; Lucio-cazana, J.; Furusu, A.; Nakayama, K. Intervention by retinoic acid in oxidative stress-induced apoptosis. *Nephrol. Dial. Transplant.* **2002**, *9*, 84–87. [CrossRef] [PubMed]
27. Liu, J.; Yan, Y.Y.; Zheng, D.D.; Zhang, J.F.; Wang, J.N. Inhibiting microRNA-200a-3p attenuates pyroptosis via targeting the SIRT1/NF-κB/NLRP3 pathway in H_2O_2-induced HAEC. *Aging* **2023**, *15*, 11184–11200. [CrossRef] [PubMed]
28. Yang, J.Y.; Han, J.; Li, Y.; Dong, B. Esculetin inhibits the apoptosis in H9c2 cardiomyocytes via the MAPK signaling pathway following hypoxia/reoxygenation injury. *Biomed. Pharmacother.* **2017**, *88*, 1206–1210. [CrossRef]
29. Zhen, A.X.; Piao, M.J.; Kang, K.A.; Fernando, P.D.S.M.; Kang, H.K.; Koh, Y.S.; Hyun, J.W. Esculetin prevents the induction of matrix metalloproteinase-1 by hydrogen peroxide in skin keratinocytes. *J. Cancer Prev.* **2019**, *24*, 123–128. [CrossRef] [PubMed]

molecules

MDPI

Review

Plant Polyphenols and Their Potential Benefits on Cardiovascular Health: A Review

Iram Iqbal [1,†], Polrat Wilairatana [2,*,†], Fatima Saqib [2,†], Bushra Nasir [3], Muqeet Wahid [1], Muhammad Farhaj Latif [1], Ahmar Iqbal [4], Rabia Naz [1] and Mohammad S. Mubarak [5,*]

1 Department of Pharmacology, Faculty of Pharmacy, Bahauddin Zakariya University, Multan 60800, Pakistan; iramiqbal.bzu@gmail.com (I.I.); muqeetsoomro@msn.com (M.W.); farhajlatif@gmail.com (M.F.L.); rabianazbhutta@gmail.com (R.N.)

2 Department of Clinical Tropical Medicine, Faculty of Tropical Medicine, Mahidol University, Bangkok 10400, Thailand; fatima.saqib@bzu.edu.pk

3 Department of Pharmaceutics, Faculty of Pharmacy, Bahauddin Zakariya University, Multan 60800, Pakistan; bushranasir@bzu.edu.pk

4 Department of General Surgery, Shanxi Medical University, Jinzhong 030600, China; ahmar204@yahoo.com

5 Department of Chemistry, The University of Jordan, Amman 11942, Jordan

* Correspondence: polrat.wil@mahidol.ac.th (P.W.); mmubarak@ju.edu.jo (M.S.M.)

† These authors contributed equally to this work.

Abstract: Fruits, vegetables, and other food items contain phytochemicals or secondary metabolites which may be considered non-essential nutrients but have medicinal importance. These dietary phytochemicals exhibit chemopreventive and therapeutic effects against numerous diseases. Polyphenols are secondary metabolites found in vegetables, fruits, and grains. These compounds exhibit several health benefits such as immune modulators, vasodilators, and antioxidants. This review focuses on recent studies on using dietary polyphenols to treat cardiovascular disorders, atherosclerosis, and vascular endothelium deficits. We focus on exploring the safety of highly effective polyphenols to ensure their maximum impact on cardiac abnormalities and discuss recent epidemiological evidence and intervention trials related to these properties. Kaempferol, quercetin, and resveratrol prevent oxidative stress by regulating proteins that induce oxidation in heart tissues. In addition, polyphenols modulate the tone of the endothelium of vessels by releasing nitric oxide (NO) and reducing low-density lipoprotein (LDL) oxidation to prevent atherosclerosis. In cardiomyocytes, polyphenols suppress the expression of inflammatory markers and inhibit the production of inflammation markers to exert an anti-inflammatory response. Consequently, heart diseases such as strokes, hypertension, heart failure, and ischemic heart disease could be prevented by dietary polyphenols.

Keywords: polyphenols; cardiovascular; atherosclerosis; oxidative stress

1. Introduction

Cardiovascular disease (CVD), encompassing conditions such as atherosclerosis, hypertension, myocardial infarction, cardiomyopathy, arrhythmia, and heart failure (HF), is a major contributor to global mortality. The incidence of CVD has experienced a notable increase [1–4]. Despite the wide range of pharmaceuticals currently utilized for the management of CVD, such as statins, angiotensin-converting enzyme inhibitors (ACEIs), angiotensin receptor blockers (ARBs), calcium channel blockers (CCBs), fibrates, and β-blockers, it is important to acknowledge that a significant number of these medications are associated with adverse effects in the human population [4]. Hence, there exists a significant clinical requirement to discover and cultivate innovative therapeutic strategies for CVD [2]. The transport of oxygen and nutrients in the human body is carried out via blood circulation along with the removal of metabolic by-products and carbon dioxide through the cardiovascular system (CVS). Coronary artery disease (CAD), cerebrovascular disease

(CVD), peripheral artery disease (PAD), congenital heart disease (CHD), hypertension, heart failure, and stroke are all disorders that affect the heart and blood arteries [1,2]. Within this context, cardiovascular diseases (CVDs) are among the leading causes of mortality throughout the world, claiming 17.9 million individual lives worldwide in 2019, which is approximately 32% of total fatalities. Approximately, 85% of these mortality rates were due to heart attacks and strokes (WHO site). Strokes kill 6.7 million people each year, and coronary heart disease claims 7.4 million lives [3,4].

Several pathologies can affect the cardiovascular system. Some of these pathologies include primary heart ailments, including cardiomyopathy and cardiac malignancies. Infectious and infectious-allergic damage to heart tissue, metabolic and systemic disorders, and diseases of other organs are also covered in this category [5,6]. CHD starts with inflammation of the blood artery walls, which narrows and causes angina pectoris [7]. In this respect, blood clots restrict arteries later in the disease's progression, resulting in severe myocardial ischemia and myocardial infarction (heart attack). Heart failure can occur in severe cases of CHD when the heart muscle's ability to pump blood around the body deteriorates [2]. Because these disorders are generally caused by arterial injury, symptoms and treatments vary depending on which arteries are afflicted [2].

Age and gender are among the most reported non-modifiable cardiovascular risk factors. Cardiovascular disorders become more common as people become older due to a rise in plasma cholesterol on one hand and the augmentation of arterial rigidity and peripheral vascular resistance on the other hand [8]. Although the risk of cardiovascular disorder varies with gender and age, the incidence is three to five times greater in men < 50 years of age compared to women. On the other hand, a considerable increase in the occurrence of CVD has been observed in women over the age of 50 years. Genetic factors, inactivity, hypertension, obesity, diabetes, smoking, and dyslipidemia are the prominent risk factors for cardiovascular disorders, as described in published reports [9–12].

Research findings have shown that a balanced diet is beneficial for preventing CVD [13]. The consumption of a high percentage of fruits and vegetables in a diet, such as the Vegan diet, is strongly correlated with a long life expectancy. In addition, it decreases the incidence of cardiovascular diseases [14]. According to epidemiological research, people who consume a diet rich in polyphenols experience a 46 percent reduction in their risk of developing CVD [15]; these food items are rich in polyphenols. Polyphenols are present in vegetables and in many fruits and seeds that we routinely consume as secondary metabolites [16]. In this regard, there is a connection between the consumption of fruits, vegetables, seeds, and nuts and a decreased incidence of chronic and age-related degenerative illnesses [17].

Polyphenols have also been found useful in enhancing endothelial function, preventing aberrant platelet aggregation, decreasing inflammation, and improving plasma lipid profile, all of which benefit cardiovascular health. Because the processes by which these chemicals exert cardioprotective activities are not entirely known, although not conclusively shown, there may be a connection between the cardiovascular advantages of some diets and their polyphenol levels [18]. Based on the preceding remarks, this work focuses on summarizing the literature dealing with the pharmacological effects of dietary polyphenols and presents an overview of the recent developments regarding their use in the prevention and treatment of different diseases. We hope that this paper can be beneficial to future research and the development of new therapeutic strategies. Depicted in Figure 1 is a sketch that shows how nutrition can help in preventing atherosclerosis, which contributes to CVD.

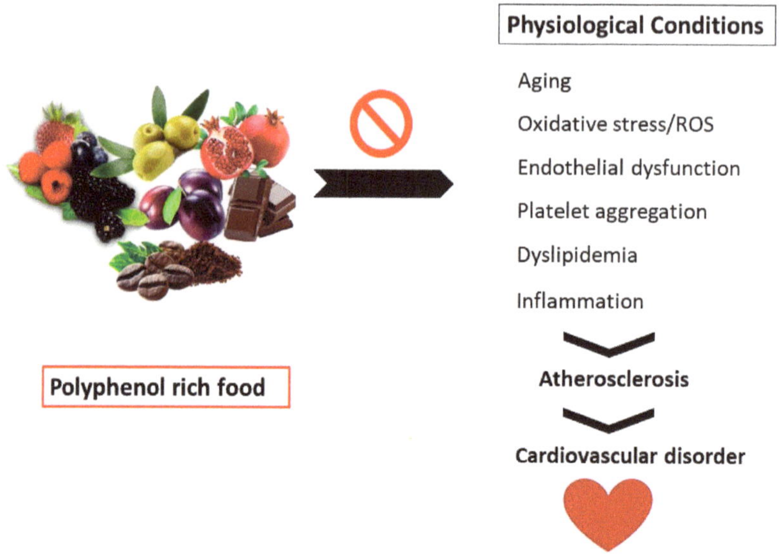

Figure 1. Nutrition can help prevent atherosclerosis, which is a pathophysiological process that contributes to the development of cardiovascular disease (CVD).

2. Polyphenols

Polyphenols are phytochemicals or secondary plant compounds that are considered non-essential nutrients in plants [19]. They are a rich collection of chemicals present in plants and algae, where their natural role is to defend the organism against UV radiation, infection, and herbivore ingestion. Polyphenols come in a variety of structural forms, from basic monomers to complex polymerized structures. Seaweed polyphenols may help lower hyperglycemia, hyperlipidemia, oxidative stress, chronic inflammation, metabolic abnormalities linked to CVDs, and diabetes sequelae. On the other hand, polyphenols from plants have been related to improved health in terms of obesity, diabetes, and CVD. A recent study has focused on marine macroalgae, presumably because of epidemiological evidence from Asian nations that suggests a diet high in seaweed lowers the occurrence of CVD, cancer, and other chronic disorders [1,20].

Polyphenols are also outstanding plant-derived secondary metabolites that exhibit anticancer, anti-cardiovascular, antidiabetic, and anti-neurodegenerative properties. Compounds like phenicic acid, stilbenes, flavonoids, coumarins, tannins, and lignins are present in numerous plants, including tonka bean (*Dipteryx odorata*), sweet grass (*Hierochloe odorata*), sweet woodruff (*Galium odoratum*), deer-tongue grass (*Dichanthelium clandestinum*), sweet clover (*Verbascum* spp.), and vanilla grass (*Anthoxanthum odoratum*). In addition to its antioxidant characteristics, resveratrol also exerts ameliorating effects against inflammation, cancer, aging, obesity, and diabetes, along with cardioprotective and neurological benefits [21]. A scientific literature analysis on PubMed with the keywords "Cardiovascular Polyphenols" showed that around 4,000 papers have been published between 2010 and 2023. Figure 2 represents our keyword occurrence analysis, in which the most focused research keywords were oxidative stress, inflammation, resveratrol, atherosclerosis, endothelium, dietary supplement, blood pressure, and apoptosis.

Phenolic compounds are the most copious non-energetic components in plant-based meals. The aptitude of polyphenols to alter enzymatic activity and, consequently, the signal-transmitting mechanisms of several processes occurring in cells may be attributed to their physicochemical properties, which allow polyphenols to participate in numerous metabolic cellular redox processes. Thus, the antioxidant scavenging properties of polyphenols make them advantageous. These are the most predominant antioxidants in the diet; they are 20 times more prevalent than vitamin E and carotenoids and 10 times more prevalent than vitamin C. Nicotinamide-adenine-dinucleotide phosphate (NADP) oxidase and xanthine oxidase are two ROS-producing enzymes that polyphenols can inhibit [22]. Listed in Table 1 are polyphenol-rich plant foods.

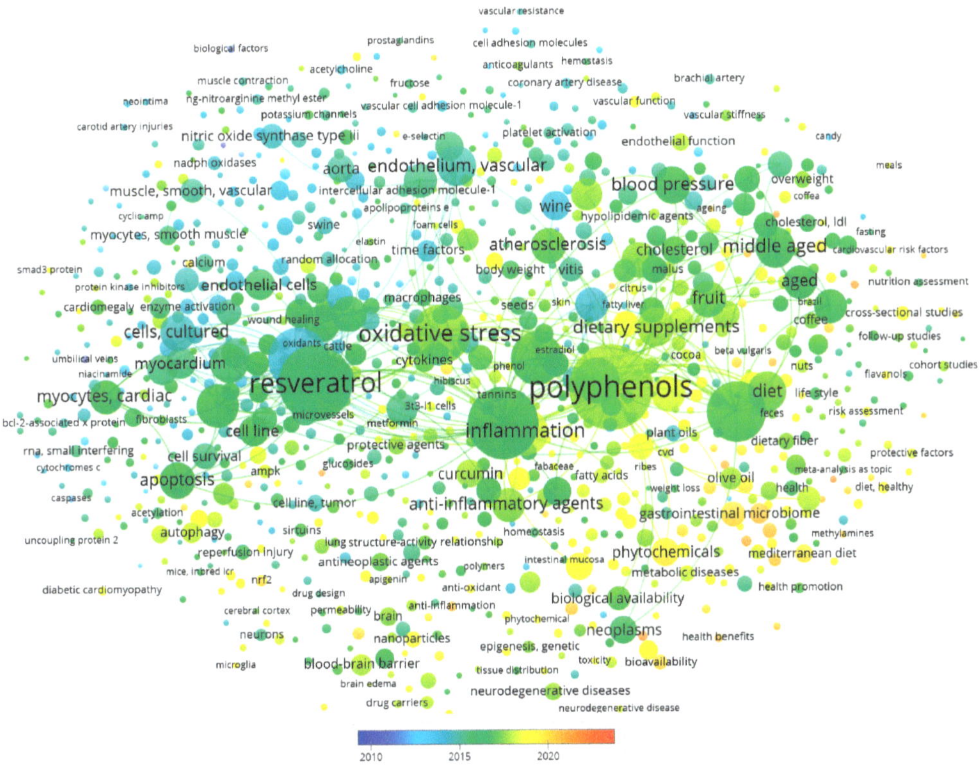

Figure 2. Scientific Literature analysis for the effect of polyphenols on cardiovascular diseases. This bibliography data were extracted from PubMed with the year range set to 2010–2023.

Table 1. Polyphenol-rich plant foods.

Plant Food	Latin Name	Edible Part	Concentration mg/100 g	Major Polyphenols	References
Apple	*Malus domestica*	Peel Flesh Total	50–120 ^{y} 0.2–0.9 5–50	Phlorizin, quercetin, phenolic acids (chlorogenic acid)	[23,24]
Blackberry	*Rubus fruticosus*	Whole	130–405	Anthocyanins, flavanols (EC), phenolic acid (ellagic acid)	[25]
Blueberry	*Vaccinium corymbosum*	Whole	160–480	Anthocyanins, flavonols (quercetin), phenolic acids (chlorogenic acid)	[25]
Coffee	*Coffea arabica*	Beverage, filtered	90	Phenolic acids (chlorogenic acid)	[25]
Chestnut (raw)	*Castanea sativa*	Whole nut	547–1960	Hydroxybenzoic acids (gallic acid, ellagic acid), tannins	[25]
Cacao	*Theabroma cacao*	Beans, powder	300–1100 ^{x}	Flavanols (EC)	[25]
Green tea	*Camellia sinensis*	Extract	29–103 ^{x}	Flavanols (EC, EGCG)	[25]
Grapefruit	*Citrus x paradisi*	Flesh	15–115	Flavonoids, phenolic acids	[25]

Table 1. *Cont.*

Plant Food	Latin Name	Edible Part	Concentration mg/100 g	Major Polyphenols	References
Olive oil, extra virgin	*Olea europaea*	Whole oil	4–200	Tyrosols, lignans (pinoresinol), phenolic acids, hydrolyzable tannins	[25]
Potato	*Solanum tuberosum*	Peel Flesh Total	180–5000 1–1000 10–50	Phenolic acids (chlorogenic acid)	[25,26]
Plum	*Prunus domestica*	Total	130–240	Phenolic acids (chlorogenic acid), procyanidins, anthocyanins	[25]
Pomegranate	*Punica granatum*	Juice	240 [x]	Punicalagin (and ellagitannin)	[27]
Grapes, Red wine	*Vitis vinifera*	Final product	25–300 [x]	Phenolic acids, anthocyanins, tannins, stilbenes (resveratrol)	[25]
Wheat	*Triticum aestivum*	Whole grain	85–220	Phenolic acids (hydroxybenzoic acids, hydroxycinnamic acids)	[25]
Spinach	*Spinacia oleracea*	Leaf	30–290	Flavonols	[25]

Abbreviations: EC = epicatechin; EGCG = epigallocatechin gallate. [x] = In juices, wine, and other beverages: mg/100 mL. [y] = Concentration in mg/cm^2. Note that the polyphenol content in purple potatoes is approximately five times higher than that in other varieties.

3. Classification of Polyphenols

Polyphenols are divided into several categories, including phenolic acids (hydroxybenzoic and hydroxycinnamic acids), flavonoids (flavones, flavonols, isoflavones, flavanones, and anthocyanins), stilbenes (resveratrol, piceatannol), lignans (sesamol, pinoresinol, sinol, enterodiol), and others, including tannins (hydrolyzable, non-hydrolyzable, and condensed tannins), lignins, xanthones, chromones, and anthraquinones, as shown in Figure 3A [28–30].

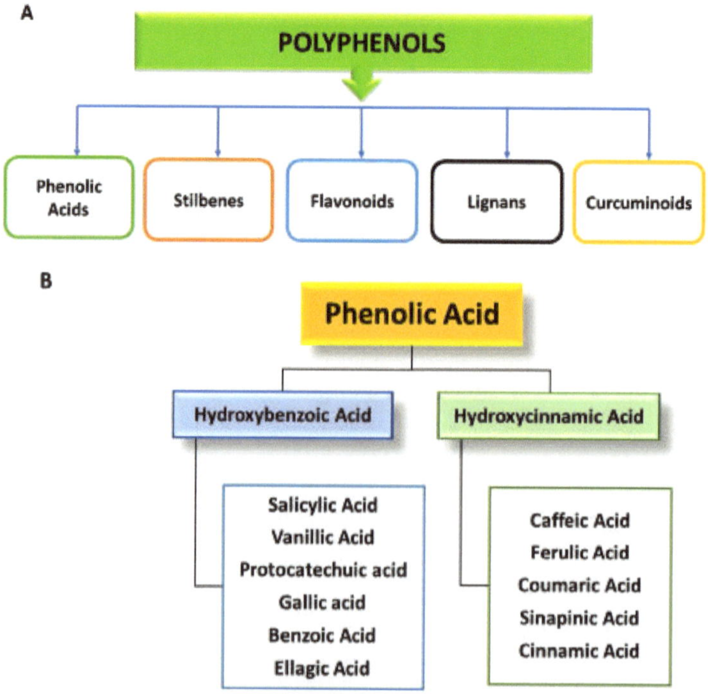

Figure 3. (**A**) Classification of polyphenols. (**B**) Classification and food sources of phenolic acids.

3.1. *Phenolic Acids*

Phenolic acids are a class of organic compounds composed of aromatic rings connected to a carboxylic acid group. The antioxidant properties of phenolic acids make them protective against CVD [28,30,31]. Vegetarian food, including seeds, fruits, and green-colored vegetables, are known to be good sources of phenolic acids. In addition to their health-promoting benefits, phenolic acids are widely used in a variety of things, including cosmetics, food, and medicine [32]. Phenolic acids are divided into hydroxybenzoic and hydroxycinnamic acids.

3.1.1. Hydroxybenzoic Acids

Hydroxybenzoic acids are benzoic acid ($C_7H_6O_2$) derivatives. The Hydroxybenzoic acids sub-category includes salicylic acid, protocatechuic acid, vanillic acid, benzoic acid, gallic acid, and ellagic acid [30,33]; gallic acid may be found in large quantities in tea and grape seeds [34]. Included within olive products are hydroxybenzoic acids, which exert anti-inflammatory, antioxidant, and cardioprotective properties, among others [29–31,35].

3.1.2. Hydroxycinnamic Acids

Aromatic acids, which are derived from cinnamic acid are represented by the family of hydroxycinnamic acids (C6-C3) [29,36]. Adequate sources of hydroxycinnamic acids include coffee, berries, apples, grains, and kiwi fruit [37,38]. Specifically, coffee contains chlorogenic acid, and caffeine, berries, and apples contain caffeic acid, while cereals contain ferulic acid. Most citrus fruits contain caffeine and chlorogenic acid (cinnamic acid) [34]. In addition to having anti-inflammatory properties, phenolic acids can also protect the body from cell damage, ROS, oxidative stress, and cardiovascular health issues like heart diseases and diabetes; they have neuroprotective and food-preservative properties that can help keep food fresh longer [30,34]. Figure 3B depicts the classification of phenolic acids.

3.2. *Lignin*

Lignin is a category of complex chemical compounds that may be found in a variety of plant tissues. Lignin is particularly important in plants and trees because it assists in cell wall formation [39]. Flaxseeds, tomatoes, peaches, apples, and some berries are examples of foods high in lignin [30].

Silymarin

Lignin silymarin, a kind of flavonolignan, has been shown to possess antioxidant properties. This kind of lignin may be found in the seeds of milk thistle and other herbaceous plants [40].

3.3. *Stilbenes*

Stilbenes are phenol-derived metabolites ($C_{14}H_{12}$). Their biological activity and health-promoting benefits are a source of research interest and a focal point for many studies. Studies have focused on their bioavailability, metabolism, and absorption rates, as well as their overall health benefits [41].

Resveratrol

The most well-known kind of stilbene is resveratrol, which has anti-inflammatory effects [42]. Resveratrol is mostly found in grapes and red wine. This compound has also been demonstrated to lower blood pressure. Additionally, studies have indicated that taking resveratrol as a supplement rather than getting it naturally is more beneficial. Moreover, findings have indicated that resveratrol may be beneficial to humans, although its specific mechanism of action is currently being researched [43]. Despite its limited bioavailability in the body, resveratrol has been shown to protect against CVDs and have a sun-protective impact, thus protecting against skin cancer [44].

3.4. *Flavonoids*

Flavonoids are naturally occurring polyphenolic compounds that are divided into six primary categories: flavanones, flavones, flavanols, isoflavones, flavan-3-ols, and anthocyanidins [44]. Figure 4 depicts the chemical structures of important flavonoids.

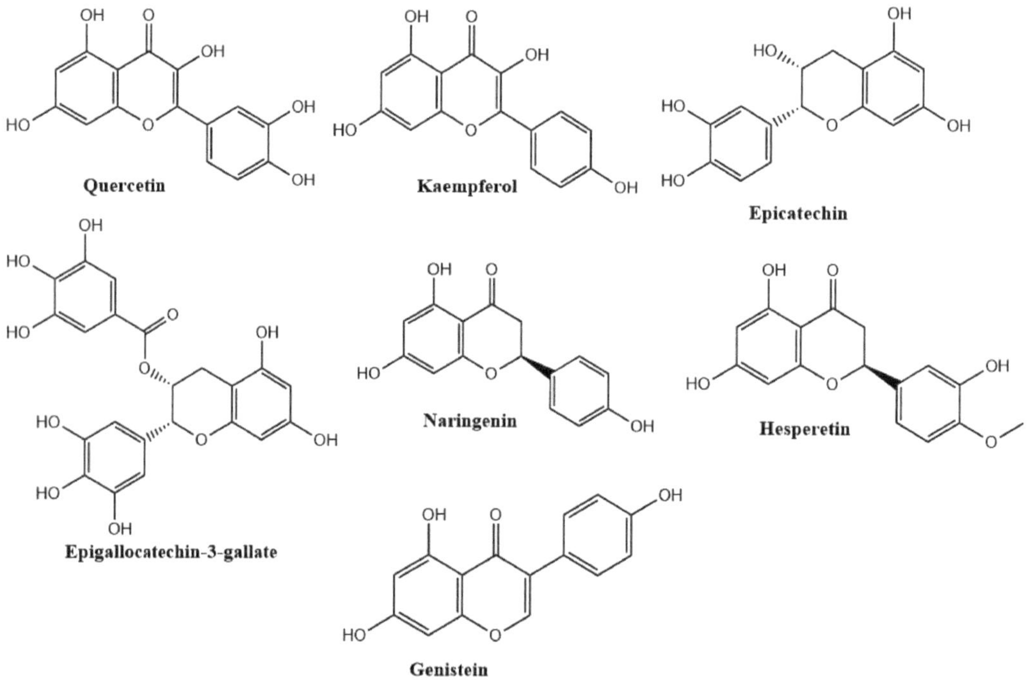

Figure 4. Chemical structures of important phytocompounds.

3.4.1. Flavones

Flavones are present in foods like garlic, chamomile tea, and celery, which are rich in luteolin [45]. The advantageous effects of luteolin that have been observed in various studies include blood pressure reduction in hypertensive rats, improving vasodilation of the aortic rings, and increasing cAMP accumulation due to the inhibition of cAMP-specific phosphodiesterase [46]. The activation of the cAMP/PKA cascade increases endothelial cell nitric oxide levels through the activation of endothelial enzyme nitric oxide synthase. This encourages the relaxation of the vasculature through nitric oxide, a mechanism carried out by potassium and calcium channels [47].

3.4.2. Flavonols

Onions, tea, broccoli, and fruit are rich sources of flavonols, which are represented by kaempferol and quercetin, which are glycosides [48].

Quercetin

Quercetin exerts its antihypertensive effect by improving endothelial function, modulating the renin–angiotensin–aldosterone system (RAAS) by modulating the mechanism of contraction of smooth muscles in blood vessels [48], producing vasodilation at a renal level that is protein kinase C-dependent, and lowering blood pressure in patients with diabetes or metabolic syndrome [49,50]. Quercetin is reported to cause oxidative stress reduction in the heart and kidneys [51].

Kaempferol

Kaempferol is present in foods such as broccoli, strawberries, green tea, and beans [52]; its antihypertensive effects are manifested by the activation of endothelial nitric oxide [53]. Besides its antihypertensive effect, kaempferol is reported to reduce proteinuria and albuminuria, and it is considered a potential contender for the improvement of these two conditions [54].

3.4.3. Flavan-3-ols

Flavan-3-ols comprise monomers such as epicatechin, gallocatechin, catechin, and oligomers (proanthocyanidin) [52]. Catechin monomers are found in apple-, tea-, cocoa-, pear-, and grape-based products in the form of aglycones (part of non-carbohydrate glycosides). Catechins have been shown to have advantageous effects on vascular function and to be cardioprotective. Additionally, studies have revealed that they can lower both systolic and diastolic blood pressure [55,56].

Epicatechin

Epicatechin-rich food results in decreasing both systolic blood and diastolic blood pressure by 4.2 mmHg and 2.1 mmHg, respectively, showing its antihypertensive effect. It has also been found to decrease myocardial rigidity in hypertrophic cardiomyopathic rats [57].

Epigallocatechin-3-gallate

Epigallocatechin-3-gallate is abundant in green tea and has been found to have antioxidant, anti-inflammatory, and antiatherogenic properties [58].

3.4.4. Flavanones

Naringenin and hesperetin are the main representatives of this class, which is predominantly found in citrus fruits, especially in their peels [59]. Their antioxidant properties are due to their free radical scavenging activity [60].

Naringenin

Naringenin has some promising effects; e.g., it reduces mean blood pressure, regulates nitric oxide levels, and provides a shield against endothelial dysfunction [61,62].

Hesperetin

Hesperetin is one of the dietary flavanones found in citrus fruits [52] which is rapidly absorbed in the intestine, and the resulting metabolites are responsible for the antihypertensive effect. They can also reduce the progression of atheroma plaque through their anti-inflammatory activity. In addition to this, the antioxidant effect of hesperetin helps to increase the amount of nitric oxide and reduces the amount of calcium ions, thus producing smooth muscle relaxation in the blood vessels [63,64].

3.4.5. Anthocyanidins

Anthocyanidins are key soluble pigments that provide color to fruits and vegetables like blue, red, or purple fruits, e.g., forest fruits and black currants, etc. [52]. The endothelium-dependent vasodilatory property has beneficial effects on the cardiovascular system, thereby reducing the risk of acute myocardial infarction [65].

3.4.6. Isoflavones

Soy is a source of isoflavones, which are structurally similar to mammalian estrogens. Thus, they can produce their agonistic effect by binding to estrogen receptor agonists. Diadzein and genistein are the two main isoflavones in this class [66].

Diadzein

Diadzein has been found to have an anti-damaging effect by reducing oxidative stress, increasing nitric oxide synthesis, reducing LDL oxidation, and increasing prostaglandin production [67].

Genistein

Genistein has the property of reducing hypertension [68].

4. Bioavailability of Polyphenols

The bioavailability of phenolic compounds in our food is critical because only the most bioavailable phenolic compounds in our diet will have the most beneficial effects on the human body [69]. These will be different for each person depending on their relationship with food, how cell walls are made, and where glycosides are found. Many epidemiological studies have shown that phenolic compounds have a lot of health benefits, such as protecting against the buildup of fat, preventing microorganisms from decaying, lowering cardiovascular diseases, preventing diabetes, stroke, and cancer, and exerting anti-inflammatory effects [30].

Recent research has placed a strong emphasis on identifying the processes governing polyphenol metabolism and bioavailability in humans [70]. A wide array of fruits and vegetables contain compounds known as phenolics. Some plants contain as much as 750 mg/100 g of fruit, which is a significant amount [19,71]. The highest dietary sources of polyphenols are dark-colored fruits (especially small berries), chocolate, cereals made entirely of whole grains, coffee, and red wine, with the latter three accounting for the lion's share of overall dietary polyphenol consumption [72]. Among the food groups consumed, polyphenols are primarily associated with carbohydrates, organic acids, and other food groups. They combine with arabinose to generate ester linkages in hemicellulose or core lignin, which allows them to form covalent connections with polysaccharides in the cell wall of the plant. While flavonoids can be found in the cytosol and endoplasmic reticulum, where they are formed, they are mostly found in free form in the cytosol and endoplasmic reticulum. Cell barriers and intracellular compartments must be damaged for the drug to be bioavailable [71]. Flavonoids found in nature are housed within plants as glycoside and non-glycosylated conjugate compounds, and as a result, the moiety's type might affect their subsequent human bioavailability [73,74]. A summary of the comparative bioavailability of different polyphenols is shown in Figure 5 [75].

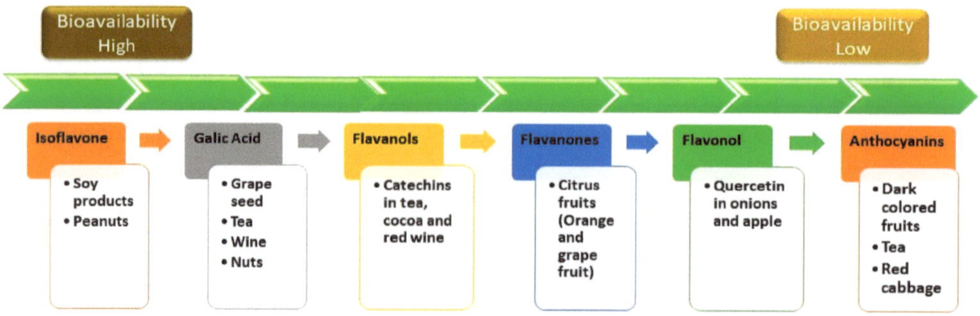

Figure 5. Comparative bioavailability of some common dietary polyphenols.

4.1. Metabolism of Polyphenols

4.1.1. Oral and Gastric Absorption

The most prevalent enzyme in the mouth cavity is amylase, which starts the digesting process. Due to the brief contact duration, the consequence of enzyme action on polyphenol release from food is predicted to be negligible [76]. On the other hand, particle size decrease occurs, resulting in particles with a diameter ranging from a few hundred microns to several thousand microns [77], allowing for increased enzyme access throughout the succeeding stages of digestion due to an increase in the digest volume fraction. Most polyphenols seem to be formed in the stomach during the digestion process. In the gastric phase digestion with pepsin, peristaltic movements, and a low pH cause formation of finely powdered digestible polyphenols with even smaller particle sizes, often less than 500 microns in diameter [78]. Another factor that might contribute to polyphenols remaining in an undissociated form is the low pH, which may accelerate the movement of polyphenols from the food matrix to the aqueous phase due to decreased interactions between ionic groups. The pH of the digestive fluids normally rises from approximately 2–4 to around 7 when digested food exits the stomach and enters the small intestine. It is possible that the pancreatic and biliary enzymes can be activated in this manner [79]. A brush border enzyme called lactase-phlorizin hydrolase (LPH) is thought to be responsible for cleaving polyphenols from their sugar moiety before cellular absorption [80].

4.1.2. Uptake in Enterocytes

The gastric absorption of some polyphenols has been strongly suggested because of how soon these polyphenols enter plasma after consumption [81]. After breaking into their corresponding aglycons, polyphenols can penetrate the intestinal epithelium via passive transport, active transport, or facilitated transport. Polyphenols with low molecular weight, like phenolic acids, flavonoid aglycon, herbal tea polyphenols, and the polyphenols in cocoa (epicatechin, procyanidin B2, catechin), are believed to be absorbed mostly by passive diffusion (based on tests in Caco-2 cells) [82–84].

The hypothesis that sodium-glucose transport, notably the protein known as "sodium-glucose-linked transporter 1 (SGLT1)", actively takes up specific glycoside polyphenols has been proposed [85]. According to these findings, glycosides may be absorbed by SGLT1 to a small amount before being re-secreted into the digestive system, or they may be further broken down by cytosolic glucosidase in the body [86]. Another approach for polyphenol absorption is the use of monocarboxylic acid transporters to facilitate the transfer of polyphenolic substances into enterocytes (MCTs). To be recognized as an absorption substrate, a polyphenol must contain an attached carboxylic acid group of a monoanionic nature as well as a side chain with a nonpolar nature or an aromatic group with a hydrophobic nature [87,88]. MCTs have been shown to take up a range of polyphenols, such as caffeic and ferulic acid (usually utilizing Caco-2 cell models) [89].

4.1.3. Effect of Microbial Fermentation in the Colon

While some nutritional flavonoids are taken in the small intestine, the majority are taken in the intestinal tract, where the intestinal microbiota additionally degrades the deconjugated type metabolites and related compounds into freely accessible compounds such as phenolic acids, which are subsequently absorbed [90]. Bacteria may have an essential role in the breakdown of plant polyphenols and phase I/II metabolism conjugates, which are usually discharged by enterohepatic recirculation. Glycosylation, hydroxylation, demethylation, deconjugation, ring cleavage (typical of the C-ring), hydrolysis, epimerization, and chain-shortening processes are among the routine occurrences [88,91].

4.1.4. Metabolism in the Enterocytes

Most polyphenols are supposed to be absorbed in the small intestine, where they are commonly processed by phase II enzymes before they enter systemic circulation [74]. After they have been extensively captivated by the intestinal epithelial cells, before inflowing into the systemic flow, they are subjected to phase II enzymatic detoxification, which results in the creation of various conjugated harvests. These include sulfates produced by the action of sulfotransferases (SULTs), glucuronides produced by the action of uridine-5′-diphosphate glucuronosyltransferases (UGT), and some methylated derivatives produced by catechol-O-methyltransferase (COMTs) action [19,92]. Aspects of polyphenol bioavailability and accumulation in tissues are also strongly related to the action of some proteins, mainly "multidrug resistance-associated proteins" such as MRP-1/MRP-2. These are efflux transporters that are ATP-dependent for their action, and together, they are referred to as phase III metabolism [93]. Aglycones undergo further phase II metabolism after being oxidatively degraded in hepatocytes, particularly in the Golgi apparatus and the peroxisome [92,94].

4.1.5. Distribution in Body and Excretion

Polyphenolics can be carried in the bloodstream in three different ways: (1) in free form, (2) when coupled to proteins, and (3) when lipoprotein (lipids)-bound. Polyphenols appear to be coupled to proteins in most cases [95,96]. In the end, via the portal blood flow, they make their way into the liver, where they move through a second phase of metabolism (phase II) before entering the systemic blood flow and peripheral bodily tissues, where they are eliminated by the kidneys [97].

Polyphenols are believed to mostly be excreted through urinary excretion, especially those with greater hydrophilicity. The percentage of polyphenols retrieved in urination (including conjugates) varies greatly, with gallic acid and isoflavones having the largest amounts (over 60%) of polyphenols recovered [19,98,99]. Figure 6 shows a schematic diagram of the route of absorption, metabolism, and excretion of polyphenols, and Table 2 shows a summary of the most prevalent dietary polyphenols and their major colonic metabolites.

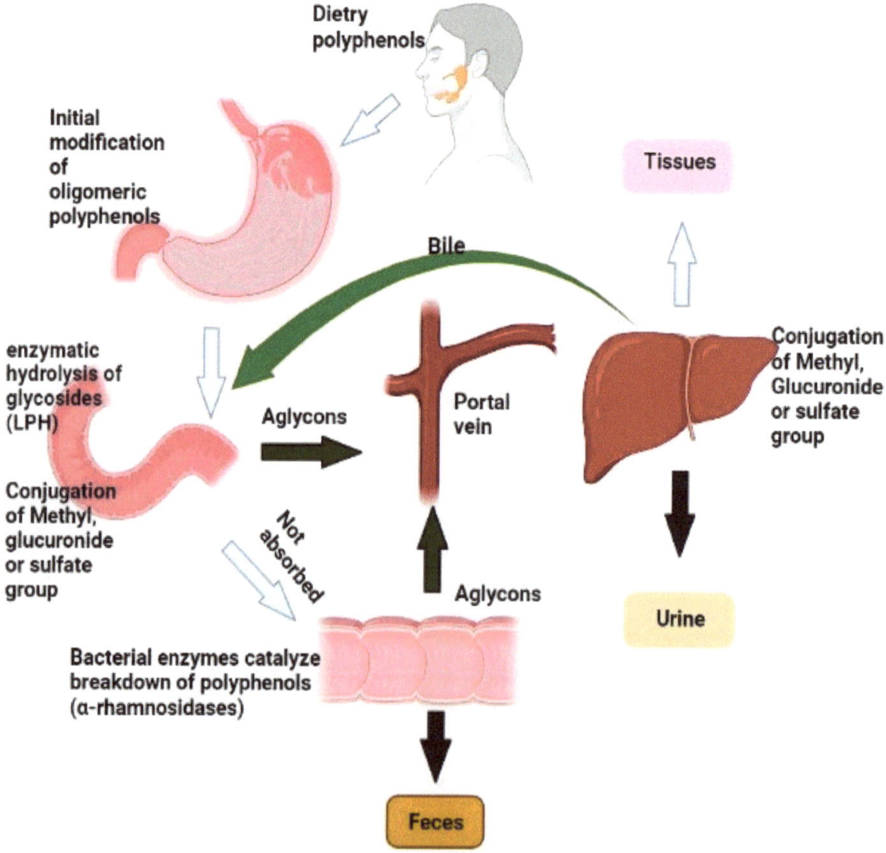

Figure 6. A schematic diagram of the route of absorption, metabolism, and excretion of polyphenols.

Table 2. The most prevalent dietary polyphenols and their major colonic metabolites.

Polyphenol Class	Metabolites	Bioavailability	References
Anthocyanins	Catechol Glucunoride conjugates Hydroxyhippuric acid Methyl conjugates Propionic acid Protocatechuic acid Pyrogallol Sulphate conjugates Syringic acid Vanillic acid	**Absorption:** A minor amount of glycosylated anthocyanin product is immediately absorbed in the gut, resulting in maximal plasma concentrations ranging from 14 to 592 nmol/L at 4–5 h after intake (doses: 68–1300 mg). **Metabolism:** Through glucosidase activity, the gut bacteria hydrolyze anthocyanins. Cleavage of the C3-ring breaks down the aglycones, which are then metabolized into various phenolic and aldehydic components. **Excretion:** Urinary excretion is estimated to be between 0.03% and 4% of the ingested dosage, with elimination half-lives of 15–3 h.	[92,100–103]
Phenolic acids	Dihydrocaffeic acid Feruloylglycine Dihydrofeluric acid Hydroxybenzoic acid Vanillic acid Hippuric acid	**Absorption:** Approximately 30 min after consumption, the maximal plasma concentration level is attained, and this is because its maximum absorption occurs in the small intestine. **Metabolism:** These chemicals are metabolized and circulated in the body as glucuronate, sulfate, and methylated metabolites with varying degrees of bioactivity. **Excretion:** Urinary excretions account for roughly 40% of total consumption, with excretion peaking after 8 h.	[92,100,104,105]
Flavonols	Hydroxyphenylacetic derivates Protocatechuric acid Propionic acid	**Absorption:** Small intestine absorption is poor. **Metabolism:** The flavonol skeleton is broken down by gut microbiota microbial enzymes, resulting in the production of low-molecular-weight polar metabolites. **Excretion:** The clearance of epicatechin metabolites relies heavily on urine excretion.	[92,100,106]
Flavan-3-ols and Proanthocyanidins	Benzoic acids Hippuric acids Phenilvalerolactones Phenylacetic acids Phenylpropionic acids Phenylvaleric acids	**Absorption:** The small intestine absorbs between 8 and 17 percent of monomeric-3-ols. **Metabolism:** The leftover unabsorbed portion reaches the other end of the large intestine practically intact, and their gut bacteria cause the breakdown of the flavonoid skeleton, producing several low-molecular-mass metabolites.	[19,92,100,107]
Ellagitanninis	Dimethyl-ellagic acid Urolithin A and B Urolithin D	**Absorption:** Ellagitannins are hydrolyzed in the gastrointestinal lumen after intake, yielding a free form of ellagic acid. **Metabolism:** The gut microbiota degrades ellagic acid in the large intestine, resulting in a variety of derivative chemicals known as urolithins, all of which have the same nucleus. Urolithins are substantially absorbed and metabolized as glucuronidated and sulfated products by hepatic and intestinal cells.	[92,108,109]
Stilbenes	3,4′-dihydro-trans-stilbene Dihydroresveratrol 3,4′-dihydroxybibenzyl	**Absorption:** The upper gastrointestinal tract absorbs resveratrol. **Metabolism:** Enterocytes and hepatocytes both metabolize it, producing glucuronide and sulfate forms.	[92,100,110–114]

5. Role of Vascular Endothelium in the Regulation of Vascular Homeostasis

A monolayer of cells produces the endothelium, which makes up the interior of blood vessels. Vascular endothelium regulates the tone and homeostasis of the vasculature, as well as the morphological changes that occur in pathological circumstances. The endothelium regulates the balance of opposing processes such as vasodilation and constriction, procoagulant and antithrombotic actions, and cell proliferation and apoptosis [115,116].

Endothelial cells minimize the interaction of the bloodstream with the basal prothrombotic arterial wall due to their selective position [117]. The primary function of Endothelial cells is to regulate vascular tone by producing vasodilator and vasoconstrictor chemicals. The endothelial NO synthase (eNOS) enzyme produces NO from L-arginine, exerting a vasodilatory effect. NO can easily diffuse into the cells of vascular smooth muscle, where it triggers guanyl cyclase, thus accumulating cyclic guanosine monophosphate (cGMP), which ultimately activates the protein kinase G and causes endothelial vasorelaxation (Figure 7). The endothelium-derived hyperpolarizing factor (EDHF) plays its role in vasodilation by targeting the K+ channels in the blood vessels. Furthermore, prostacyclin (PGI2) produced during the cyclooxygenase (COX) pathway has vasodilatory effects. Some other factors can be produced by the endothelium having vasoconstrictive effects on blood vessels such as angiotensin II (Ang II), endothelin-1 (ET-1), and thromboxane A2 (TXA2) [15].

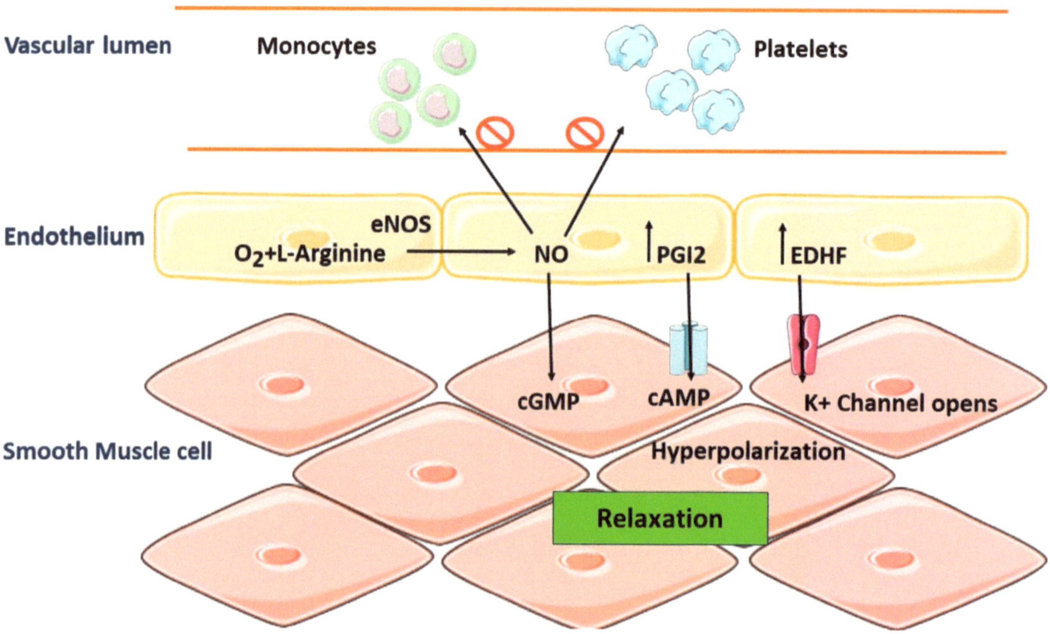

Figure 7. Normal regulation of vascular homeostasis. Vascular homeostasis is regulated in part by endothelium-derived NO. The endothelial NO synthase (eNOS) enzyme produces NO from L-arginine, exerting a vasodilatory effect. NO can easily diffuse into the cells of vascular smooth muscle, where it triggers guanyl cyclase, thus accumulating cyclic guanosine monophosphate (cGMP), which ultimately activates the protein kinase G and causes vasorelaxation in endothelial.

Regarding blood–tissue contact, the endothelium does play a crucial role, interacting directly with a variety of circulating substances, including antioxidants, oxidized LDLs, and pro-inflammatory cytokines such as tumor necrosis factor (TNF) and interleukins (IL) [115,116]. These variables can cause vasomotricity or the manufacturing of endothelial agents like nitric oxide (NO). Their role in a variety of physiological processes has been reported to influence biological processes such as the apoptosis, proliferation, and migration of endothelial cells [15,118,119]. Thus, endothelial dysfunction can have several detrimental effects on vascular cells and surrounding tissue, resulting in the development of cardiovascular disorders such as atherosclerosis and hypertension [120].

Diets like the Mediterranean diet have been linked to better cardiovascular health [121], which might be due to the high consumption of polyphenol-rich drinks and foods, as well as fruits and vegetables. Polyphenol-rich foods, including red wine, chocolate, green tea, and berries, also help to promote cardiovascular health [122,123]. Polyphenols have been linked to improving cardiovascular health in various ways. Their advantages include an improvement in lipid profiles. They also have direct effects on endothelial cells and have anti-atherosclerotic, anti-hypertensive, and anti-inflammatory properties (Figure 8).

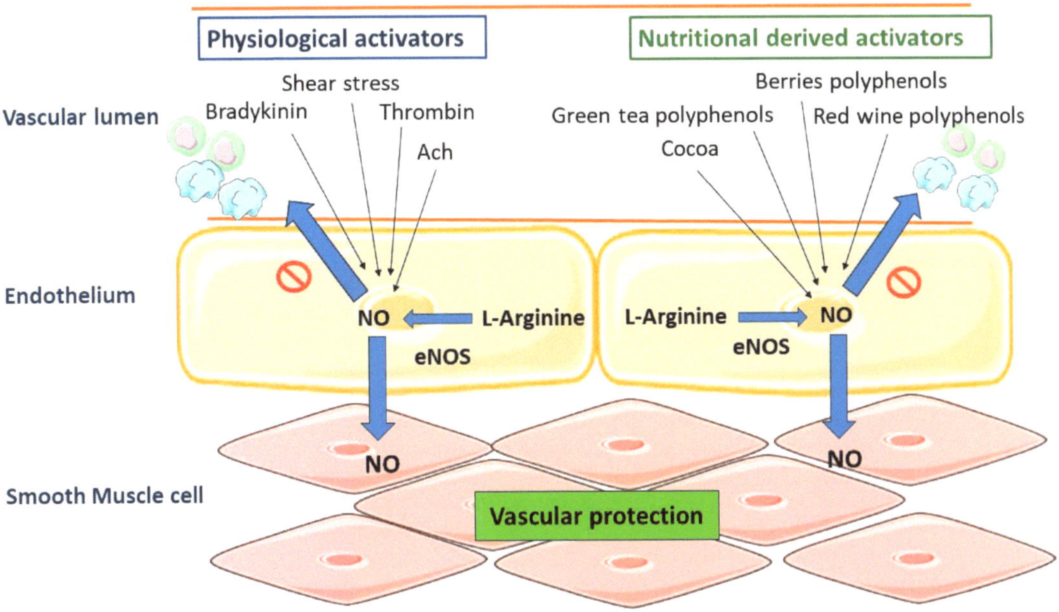

Figure 8. Numerous physiological activators on the endothelium surface can boost endothelial NO production to produce vascular protection. Also, polyphenol-rich dietary items, including chocolate, berries, red wine, and green tea, can enhance endothelial NO production. The endothelial NO synthase (eNOS) enzyme produces NO from L-arginine, exerting a vasodilatory effect. NO can easily diffuse into the cells of vascular smooth muscle, where it triggers guanyl cyclase, thus accumulating cyclic guanosine monophosphate (cGMP), which ultimately activates the protein kinase G and causes vasorelaxation in endothelial.

6. Pathophysiology: Oxidative Stress and CVD

In healthy cells, antioxidant defense systems such as superoxide dismutase (SOD), catalase (CAT), and glutathione reductase (GSR) limit the generation of radicals during various physiological activities such as metabolism and cellular respiration [1,124,125]. Long-term exposure to stress [126], pollution [127], smoking, and excessive drinking [128,129], as well as aging [130], can cause an imbalance of oxidative species (also known as reactive oxygen species; ROS) in comparison to endogenous defenses, resulting in oxidative stress [1,124]. ROS can bind to proteins, lipids, and DNA, thus oxidizing them and changing a healthy state into a diseased one. An increase in the level of ROS results in oxidative stress, and the cell's antioxidant system may become overburdened, endangering the health and integrity of the cell [131].

A similar pathological mechanism, atherosclerosis, underpins cardiovascular illnesses such as coronary artery disease, ischemic stroke, and peripheral artery disease [132]. Atherosclerosis is a multifactorial, degenerative ailment of the medium and great conduit arteries that is fueled by lipid buildup in the artery wall [133,134]. The risk factors of this disease include old age, chronic smoking, hyperlipidemia, hypertension, and a history of diabetes. The atherogenic process is tightly linked to inflammation and endothelial dysfunction [132,135]. Endothelial damage due to ROS leads to the development of atherosclerosis, which may result in myocardial infarction and ischemic reperfusion [136–138]. Oxidative stress and ROS target all body cells, especially smooth muscle cells and endothelial cells, with the help of neutrophils, macrophages, and platelets [139].

ROS has an impact on a variety of endothelium-related processes [2]. The most well-known is endothelium-dependent vasorelaxation, which has long been marked as a key component in the prognosis of cardiovascular health, which is harmed by a decrease in NO bioactivity and/or bioavailability [140,141]. There are some mechanisms responsible for reduced NO bioavailability. This can either be caused by a decreased expression of the enzyme responsible for the NO production in endothelial cells, i.e., endothelial NOS (eNOS), or a decrease in the existing NO owing to ROS destruction, among other things [141]. NO is a powerful vasodilator that also inhibits the activation and adherence of inflammatory cells [142].

In the vasculature, there are several sources of ROS, including mitochondrial enzymes such as NADH/NADPH oxidase and xanthine oxidase [141,143]. Endothelium-derived NO reacts quickly with the superoxide radical (O_2^-) to create peroxynitrite (ONOO), a potent oxidant that is equally damaging to endothelial cells [144]. In this respect, studies have shown that the continued contact of endothelial cells with oxygen and different oxidants like hydrogen peroxide, ONOO, and/or oxidized LDL (ox-LDL) causes epithelial damage by promoting apoptosis, leading to cell damage and endothelial cell dysfunction, which has been reported as a critical early step in atherogenesis. Atherosclerotic lesions can arise from leaky and dysfunctional endothelium [132,133].

LDL normally diffuses easily in both directions across the compromised endothelium. Oxidative stress converts LDL to ox-LDL by peroxidation, which has cytotoxic effects and can cause inflammation [145]. The first step in the formation of atherosclerotic plaques is the oxidation of LDL and its subsequent passage through the endothelial barrier. Furthermore, the interaction of hypercholesterolemia, oxidative stress radicals, and inflammatory molecules creates an environment conducive to severe endothelial damage, which is a characteristic of atherosclerosis development [124,146].

A particular adhesion molecule called VCAM-1, which is crucial for binding monocytes and T cells before they transmigrate into the arterial wall, is produced by wounded or activated endothelial cells [145]. Reportedly VCAM-1, ICAM-1, and E-selectin enhance the adherence of leukocytes to the vascular endothelium at the sites of atherosclerotic lesions, consequently boosting signal transduction cascades [146]. These monocytes develop into macrophages after activation, which then become puffy with the uptake of ox LDL by scavenger receptor-mediated phagocytosis, resulting in fatty bands in the artery wall [132,140,147]. Furthermore, lipid-engorged macrophages (foam cells) eventually die in situ because of necrotic cell death, resulting in the creation of a tender and unstable core inside the atherosclerotic plaques, which has a high consistency of lipids [148]. This plaque is stabilized by a protective cap secreted by smooth muscle cells. It consists of a collagen-rich matrix comprising fibroblasts, which can stop the disease from progressing. Prolonged inflammation, on the other hand, might result in plaques that are unstable and prone to rupture [1,148]. Such ruptured plaques cause a fast thrombotic reaction, resulting in arterial blockage and, depending on the location of the atherosclerotic lesion, potentially causing heart attacks, ischemic strokes, or peripheral ischemia [133]. Research findings have linked the plaques in the walls of coronary arteries to chronic atherosclerotic lesions, which can limit the channel lumen. The current preferred hypothesis is that acute coronary syndrome (ACS) is caused by a rupture of the fibrous cap of an atherosclerotic plaque, which facilitates blood contact with extracellular matrix collagen and tissue factors previously deposited in the plaque, resulting in the formation of a thrombus [149–152]. On the other hand, increased platelet activation, including the adhesion, secretion, and aggregation at the site of arterial injury or in atherosclerotic arteries, plays a key part in the etiology of CVD [153,154]. When an atherosclerotic plaque ruptures, activated platelets can bind to the endothelium, causing fastening, aggregation, and thrombus development, which leads to embolism and the constriction of vessels, two features of myocardial infarction [155].

Physiological hemostasis is a natural defense against excessive blood loss that is based on the creation of a regulated thrombus at the site of the blood vessel injury. Platelets, the smallest (2–4 m) blood corpuscles, are formed at a rate of 40×10^3/mL/day in the bone marrow from megakaryocytes and play a key role in hemostasis. Platelets play a function in hemostasis that extends beyond forming the platelet plug, which also serves as the site of fibrin production (at the site of the vessel wall injury), to include a beneficial influence on vessel wall contraction and participation in clotting responses [8]. Collagen and tissue factor (TF) located in the sub-endothelial matrix come into contact with the lowing blood when the endothelium is injured, causing a clot to develop [156]. Denuded collagen directly promotes platelet pooling and activation, and the denuded tissue factor starts the synthesis of thrombin, which not only transforms fibrinogen to fibrin but also activates platelets [150]. The presence of collagen receptors (including integrin a2b1 and glycoprotein complex GPIb/IX/V) on the platelet surface allows platelets to connect with the subendothelial layer. Platelet attachment leads them to change shape from discoid to spherical, resulting in the creation of pseudopodia and the release of chemicals held in granules (e.g., ADP, P-selectin, von Willebrand factor [vWF], thrombospondin) and, as a result, platelet aggregation [149,157].

7. Beneficial Effects of Polyphenols on Cardiovascular Disorders

7.1. Polyphenols as Antioxidant Therapy

Antioxidant therapy is becoming better recognized as a strategy for reducing ROS in the vasculature and, as a result, reducing their harmful effects [158]. Blockers of the angiotensin-converting enzyme (ACE) that lower circulatory Ang II detoxify in addition to exhibiting their antihypertensive attributes. In this regard, statins (cholesterol-lowering drugs) are used for the same intent in addition to their cholesterol-lowering properties by regulating HMG CoA reductase. Similarly, vitamins E and C are widely used as dietary supplements in combination with other drugs to reduce oxidative stress [158]. Polyphenols, on the other hand, are gaining attention as possible therapeutic agents for reducing oxidative stress and thereby protecting people from heart diseases [159,160]. In the diet, polyphenols are the most prevalent antioxidants, and their consumption is ten times that of water-soluble vitamin C and one hundred times that of lipid-soluble vitamin E and carotenoids [15].

Polyphenols have enormous antioxidant features. The presence of catechol groups, as well as hydroxylation patterns, such as the 3-hydroxy group in flavanols or electron shortage in anthocyanins, is essential in the antioxidant actions [161,162]. The presence of a catechol ring in the structure of various polyphenols has been positively linked with their antioxidant activity, as demonstrated by the ferric-reducing ability power (FRAP). In one study, further increases in FRAP were achieved by using aliphatic substitution or a double bond in the aliphatic group conjugated with the catechol ring; there were no benefits from adding more OH groups [163]. Polyphenols may serve as antioxidants by scavenging free radicals in a variety of ways. They exert their antioxidant potential either by inhibiting or potentiating the activity of various enzymes or by direct interaction with free radicals [164]. ROS that can be extremely toxic to lipids, proteins, and DNA include superoxide (O_2^-), hydrogen peroxide (H_2O_2) [165], and hypochlorous acid (HOCl) [166], which are all immediately scavenged by polyphenols like quercetin and catechin. In this respect, ROS can be made less reactive by the phenolic core acting as a buffer and collecting electrons [167]. Polyphenols can have indirect effects on cellular detoxification systems such as catalase (CAT), superoxide dismutases (SODs), and glutathione peroxidases [168,169]. Polyphenols can also inhibit enzymes that produce ROS, such as xanthine oxidase and nicotinamide adenine dinucleotide phosphate (NADPH) oxidase [170,171]. In addition to the production of ROS, there is an increase in the quantities of free metal ions. Due to their low redox potentials, flavonoids can chelate these metal ions, which prevents the production of free radicals. Research findings have indicated that quercetin is the flavonoid with the best capacity to chelate metal ions [171–173].

Polyphenols are known to be potent antioxidants due to their health advantages. Phenolic compounds may easily donate an electron or H atom from an aromatic hydroxyl group to a free radical, effectively neutralizing its effects. It all comes down to how the functional groups are arranged in the polyphenol's core structure [174]. Although polyphenols have been shown to have good antioxidant activity in vitro, their antioxidant capacity in vivo is lower than it is in vitro. Several variables play a role, including the metabolism of polyphenols into compounds with lower antioxidant activity. By inhibiting the –OH group, metabolism reduces polyphenols' ability to scavenge radicals [1,175]. Because proteins, uric acid, vitamin C, and thiols create an antioxidant barrier strong enough to overlook phenolic contribution in plasma, the polyphenolic antioxidant contribution is low [176]. Taking this into account, the theory that eating polyphenol-rich foods boosts plasma antioxidant capacity is debunked. Other dietary components absorbed alongside polyphenols, such as vitamins C and E, may be to blame for this increase [177]. The recognized interaction between fructose and uric acid is more likely to cause the antioxidant effect of a fruit- and vegetable-rich diet [178].

7.2. Polyphenols and Vascular Tone

The significance of endothelium-produced nitric oxide (NO) in controlling vascular tone and blood pressure is well understood. The central mechanism of NO action is the activation of the cGMP-protein kinase G cascade in artery smooth muscle cells. The potassium channels are triggered when the cascade is activated, resulting in membrane hyperpolarization and preventing intracellular calcium influx, which induces vasodilation. On the other hand, protein kinase G reduces smooth muscle vasoconstriction in arteries by phosphorylating myosin light chains [179,180]. NO generation is primarily responsible for the polyphenols' effect on the endothelium [181–183].

After ingesting red wine or polyphenols (1 g/kg body weight) circulating NO concentrations reach 30 and 40 nM after 30 min in adults. A decrease in blood pressure (11 mmHg) and an increase in heart rate have also been observed [184]. Research findings have shown that olive oil can help hypertensive people lower their blood pressure [185], whereas red wine polyphenolic compounds (RWPC) can produce the endothelium-dependent relaxation of isolated arteries such as the rat's mesenteric artery or aorta [181]. In addition, red wine polyphenols, polyphenols from grape skin, and quercetin exhibit antihypertensive effects. In this respect, short-term oral treatment with RWPC lowers blood pressure in normotensive rats. This hemodynamic effect was correlated with enhanced endothelium-dependent relaxation and the induction of the genes responsible for inducible NO synthase and COX-2 inside the artery wall, thus contributing to the maintenance of agonist-induced contractility [186]. The higher synthesis of NO in consequence to the impact of polyphenols found in wine extract is linked to the calcium ion-dependent pathway, among several other things [187]. Resveratrol and quercetin cause an increase in the intracellular ion concentration of (Ca^{2+}) ions through the opening of potassium channels or the inhibition of Ca^{2+} ATP-ase within the endoplasmic reticulum of endothelial cells [188,189]. Similarly, delphinidin, an anthocyanin present in natural foods like red wine, can activate endothelial cells. This anthocyanin raises intracellular protein-Ca^{2+} and tyrosin phosphorylation, which controls eNOS. Tyrosine kinases and phospholipase C are both involved in Ca^{2+} signaling [190]. Furthermore, RWPC might even enhance endothelial NO production via the redox-responsive PI3/Akt channel, according to another report [191].

In addition, the effect of polyphenolic compounds on endothelial cells in preventing cardiovascular diseases is not limited to the stimulation of NO production. Because of the increased production of PGI2, the vasodilating effect is also boosted. In vitro studies on human endothelial cells exposed to the action of cocoa extract rich in procyanidins at a concentration of 2 mg/L and in vivo studies on procyanidins contained in chocolate administered to healthy volunteers showed that the ratio of cysteinyl leukotrienes (LTC4, LTD4, LTE4) to PGI2 can be reduced by 58 and 52%, respectively [192]. In contrast, isolavonoids, particularly genistein, limit the procoagulant action of vascular endothelium by, for example, lowering ET-1 expression [193]. Finally, polyphenols can affect endothelial cells' NO levels by affecting PDE-2 and PDE-4, two phosphodiesterases [194,195]. Taken together, plant polyphenols may have complex effects on the circulatory system's NO balance, which could account for their antihypertensive effects [196].

7.3. Polyphenols and Atherosclerosis

Atherosclerosis is the hardening and narrowing of the arteries, which is triggered by the buildup of lipids, cholesterol, and other substances in and on the artery walls over time. This then progresses into the endothelium, where they are oxidized by endothelial smooth muscle cells and activated macrophages [197,198]. ROS and reactive nitrogen species (RNS) production can enhance LDL oxidation. This causes a buildup of macrophages in this area, which clear oxidized LDL and transform them into foam cells. Endothelial dysfunction, as well as the concentration of monocytes/macrophages in the vascular intima under the influence of chemokines and adhesion molecules, foam cell development, and vascular smooth muscle proliferation, are all linked to the inflammatory backdrop of atherosclerotic lesions [199]. There is also a rise in extracellular matrix buildup surrounding the spot of inflammation, leading to plaque development which blocks the vessel, resulting in the loss of the blood artery's natural capacity to relax [198,199].

Published research has dealt with the potential benefits of polyphenols on atherosclerosis. Within this context, various studies have shown that RWPC and purple grape juice slow down atherosclerosis onset and progression through their anti-LDL oxidation, antioxidant properties, and inhibition of platelet aggregation properties, as well as an increase in HDL concentration and delay of vascular smooth muscle cell (SMC) propagation. In conclusion, polyphenols may be able to maintain "healthy blood vessels" by producing NO, which is important for vascular tone [200–202]. Recently, researchers found that giving rabbits red wine polyphenolic compounds (RWPC) orally reduces neointimal growth, lipid buildup, and inflammation in their iliac arteries. This is because RWPC has an anti-inflammatory effect [203]. In addition, when hamsters are given red wine, they have less neointimal hyperplasia, which is caused by a decrease in a protein that helps monocytes enter the artery wall. This is one of the ways that the artery reopens [204].

7.4. Polyphenols and Anti-Platelet Action

The excessive activation of platelets is linked to several long-term vascular diseases. This is due to the many adhesion proteins in the granules that, when highly activated, can lead to different types of thrombotic diseases [205,206]. In this respect, numerous important things happen in the process of platelet activation. One of them is the conversion of arachidonic acid to thromboxane A2, an arachidonate metabolite, through the cyclooxygenase pathway [207]. In this context, polyphenols are valuable from the perspective of platelet activation, which includes the adhesion and aggregation of platelets, due to the antioxidant effect of polyphenols. The first step in platelet activation involves platelets sticking to the collagen in the body; as a result, the platelets become activated. In this respect, proteins like fibrinogen and thrombospondin act as adhesion proteins, and platelet receptors work together to help platelets stick together, leading to the start of a signaling process inside cells and the activation of platelets [208]. However, it has not been fully explained how polyphenols make platelets less likely to stick together. It turns out that extracts rich in polyphenolic compounds, like grape seed and Yucca schidigera extracts, can help stop platelets from sticking to collagen. These extracts contain resveratrol and its derivatives, which make platelets less likely to stick together when they are stimulated by thrombin [153].

Thromboxane A2 (TXA2) is the key compound that is formed from the breakdown of arachidonic acid (ARA). It has some surface receptors that make platelets clump together. Evidence from the literature has indicated that the anti-aggregative effect of polyphenols is linked to numerous complicated molecular processes [209]. The capacity of polyphenols to hinder the enzymes involved in the formation of TXA2, COX, and LOX is the primary method by which they exert their anti-platelet aggregate effects on platelets [209,210]. However, they are also antagonists of the thromboxane A2 receptor, which suggests that flavonoids, through their indirectly suppressive effect on COX1, can lower TXA2 levels in the blood [211]. In an in vivo dog model, researchers investigated the effects of grape juice and red and white wine on platelet aggregation activity. The results revealed the antiplatelet effects of red wine and grape juice, while white wine does not yet have this impact [212].

Flavonoids have been shown to lower platelet aggregation because collagen metabolism is altered by these compounds, in addition to their interference in arachidonic acid metabolism. This is expressed as the antiplatelet action of collagen in the early stages of the aggregation of platelets. Moreover, the oxidative stress results in the aggregation of platelet in response to collagen via the activation of the inositol pathway, boosting intracellular calcium levels in the process. Flavonoids such as quercetin, catechin, and kaempferol, among others, have been shown to decrease oxidative stress by impeding the enzyme NADPH-oxidase [205].

7.5. Polyphenols as Anti-Inflammatory Agents

Inflammatory response to injury is a complicated biotic process that happens in response to a damaging stimulus. Different enzymes, including cyclooxygenase (COX), lipoxygenase (LOX), tyrosine kinase (TK), phospholipase A2 (PLA2s), and protein kinase C), are responsible for the proper function of an inflammatory response. Certain flavonoids have been demonstrated to act directly on several such enzymes, blocking them and therefore directly affecting inflammation [213,214]. One of the most important elements in preventing and treating chronic inflammation, according to epidemiological research, is nutrition. Through ex vivo and in vivo models, researchers have discovered that some flavonoids exert anti-inflammatory effects. One of the key bodily functions that flavonoids have an impact on is the synthesis of prostaglandins. Hesperidin and diosmin can reduce the generation of prostaglandins, according to several in vivo studies [215].

The mobilization of leukocytes is known as a critical stage in the progression of inflammation that occurs in cardiovascular illnesses and other conditions. The production of arachidonic acid ultimately results in the generation of cytokines (IL-1) and chemokines (IL-8) by neutrophils, which is mediated by both COX and LOX. In this regard, quercetin, a polyphenol, is especially effective in suppressing the formation of prostaglandins (PGs), leukotrienes (LT), and thromboxanes (TXA) by preventing the enzymes COX and LOX, respectively [216–218]. Evidence from numerous ex vivo experiments shows that some flavonoids, for example, bilobetine, morelloflavone, amentoflavone, and those found in *Sophora flavescens*, exert their effect by inhibiting the production of arachidonic acid [219]. Furthermore, resveratrol is regarded as a molecule with anti-inflammatory properties, as it inhibits the production of PGs [220]. Table 3 lists the key cardioprotective mechanisms of action behind the beneficial effects dietary polyphenols have on human health.

Table 3. Cardioprotective mechanisms of action of polyphenols.

Beneficial Effect	Specific Mechanism
Antioxidant	Generation of stable flavonoid radicals, increasing the protection of antioxidant systems and the elimination of ROS [83–103].
Antihypertensive	Modulate the RAAS and prompt an increase in the endothelium-derived nitric oxide concentration [125–143].
Anti-atherogenic	Through their antioxidant action, the diminished oxidation of LDL, and antiplatelet clumping action, they inhibit the formation and progression of atherosclerosis. They also, inhibit the oxidative degradation of lipoproteins and decrease the circulatory lipid levels [123–129].
Antiplatelet	Inhibitory effect on excessive platelet activation and produce decreased platelet adhesion [130–137].
Anti-inflammatory	Blocking inflammatory enzymes (COX, LOX, TK, PLA2s, protein kinase C), interfering with the production of prostaglandins, and suppressing the formation of PGs, LT, and TXA [159–166].

8. Interactions between Polyphenols and Nutrients and Drugs

Flavonoids form protein complexes through several nonspecific ways, including hydrogen bonds, hydrophobic interactions, and covalent bonds [221–223]. By forming complexes with proteins, polyphenols can alter their function, structure, solubility, hydrophobicity, thermal stability, isoelectric point, and susceptibility to digestive enzymes [223]. These modifications can influence digestion and the utilization of dietary proteins. In addition, polyphenols may affect the structure of digestive enzymes such as amylases, proteases, and lipases, disrupting their function and causing biochemical processes to malfunction. Furthermore, polyphenols can modify the processes of drug absorption, distribution, and metabolism. This is achieved via two mechanisms: the inhibition of P450 activity, which can occur through competitive, non-competitive, or uncompetitive enzyme inhibition, and the reduction of P450 activity. These alterations in P450 activity directly impact the clinical outcomes of drugs [221].

Similarly, polyphenols interact with reactive oxygen species, free radicals, and many other chemical compounds within their immediate environment due to their high concentration of active functional groups. In some cases, specific interactions can have adverse effects on human health [224]. It is critical to consider the interaction between polyphenols and pharmacological agents, namely the iron-containing preparations used to treat anemia. This interaction can significantly affect drug metabolism and pharmacokinetics and may modify pharmaceutical therapeutic effects [225]. For example, consumers acknowledge that grapefruit juice and herbal infusions are contraindicated for consuming pharmaceutical medications concurrently. These modifications can cause an increase in treatment effectiveness or a decrease in effectiveness. Furthermore, polyphenols have been demonstrated to have a significant effect on the drug-metabolizing enzymes involved in phases I and II of drug metabolism, although the fundamental rationale for such recommendations remains undisclosed; these enzymes include cytochrome P450, glutathione S-transferase, UDP-glucuronosyltransferase, sulfotransferase, N-acetyltransferase, methyltransferase, epoxide hydrolase, and NAD(P)H, a quinone oxidase transporter [223–225].

9. Adverse Effects of Polyphenols

The human diet contains a vast array of polyphenolic chemicals, about 8000 [226]. However, it has been found that consuming excessive amounts of polyphenols can lead to side effects [227]. In this regard, green tea extracts that include the well-known catechin (−)-epigallocatechin-3-gallate (EGCG) are marketed with the intention of facilitating weight loss; however, hepatotoxicity has been reported in a subset of individuals who consumed the product [228]. Similarly, chlorogenic acid, a prevalent polyphenol found in coffee, has been associated with numerous health advantages, as well as cytotoxic and genotoxic effects [229]. Polyphenols have been found to elicit mutagenic effects, promote the development of cancer, and produce genotoxicity. Within this context, multiple investigations have demonstrated that flavonoids engage in interactions with topoisomerase II and II, resulting in the induction of DNA cleavage. The present paper provides evidence that genistein exhibits an augmenting effect on the DNA cleaving activity of both human topoisomerase II and II, as reported in a previous study [230]. According to prior studies, it has been observed that (−)-epigallocatechin gallate (EGCG) [231] exhibits redox-dependent characteristics as a topoisomerase II toxin through the creation of covalent linkages with the enzyme. Moreover, numerous studies have shown that polyphenols exert inhibitory effects on topoisomerase via multiple pathways. Research findings have indicated that topoisomerase II poisons have an impact on EGCG, a compound that is influenced by redox reactions. The toxins belonging to the kaempferol and quercetin classes have mechanisms of action that are dependent on redox processes, as well as additional action mechanisms. Research findings have shown that the effects of (−)-epicatechin gallate (ECG) and (−)-epicatechin/EC/ are statistically insignificant [232].

Although polyphenols exhibit potent antioxidant effects, they may demonstrate prooxidant characteristics in the presence of elevated amounts of metals, pH, and oxygen. Specifically, copper and iron contribute to the heightened prooxidant activity of EGCG. The chemical exhibits prooxidant characteristics through the formation of a redox complex with either a transition metal ion or a phenoxyl radical [233]. Phenoxyl radicals produce reactive oxygen species (ROS) such as $O_2\bullet$ and H_2O_2 in response to the presence of oxygen.

Consequently, these chemical substances induce DNA damage, lipid peroxidation, and various signs of molecular oxidation. It has been proposed that the potential pro-oxidant impacts of polyphenols, specifically regarding EGCG, elicit noteworthy responses [234]. Previous studies have noted that the oxidation of polyphenols possessing small molecular structures, such as dihydroxycinnamic acids, leads to DNA incision or lipid peroxidation [233]. Zeng et al. [235] conducted a study that revealed that compounds possessing dihydroxyl groups in the ortho-conformation, such as caffeic acid and chlorogenic acid, as well as those containing 4-hydroxy-3-methoxyl groups, such as sinapic acid and ferulic acid, induced a significantly higher level of DNA damage compared to compounds lacking these functional groups.

10. Materials and Methods

10.1. Literature Search and Methodology

In this current review on Plant polyphenols and their potential benefits on cardiovascular health, different databases, including PubMed, Google Scholar, NIH National Library of Medicine, Scopus, and Web of Sciences were surveyed to retrieve data using a series of search terms, namely "cardiovascular system", "Coronary artery disease", "Polyphenols", "vascular Homeostasis", "Vascular endothelium", "Oxidative stress", "polyphenols as antioxidants", and "Atherosclerosis", without any publication date limit. Research articles and reviews were included, whereas conference abstracts and non-English publications were excluded.

10.2. Illustrations and Figures

The chemical assemblies were drawn using ChemDraw 22.0.0 with the assistance of PubChem. The figures demonstrating mechanisms of action were drawn in Microsoft PowerPoint 2019 and Biorender (https://biorender.com/, accessed on 25 May 2023).

11. Conclusions

Food and plant-derived natural substances are attractive because they are relatively safe compared to synthetic drugs, affordable, and widely accessible. At present, people consume natural food products obtained from fruits and vegetables and other food items to treat and manage cardiovascular diseases. These food items contain natural compounds, especially polyphenols, which exhibit antioxidant properties among other things. There is uncertainty regarding the precise physiological actions of these substances, particularly regarding their effects on the cardiovascular system. In summary, the findings presented in this work show that significant amounts of polyphenols can be found in fruits, especially berries, and drinks such as tea and coffee. Similarly, substantial amounts of these potential cardioprotective compounds are present in vegetables, leguminous plants, and grains. In addition, the findings presented in this paper demonstrate the anti-atherosclerotic properties of dietary polyphenols, including improvements in endothelium and vascular function, hemostasis and platelet function, and inflammatory biomarkers. Long-term dietary intervention research comparing the effectiveness of various dosages of conventional pharmaceuticals to polyphenol treatments in a range of clinical populations will increase the clinical acceptance of polyphenol therapies. The scientific community would benefit greatly from more research into the mechanisms underlying in vivo bioavailability in humans and the safety implications of consuming foods high in polyphenols. Although there have been some debates over its absorption, oral polyphenol consumption has shown encouraging results as a supplemental treatment option in lowering atherosclerosis development in at-risk patients. However, more extensive studies involving human subjects are required to establish the efficacy and safety of dietary polyphenols for long-term use in treating human diseases such as cancer and cardiovascular diseases.

There are numerous mechanisms through which polyphenols can influence the complex pathophysiology of cardiovascular disease (CVD); these include decreasing blood pressure, reducing cholesterol levels, acting as antioxidants, mitigating inflammation, inhibiting cell proliferation and angiogenesis, promoting endothelial function recovery, preventing thrombosis, and providing protection for the myocardium, among other functions. Nevertheless, there are several significant challenges that prevent the clinical utilization of polyphenols. These include dosage, specificity, potency, practicality, and the potential short- or long-term side effects on human subjects. Although natural polyphenols are generally considered to be safe, their potential adverse effects depend on their distribution and target cells within the body. It is possible for certain polyphenols to interact with conventional drugs as well as nutrition, thereby posing potential safety risks. Future animal experiments, large-scale cohort studies, and human intervention trials are required to address these challenges.

Author Contributions: Conceptualization, I.I., F.S. and M.W.; Methodology, I.I. and B.N.; Writing—original draft preparation, A.I., I.I. and M.W.; Writing—review and editing, R.N., M.F.L., P.W. and M.S.M.; Supervision, F.S. and M.S.M. All authors have read and agreed to the published version of the manuscript.

Funding: This research received no external funding.

Institutional Review Board Statement: Not applicable.

Informed Consent Statement: Not applicable.

Data Availability Statement: Not applicable.

Conflicts of Interest: The authors declare no conflict of interest.

References

1. Goszcz, K.; Duthie, G.G.; Stewart, D.; Leslie, S.J.; Megson, I.L. Bioactive polyphenols and cardiovascular disease: Chemical antagonists, pharmacological agents or xenobiotics that drive an adaptive response? *Br. J. Pharmacol.* **2017**, *174*, 1209–1225. [CrossRef]
2. Alam, M.A. Anti-hypertensive effect of cereal antioxidant ferulic acid and its mechanism of action. *Front. Nutr.* **2019**, *6*, 121. [CrossRef]
3. Jamee Shahwan, A.; Abed, Y.; Desormais, I.; Magne, J.; Preux, P.M.; Aboyans, V.; Lacroix, P. Epidemiology of coronary artery disease and stroke and associated risk factors in Gaza community-Palestine. *PLoS ONE* **2019**, *14*, e0211131. [CrossRef]
4. Li, J.; Liao, R.; Zhang, S.; Weng, H.; Liu, Y.; Tao, T.; Yu, F.; Li, G.; Wu, J. Promising remedies for cardiovascular disease: Natural polyphenol ellagic acid and its metabolite urolithins. *Phytomedicine* **2023**, *18*, 154867. [CrossRef]
5. Sharifi-Rad, J.; Rodrigues, C.F.; Sharopov, F.; Docea, A.O.; Can Karaca, A.; Sharifi-Rad, M.; Kahveci Karıncaoglu, D.; Gülseren, G.; Şenol, E.; Demircan, E.; et al. Diet, lifestyle and cardiovascular diseases: Linking pathophysiology to cardioprotective effects of natural bioactive compounds. *Int. J. Environ. Res. Public Health* **2020**, *17*, 2326. [CrossRef]
6. Blauwet, L.A.; Cooper, L.T. Myocarditis. *Prog. Cardiovasc. Dis.* **2010**, *52*, 274–288. [CrossRef]
7. Wirtz, P.H.; von Känel, R. Psychological stress, inflammation, and coronary heart disease. *Curr. Cardiol. Rep.* **2017**, *19*, 111. [CrossRef]
8. Khan, J.; Deb, P.K.; Priya, S.; Medina, K.D.; Devi, R.; Walode, S.G.; Rudrapal, M. Dietary flavonoids: Cardioprotective potential with antioxidant effects and their pharmacokinetic, toxicological and therapeutic concerns. *Molecules* **2021**, *26*, 4021. [CrossRef]
9. Curry, S.J.; Krist, A.H.; Owens, D.K.; Barry, M.J.; Caughey, A.B.; Davidson, K.W.; Doubeni, C.A.; Epling, J.W., Jr.; Kemper, A.R.; Kubik, M.; et al. Risk assessment for cardiovascular disease with nontraditional risk factors: US preventive services task force recommendation statement. *JAMA* **2018**, *320*, 272–280.
10. Holvoet, P. Stress in obesity and associated metabolic and cardiovascular disorders. *Scientifica* **2012**, *2012*, 205027. [CrossRef]
11. Matsuda, M.; Shimomura, I. Increased oxidative stress in obesity: Implications for metabolic syndrome, diabetes, hypertension, dyslipidemia, atherosclerosis, and cancer. *Obes. Res. Clin. Pract.* **2013**, *7*, e330–e341. [CrossRef]
12. Knowles, J.W.; Ashley, E.A. Cardiovascular disease: The rise of the genetic risk score. *PLoS Med.* **2018**, *15*, e1002546. [CrossRef]
13. Santhakumar, A.B.; Battino, M.; Alvarez-Suarez, J.M. Dietary polyphenols: Structures, bioavailability and protective effects against atherosclerosis. *Food Chem. Toxicol.* **2018**, *113*, 49–65. [CrossRef]
14. Malekmohammad, K.; Sewell, R.D.; Rafieian-Kopaei, M. Antioxidants and atherosclerosis: Mechanistic aspects. *Biomolecules* **2019**, *9*, 301. [CrossRef]
15. Sanches-Silva, A.; Testai, L.; Nabavi, S.F.; Battino, M.; Devi, K.P.; Tejada, S.; Sureda, A.; Xu, S.; Yousefi, B.; Majidinia, M.; et al. Therapeutic potential of polyphenols in cardiovascular diseases: Regulation of mTOR signaling pathway. *Pharmacol. Res.* **2020**, *152*, 104626. [CrossRef]
16. Quideau, S.; Deffieux, D.; Douat-Casassus, C.; Pouységu, L. Plant polyphenols: Chemical properties, biological activities, and synthesis. *Angew. Chem.* **2011**, *50*, 586–621. [CrossRef]

17. Nayak, B.; Liu, R.H.; Tang, J. Effect of processing on phenolic antioxidants of fruits, vegetables, and grains—A review. *Crit. Rev. Food Sci. Nutr.* **2015**, *55*, 887–919. [CrossRef]
18. Murray, M.; Dordevic, A.L.; Ryan, L.; Bonham, M.P. An emerging trend in functional foods for the prevention of cardiovascular disease and diabetes: Marine algal polyphenols. *Crit. Rev. Food Sci. Nutr.* **2018**, *58*, 1342–1358. [CrossRef]
19. Manach, C.; Scalbert, A.; Morand, C.; Rémésy, C.; Jiménez, L. Polyphenols: Food sources and bioavailability. *Am. J. Clin. Nutr.* **2004**, *79*, 727–747. [CrossRef]
20. Golovinskaia, O.; Wang, C.K. The hypoglycemic potential of phenolics from functional foods and their mechanisms. *Food Sci. Hum. Wellness* **2023**, *12*, 986–1007. [CrossRef]
21. Leuci, R.; Brunetti, L.; Poliseno, V.; Laghezza, A.; Loiodice, F.; Tortorella, P.; Piemontese, L. Natural compounds for the prevention and treatment of cardiovascular and neurodegenerative diseases. *Foods* **2021**, *10*, 29. [CrossRef] [PubMed]
22. Quiñones, M.; Miguel, M.; Aleixandre, A. Beneficial effects of polyphenols on cardiovascular disease. *Pharmacol. Res.* **2013**, *68*, 125–131. [CrossRef] [PubMed]
23. Wahid, M.; Saqib, F.; Chicea, L.; Ahmedah, H.T.; Sajer, B.H.; Marc, R.A.; Pop, O.L.; Moga, M.; Gavris, C. Metabolomics analysis delineates the therapeutic effects of hydroethanolic extract of *Cucumis sativus* L. seeds on hypertension and isoproterenol-induced myocardial infarction. *Biomed. Pharmacother.* **2022**, *148*, 112704. [CrossRef] [PubMed]
24. Bouayed, J.; Deußer, H.; Hoffmann, L.; Bohn, T. Bioaccessible and dialysable polyphenols in selected apple varieties following in vitro digestion vs. their native patterns. *Food Chem.* **2012**, *131*, 1466–1472. [CrossRef]
25. Rothwell, J.A.; Perez-Jimenez, J.; Neveu, V.; Medina-Remon, A.; M'hiri, N.; García-Lobato, P.; Manach, C.; Knox, C.; Eisner, R.; Wishart, D.S. Phenol-Explorer 3.0: A major update of the Phenol-Explorer database to incorporate data on the effects of food processing on polyphenol content. *Database* **2013**, *2013*, bat070. [CrossRef] [PubMed]
26. Deußer, H.; Guignard, C.; Hoffmann, L.; Evers, D. Polyphenol and glycoalkaloid contents in potato cultivars grown in Luxembourg. *Food Chem.* **2012**, *135*, 2814–2824. [CrossRef] [PubMed]
27. Rojanathammanee, L.; Puig, K.L.; Combs, C.K. Pomegranate polyphenols and extract inhibit nuclear factor of activated T-cell activity and microglial activation in vitro and in a transgenic mouse model of Alzheimer disease. *J. Nutr.* **2013**, *143*, 597–605. [CrossRef]
28. Rasouli, H.; Farzaei, M.H.; Khodarahmi, R. Polyphenols and their benefits: A review. *Int. J. Food Prop.* **2017**, *20*, 1700–1741. [CrossRef]
29. Singla, R.K.; Dubey, A.K.; Garg, A.; Sharma, R.K.; Fiorino, M.; Ameen, S.M.; Haddad, M.A.; Al-Hiary, M. *Natural Polyphenols: Chemical Classification, Definition of Classes, Subcategories, and Structures*; Oxford University Press: Oxford, UK, 2019; Volume 102, pp. 1397–1400.
30. Prabhu, S.; Molath, A.; Choksi, H.; Kumar, S.; Mehra, R. Classifications of polyphenols and their potential application in human health and diseases. *Int. J. Physiol, Nutr. Phys. Educ.* **2021**, *6*, 293–301. [CrossRef]
31. Shahidi, F.; Yeo, J. Bioactivities of phenolics by focusing on suppression of chronic diseases: A review. *Inter. J. Mol. Sci.* **2018**, *19*, 1573. [CrossRef]
32. Rashmi, H.B.; Negi, P.S. Phenolic acids from vegetables: A review on processing stability and health benefits. *Food Res. Int.* **2020**, *136*, 109298. [CrossRef] [PubMed]
33. Ozcan, T.; Akpinar-Bayizit, A.; Yilmaz-Ersan, L.; Delikanli, B. Phenolics in human health. *Int. J. Chem. Eng. Appl.* **2014**, *5*, 393. [CrossRef]
34. Teixeira, J.; Gaspar, A.; Garrido, E.M.; Garrido, J.; Borges, F. Hydroxycinnamic acid antioxidants: An electrochemical overview. *BioMed. Res. Int.* **2013**, *2013*, 251754. [CrossRef] [PubMed]
35. Gupta, S.; Bishnoi, J.P.; Kumar, N.; Kumar, H.; Nidheesh, T. *Terminalia arjuna* (Roxb.) Wight & Arn.: Competent source of bioactive components in functional food and drugs. *Pharma Innov. J.* **2018**, *7*, 223–231.
36. Hordiei, K.; Gontova, T.; Trumbeckaite, S.; Yaremenko, M.; Raudone, L. Phenolic Composition and Antioxidant Activity of *Tanacetum parthenium* Cultivated in Different Regions of Ukraine: Insights into the Flavonoids and Hydroxycinnamic Acids Profile. *Plants* **2023**, *12*, 2940. [CrossRef]
37. Yadav, D.; Kumar, H.; Kumar, A.; Jha, A.; Kumar, P.; Goyal, A. Optimization of polyphenolic fortification of grape peel extract in stirred yogurt by response surface methodology. *Indian J. Dairy Sci.* **2016**, *69*, 41–49.
38. Baniwal, P.; Mehra, R.; Kumar, N.; Sharma, S.; Kumar, S. Cereals: Functional constituents and its health benefits. *Pharma Innov. J.* **2021**, *10*, 343–349. [CrossRef]
39. Galletti, S.; Cianchetta, S.; Righini, H.; Roberti, R. A Lignin-rich extract of giant reed (*Arundo donax* L.) as a possible tool to manage soilborne pathogens in horticulture: A preliminary study on a model pathosystem. *Horticulturae* **2022**, *8*, 589. [CrossRef]
40. Soleimani, V.; Delghandi, P.S.; Moallem, S.A.; Karimi, G. Safety and toxicity of silymarin, the major constituent of milk thistle extract: An updated review. *Phytother. Res.* **2019**, *33*, 1627–1638. [CrossRef]
41. Zhang, L.X.; Li, C.X.; Kakar, M.U.; Khan, M.S.; Wu, P.F.; Amir, R.M.; Dai, D.F.; Naveed, M.; Li, Q.Y.; Saeed, M.; et al. Resveratrol (RV): A pharmacological review and call for further research. *Biomed. Pharmacother.* **2021**, *143*, 112164. [CrossRef]
42. Shaito, A.; Posadino, A.M.; Younes, N.; Hasan, H.; Halabi, S.; Alhababi, D.; Al-Mohannadi, A.; Abdel-Rahman, W.M.; Eid, A.H.; Nasrallah, G.K.; et al. Potential adverse effects of resveratrol: A literature review. *Int. J. Mol. Sci.* **2020**, *21*, 2084. [CrossRef] [PubMed]

43. Ferrazzano, G.F.; Amato, I.; Ingenito, A.; Zarrelli, A.; Pinto, G.; Pollio, A. Plant polyphenols and their anti-cariogenic properties: A review. *Molecules* **2011**, *16*, 1486–1507. [CrossRef] [PubMed]
44. Clark, J.L.; Zahradka, P.; Taylor, C.G. Efficacy of flavonoids in the management of high blood pressure. *Nutr. Rev.* **2015**, *73*, 799–822. [CrossRef]
45. Su, J.; Xu, H.T.; Yu, J.J.; Gao, J.L.; Lei, J.; Yin, Q.S.; Li, B.; Pang, M.X.; Su, M.X.; Mi, W.J.; et al. Luteolin Ameliorates Hypertensive Vascular Remodeling through Inhibiting the Proliferation and Migration of Vascular Smooth Muscle Cells. *J. Evid. -Based Complement. Altern. Med.* **2015**, *2015*, 364876. [CrossRef] [PubMed]
46. Liu, D.; Homan, L.L.; Dillon, J.S. Genistein acutely stimulates nitric oxide synthesis in vascular endothelial cells by a cyclic adenosine 5′-monophosphate-dependent mechanism. *Endocrinology* **2004**, *145*, 5532–5539. [CrossRef]
47. Kumar, S.; Pandey, A.K. Chemistry and biological activities of flavonoids: An overview. *Sci. World J.* **2013**, *2013*, 162750. [CrossRef]
48. Rivera, L.; Morón, R.; Sánchez, M.; Zarzuelo, A.; Galisteo, M. Quercetin ameliorates metabolic syndrome and improves the inflammatory status in obese Zucker rats. *Obesity* **2008**, *16*, 2081–2087. [CrossRef]
49. Olaleye, M.; Crown, O.; Akinmoladun, A.; Akindahunsi, A. Rutin and quercetin show greater efficacy than nifedipin in ameliorating hemodynamic, redox, and metabolite imbalances in sodium chloride-induced hypertensive rats. *Hum. Exp. Toxicol.* **2014**, *33*, 602–608. [CrossRef]
50. Oyagbemi, A.A.; Omobowale, T.O.; Ola-Davies, O.E.; Asenuga, E.R.; Ajibade, T.O.; Adejumobi, O.A.; Arojojoye, O.A.; Afolabi, J.M.; Ogunpolu, B.S.; Falayi, O.O. Quercetin attenuates hypertension induced by sodium fluoride via reduction in oxidative stress and modulation of HSP 70/ERK/PPARγ signaling pathways. *Biofactors* **2018**, *44*, 465–479. [CrossRef]
51. Tiwari, P.; Mishra, R.; Mazumder, A.; Mazumder, R.; Singh, A. An insight into diverse activities and targets of flavonoids. *Curr. Drug Targets* **2023**, *24*, 89–102.
52. Leeya, Y.; Mulvany, M.J.; Queiroz, E.F.; Marston, A.; Hostettmann, K.; Jansakul, C. Hypotensive activity of an n-butanol extract and their purified compounds from leaves of *Phyllanthus acidus* (L.) Skeels in rats. *Eur. J. Pharmacol.* **2010**, *649*, 301–313. [CrossRef]
53. Thamcharoen, N.; Susantitaphong, P.; Wongrakpanich, S.; Chongsathidkiet, P.; Tantrachoti, P.; Pitukweerakul, S.; Avihingsanon, Y.; Praditpornsilpa, K.; Jaber, B.L.; Eiam-Ong, S. Effect of N-and T-type calcium channel blocker on proteinuria, blood pressure and kidney function in hypertensive patients: A meta-analysis. *Hypertens. Res.* **2015**, *38*, 847–855. [CrossRef] [PubMed]
54. Ried, K.; Sullivan, T.R.; Fakler, P.; Frank, O.R.; Stocks, N.P. Effect of cocoa on blood pressure. *Cochrane Database Syst. Rev.* **2012**. [CrossRef]
55. Igarashi, K.; Honma, K.; Yoshinari, O.; Nanjo, F.; Hara, Y. Effects of dietary catechins on glucose tolerance, blood pressure and oxidative status in Goto-Kakizaki rats. *J. Nutr. Sci. Vitaminol.* **2007**, *53*, 496–500. [CrossRef] [PubMed]
56. Ellinger, S.; Reusch, A.; Stehle, P.; Helfrich, H.-P. Epicatechin ingested via cocoa products reduces blood pressure in humans: A nonlinear regression model with a Bayesian approach. *Am. J. Clin. Nutr.* **2012**, *95*, 1365–1377. [CrossRef] [PubMed]
57. DA Silva, N.D., Jr.; Fernandes, T.; Soci, U.P.; Monteiro, A.W.; Phillips, M.I.; EM, D.E.O. Swimming training in rats increases cardiac MicroRNA-126 expression and angiogenesis. *Med. Sci. Sports Exerc.* **2012**, *44*, 1453–1462. [CrossRef]
58. Mahmoud, A.M.; Hernández Bautista, R.J.; Sandhu, M.A.; Hussein, O.E. Beneficial effects of citrus flavonoids on cardiovascular and metabolic health. *Oxid. Med. Cell. Longev.* **2019**, *2019*, 5484138. [CrossRef]
59. Khan, H.; Jawad, M.; Kamal, M.A.; Baldi, A.; Xiao, J.; Nabavi, S.M.; Daglia, M. Evidence and prospective of plant derived flavonoids as antiplatelet agents: Strong candidates to be drugs of future. *Food Chem. Toxicol.* **2018**, *119*, 355–367. [CrossRef]
60. Ikemura, M.; Sasaki, Y.; Giddings, J.C.; Yamamoto, J. Preventive effects of hesperidin, glucosyl hesperidin and naringin on hypertension and cerebral thrombosis in stroke-prone spontaneously hypertensive rats. *Phytother. Res.* **2012**, *26*, 1272–1277. [CrossRef]
61. Yang, Y.; Li, P.-Y.; Cheng, J.; Mao, L.; Wen, J.; Tan, X.-Q.; Liu, Z.-F.; Zeng, X.-R. Function of BKCa channels is reduced in human vascular smooth muscle cells from Han Chinese patients with hypertension. *Hypertension* **2013**, *61*, 519–525. [CrossRef]
62. Yamamoto, M.; Jokura, H.; Hashizume, K.; Ominami, H.; Shibuya, Y.; Suzuki, A.; Hase, T.; Shimotoyodome, A. Hesperidin metabolite hesperetin-7-O-glucuronide, but not hesperetin-3′-O-glucuronide, exerts hypotensive, vasodilatory, and anti-inflammatory activities. *Food Funct.* **2013**, *4*, 1346–1351. [CrossRef] [PubMed]
63. Liu, Y.; Niu, L.; Cui, L.; Hou, X.; Li, J.; Zhang, X.; Zhang, M. Hesperetin inhibits rat coronary constriction by inhibiting Ca^{2+} influx and enhancing voltage-gated K+ channel currents of the myocytes. *Eur. J. Pharmacol.* **2014**, *735*, 193–201. [CrossRef]
64. McKay, D.L.; Chen, C.O.; Saltzman, E.; Blumberg, J.B. *Hibiscus sabdariffa* L. tea (tisane) lowers blood pressure in prehypertensive and mildly hypertensive adults. *J. Nutr.* **2010**, *140*, 298–303. [CrossRef] [PubMed]
65. Jiang, R.-W.; Lau, K.-M.; Lam, H.-M.; Yam, W.-S.; Leung, L.-K.; Choi, K.-L.; Waye, M.M.; Mak, T.C.; Woo, K.-S.; Fung, K.-P. A comparative study on aqueous root extracts of Pueraria thomsonii and Pueraria lobata by antioxidant assay and HPLC fingerprint analysis. *J. Ethnopharmacol.* **2005**, *96*, 133–138. [CrossRef]
66. Lu, X.-L.; Liu, J.-X.; Wu, Q.; Long, S.-M.; Zheng, M.-Y.; Yao, X.-L.; Ren, H.; Wang, Y.-G.; Su, W.-W.; Cheung, R.T.F. Protective effects of puerarin against Aß40-induced vascular dysfunction in zebrafish and human endothelial cells. *Eur. J. Pharmacol.* **2014**, *732*, 76–85. [CrossRef] [PubMed]
67. Mita, M.; Tanaka, H.; Yanagihara, H.; Nakagawa, J.; Hishinuma, S.; Sutherland, C.; Walsh, M.P.; Shoji, M. Membrane depolarization-induced RhoA/Rho-associated kinase activation and sustained contraction of rat caudal arterial smooth muscle involves genistein-sensitive tyrosine phosphorylation. *J. Smooth Muscle Res.* **2013**, *49*, 26–45. [CrossRef]

68. D'Archivio, M.; Filesi, C.; Varì, R.; Scazzocchio, B.; Masella, R. Bioavailability of the polyphenols: Status and controversies. *Int. J. Mol. Sci.* **2010**, *11*, 1321–1342. [CrossRef]
69. Karas, D.; Ulrichová, J.; Valentová, K. Galloylation of polyphenols alters their biological activity. *Food Chem. Toxicol.* **2017**, *105*, 223–240. [CrossRef]
70. Bohn, T. Dietary factors affecting polyphenol bioavailability. *Nutr. Rev.* **2014**, *72*, 429–452. [CrossRef]
71. Gonzalez, S.; Fernandez, M.; Cuervo, A.; Lasheras, C. Dietary intake of polyphenols and major food sources in an institutionalised elderly population. *J. Hum. Nutr. Diet.* **2014**, *27*, 176–183. [CrossRef]
72. Manach, C.; Williamson, G.; Morand, C.; Scalbert, A.; Rémésy, C. Bioavailability and bioefficacy of polyphenols in humans. I. Review of 97 bioavailability studies. *Am. J. Clin. Nutr.* **2005**, *81* (Suppl. S1), 230s–242s. [CrossRef] [PubMed]
73. Crozier, A.; Del Rio, D.; Clifford, M.N. Bioavailability of dietary flavonoids and phenolic compounds. *Mol. Aspects Med.* **2010**, *31*, 446–467. [CrossRef] [PubMed]
74. Behl, T.; Bungau, S.; Kumar, K.; Zengin, G.; Khan, F.; Kumar, A.; Kaur, R.; Venkatachalam, T.; Tit, D.M.; Vesa, C.M.; et al. Pleotropic effects of polyphenols in cardiovascular system. *Biomed. Pharmacother.* **2020**, *130*, 110714. [CrossRef] [PubMed]
75. Laurent, C.; Besançon, P.; Caporiccio, B. Flavonoids from a grape seed extract interact with digestive secretions and intestinal cells as assessed in an in vitro digestion/Caco-2 cell culture model. *Food Chem.* **2007**, *100*, 1704–1712. [CrossRef]
76. Lemmens, L.; Van Buggenhout, S.; Van Loey, A.M.; Hendrickx, M.E. Particle size reduction leading to cell wall rupture is more important for the β-carotene bioaccessibility of raw compared to thermally processed carrots. *J. Agric. Food Chem.* **2010**, *58*, 12769–12776. [CrossRef]
77. Grgić, J.; Šelo, G.; Planinić, M.; Tišma, M.; Bucić-Kojić, A. Role of the encapsulation in bioavailability of phenolic compounds. *Antioxidants* **2020**, *9*, 923. [CrossRef]
78. Bohn, T. Bioavailability of non-provitamin A carotenoids. *Curr. Nutr. Food Sci.* **2008**, *4*, 240–258. [CrossRef]
79. Day, A.J.; Cañada, F.J.; Díaz, J.C.; Kroon, P.A.; Mclauchlan, R.; Faulds, C.B.; Plumb, G.W.; Morgan, M.R.; Williamson, G. Dietary flavonoid and isoflavone glycosides are hydrolysed by the lactase site of lactase phlorizin hydrolase. *FEBS Lett.* **2000**, *468*, 166–170. [CrossRef]
80. Alotaibi, B.S.; Ijaz, M.; Buabeid, M.; Kharaba, Z.J.; Yaseen, H.S.; Murtaza, G. Therapeutic effects and safe uses of plant-derived polyphenolic compounds in cardiovascular diseases: A review. *Drug Des. Devel. Ther.* **2021**, *15*, 4713–4732. [CrossRef]
81. Bravo, L. Polyphenols: Chemistry, dietary sources, metabolism, and nutritional significance. *Nutr. Rev.* **1998**, *56*, 317–333. [CrossRef]
82. Hong, J.; Lu, H.; Meng, X.; Ryu, J.-H.; Hara, Y.; Yang, C.S. Stability, cellular uptake, biotransformation, and efflux of tea polyphenol (−)-epigallocatechin-3-gallate in HT-29 human colon adenocarcinoma cells. *Cancer Res.* **2002**, *62*, 7241–7246.
83. Kosińska, A.; Andlauer, W. Cocoa polyphenols are absorbed in Caco-2 cell model of intestinal epithelium. *Food Chem.* **2012**, *135*, 999–1005. [CrossRef]
84. Wolffram, S.; Block, M.; Ader, P. Quercetin-3-glucoside is transported by the glucose carrier SGLT1 across the brush border membrane of rat small intestine. *J. Nutr.* **2002**, *132*, 630–635. [CrossRef]
85. Henry-Vitrac, C.; Desmoulière, A.; Girard, D.; Mérillon, J.-M.; Krisa, S. Transport, deglycosylation, and metabolism of trans-piceid by small intestinal epithelial cells. *Eur. J. Nutr.* **2006**, *45*, 376–382. [CrossRef] [PubMed]
86. Rahman, B.; Schneider, H.-P.; Bröer, A.; Deitmer, J.W.; Bröer, S. Helix 8 and helix 10 are involved in substrate recognition in the rat monocarboxylate transporter MCT1. *Biochemistry* **1999**, *38*, 11577–11584. [CrossRef] [PubMed]
87. Cermak, R.; Breves, G.; Lüpke, M.; Wolffram, S. In vitro degradation of the flavonol quercetin and of quercetin glycosides in the porcine hindgut. *Arch. Anim. Nutr.* **2006**, *60*, 180–189. [CrossRef] [PubMed]
88. Konishi, Y.; Shimizu, M. Transepithelial transport of ferulic acid by monocarboxylic acid transporter in Caco-2 cell monolayers. *Biosci. Biotech. Biochem.* **2003**, *67*, 856–862. [CrossRef]
89. Selma, M.V.; Espín, J.C.; Tomás-Barberán, F.A. Interaction between phenolics and gut microbiota: Role in human health. *J. Agric. Food Chem.* **2009**, *57*, 6485–6501. [CrossRef]
90. Ferruzzi, M.G. The influence of beverage composition on delivery of phenolic compounds from coffee and tea. *Physiol. Behav.* **2010**, *100*, 33–41. [CrossRef]
91. Del Rio, D.; Rodriguez-Mateos, A.; Spencer, J.P.; Tognolini, M.; Borges, G.; Crozier, A. Dietary (poly) phenolics in human health: Structures, bioavailability, and evidence of protective effects against chronic diseases. *Antioxid. Redox Signal.* **2013**, *18*, 1818–1892. [CrossRef]
92. Leonarduzzi, G.; Testa, G.; Sottero, B.; Gamba, P.; Poli, G. Design and development of nanovehicle-based delivery systems for preventive or therapeutic supplementation with flavonoids. *Curr. Med. Chem.* **2010**, *17*, 74–95. [CrossRef]
93. Lambert, J.D.; Sang, S.; Yang, C.S. Biotransformation of Green Tea Polyphenols and the Biological Activities of Those Metabolites. *Mol. Pharm.* **2007**, *4*, 819–825. [CrossRef]
94. Boulton, D.W.; Walle, U.K.; Walle, T. Extensive binding of the bioflavonoid quercetin to human plasma proteins. *J. Pharm. Pharmacol.* **1998**, *50*, 243–249. [CrossRef] [PubMed]
95. Manach, C.; Morand, C.; Texier, O.; Favier, M.-L.; Agullo, G.; Demigné, C.; Régérat, F.; Rémésy, C. Quercetin metabolites in plasma of rats fed diets containing rutin or quercetin. *J. Nutr.* **1995**, *125*, 1911–1922. [CrossRef] [PubMed]
96. Chen, L.; Cao, H.; Huang, Q.; Xiao, J.; Teng, H. Absorption, metabolism and bioavailability of flavonoids: A review. *Crit. Rev. Food Sci. Nutr.* **2022**, *62*, 7730–7742. [CrossRef] [PubMed]

97. Bohn, T.; Blackwood, M.; Francis, D.; Tian, Q.; Schwartz, S.J.; Clinton, S.K. Bioavailability of phytochemical constituents from a novel soy fortified lycopene rich tomato juice developed for targeted cancer prevention trials. *Nutr. Cancer* **2013**, *65*, 919–929. [CrossRef] [PubMed]
98. Lippolis, T.; Cofano, M.; Caponio, G.R.; De Nunzio, V.; Notarnicola, M. Bioaccessibility and bioavailability of diet polyphenols and their modulation of gut microbiota. *Int. J. Mol. Sci.* **2023**, *24*, 3813. [CrossRef]
99. Zanotti, I.; Dall'Asta, M.; Mena, P.; Mele, L.; Bruni, R.; Ray, S.; Del Rio, D. Atheroprotective effects of (poly) phenols: A focus on cell cholesterol metabolism. *Food Funct.* **2015**, *6*, 13–31. [CrossRef]
100. Buscemi, S.; Verga, S.; Batsis, J.A.; Donatelli, M.; Tranchina, M.R.; Belmonte, S.; Mattina, A.; Re, A.; Cerasola, G. Acute effects of coffee on endothelial function in healthy subjects. *Eur. J. Clin. Nutr.* **2010**, *64*, 483–489. [CrossRef]
101. Calani, L.; Del Rio, D.; Luisa Callegari, M.; Morelli, L.; Brighenti, F. Updated bioavailability and 48 h excretion profile of flavan-3-ols from green tea in humans. *Int. J. Food Sci. Nutr.* **2012**, *63*, 513–521. [CrossRef]
102. Kay, C.D. Aspects of anthocyanin absorption, metabolism and pharmacokinetics in humans. *Nutr. Res. Rev.* **2006**, *19*, 137–146. [CrossRef] [PubMed]
103. Stalmach, A.; Mullen, W.; Barron, D.; Uchida, K.; Yokota, T.; Cavin, C.; Steiling, H.; Williamson, G.; Crozier, A. Metabolite profiling of hydroxycinnamate derivatives in plasma and urine after the ingestion of coffee by humans: Identification of biomarkers of coffee consumption. *Drug Metab. Dispos.* **2009**, *37*, 1749–1758. [CrossRef]
104. Heleno, S.A.; Martins, A.; Queiroz, M.J.R.; Ferreira, I.C. Bioactivity of phenolic acids: Metabolites versus parent compounds: A review. *Food Chem.* **2015**, *173*, 501–513. [CrossRef] [PubMed]
105. Rodriguez-Mateos, A.; Cifuentes-Gomez, T.; Gonzalez-Salvador, I.; Ottaviani, J.I.; Schroeter, H.; Kelm, M.; Heiss, C.; Spencer, J.P. Influence of age on the absorption, metabolism, and excretion of cocoa flavanols in healthy subjects. *Mol. Nutr. Food Res.* **2015**, *59*, 1504–1512. [CrossRef] [PubMed]
106. Saura-Calixto, F.; Serrano, J.; Goñi, I. Intake and bioaccessibility of total polyphenols in a whole diet. *Food Chem.* **2007**, *101*, 492–501. [CrossRef]
107. Espín, J.C.; González-Barrio, R.; Cerdá, B.; López-Bote, C.; Rey, A.I.; Tomás-Barberán, F.A. Iberian Pig as a Model To Clarify Obscure Points in the Bioavailability and Metabolism of Ellagitannins in Humans. *J. Agric. Food Chem.* **2007**, *55*, 10476–10485. [CrossRef]
108. Larrosa, M.; García-Conesa, M.T.; Espín, J.C.; Tomás-Barberán, F.A. Ellagitannins, ellagic acid and vascular health. *Mol. Aspects Med.* **2010**, *31*, 513–539. [CrossRef]
109. Hedayati, N.; Yaghoobi, A.; Salami, M.; Gholinezhad, Y.; Aghadavood, F.; Eshraghi, R.; Aarabi, M.H.; Homayoonfal, M.; Asemi, Z.; Mirzaei, H.; et al. Impact of polyphenols on heart failure and cardiac hypertrophy: Clinical effects and molecular mechanisms. *Front. Cardiovasc. Med.* **2023**, *10*, 1174816. [CrossRef]
110. Bialonska, D.; Kasimsetty, S.G.; Khan, S.I.; Ferreira, D. Urolithins, intestinal microbial metabolites of pomegranate ellagitannins, exhibit potent antioxidant activity in a cell-based assay. *J. Agric. Food Chem.* **2009**, *57*, 10181–10186. [CrossRef]
111. Borges, G.; Degeneve, A.; Mullen, W.; Crozier, A. Identification of flavonoid and phenolic antioxidants in black currants, blueberries, raspberries, red currants, and cranberries. *J. Agric. Food Chem.* **2010**, *58*, 3901–3909. [CrossRef]
112. Boesch-Saadatmandi, C.; Loboda, A.; Wagner, A.E.; Stachurska, A.; Jozkowicz, A.; Dulak, J.; Döring, F.; Wolffram, S.; Rimbach, G. Effect of quercetin and its metabolites isorhamnetin and quercetin-3-glucuronide on inflammatory gene expression: Role of miR-155. *J. Nutr. Biochem.* **2011**, *22*, 293–299. [CrossRef] [PubMed]
113. Bolling, B.W.; Chen, C.-Y.O.; McKay, D.L.; Blumberg, J.B. Tree nut phytochemicals: Composition, antioxidant capacity, bioactivity, impact factors. A systematic review of almonds, Brazils, cashews, hazelnuts, macadamias, pecans, pine nuts, pistachios and walnuts. *Nutr. Res. Rev.* **2011**, *24*, 244–275. [CrossRef] [PubMed]
114. Brown, A.A.; Hu, F.B. Dietary modulation of endothelial function: Implications for cardiovascular disease. *Am. J. Clin. Nutr.* **2001**, *73*, 673–686. [CrossRef] [PubMed]
115. Cines, D.B.; Pollak, E.S.; Buck, C.A.; Loscalzo, J.; Zimmerman, G.A.; McEver, R.P.; Pober, J.S.; Wick, T.M.; Konkle, B.A.; Schwartz, B.S.; et al. Endothelial cells in physiology and in the pathophysiology of vascular disorders. *Blood* **1998**, *91*, 3527–3561.
116. Saqib, F.; Wahid, M.; AL-Huqail, A.A.; Ahmedah, H.T.; Bigiu, N.; Irimie, M.; Moga, M.; Marc, R.A.; Pop, O.L.; Chicea, L.M. Metabolomics based mechanistic insights to vasorelaxant and cardioprotective effect of ethanolic extract of Citrullus lanatus (Thunb.) Matsum. & Nakai. seeds in isoproterenol-induced myocardial infarction. *Phytomedicine* **2022**, *100*, 154069.
117. Akimoto, S.; Mitsumata, M.; Sasaguri, T.; Yoshida, Y. Laminar shear stress inhibits vascular endothelial cell proliferation by inducing cyclin-dependent kinase inhibitor p21(Sdi1/Cip1/Waf1). *Circ. Res.* **2000**, *86*, 185–190. [CrossRef]
118. Ziche, M.; Morbidelli, L.; Masini, E.; Amerini, S.; Granger, H.J.; Maggi, C.A.; Geppetti, P.; Ledda, F. Nitric oxide mediates angiogenesis in vivo and endothelial cell growth and migration in vitro promoted by substance P. *J. Clin. Investig.* **1994**, *94*, 2036–2044. [CrossRef]
119. Matz, R.L.; Andriantsitohaina, R. Age-related endothelial dysfunction: Potential implications for pharmacotherapy. *Drugs Aging* **2003**, *20*, 527–550. [CrossRef]
120. Sofi, F.; Cesari, F.; Abbate, R.; Gensini, G.F.; Casini, A. Adherence to Mediterranean diet and health status: Meta-analysis. *BMJ* **2008**, *9*, 337. [CrossRef]
121. Rana, A.; Samtiya, M.; Dhewa, T.; Mishra, V.; Aluko, R.E. Health benefits of polyphenols: A concise review. *J. Food Biochem.* **2022**, *46*, e14264. [CrossRef]

122. Habauzit, V.; Morand, C. Evidence for a protective effect of polyphenols-containing foods on cardiovascular health: An update for clinicians. *Ther. Adv. Chronic Dis.* **2012**, *3*, 87–106. [CrossRef] [PubMed]
123. Khurana, S.; Venkataraman, K.; Hollingsworth, A.; Piche, M.; Tai, T.C. Polyphenols: Benefits to the cardiovascular system in health and in aging. *Nutrients* **2013**, *5*, 3779–3827. [CrossRef] [PubMed]
124. Brieger, K.; Schiavone, S.; Miller, F.J., Jr.; Krause, K.H. Reactive oxygen species: From health to disease. *Swiss Med. Wkly.* **2012**, *142*, w13659. [CrossRef] [PubMed]
125. Bouayed, J.; Rammal, H.; Soulimani, R. Oxidative stress and anxiety: Relationship and cellular pathways. *Oxid. Med. Cell. Longevity* **2009**, *2*, 63–67. [CrossRef]
126. Lodovici, M.; Bigagli, E. Oxidative stress and air pollution exposure. *J. Toxicol.* **2011**, *2011*, 487074. [CrossRef] [PubMed]
127. Li, S.; Tan, H.Y.; Wang, N.; Zhang, Z.J.; Lao, L.; Wong, C.W.; Feng, Y. The role of oxidative stress and antioxidants in liver diseases. *Int. J. Mol. Sci.* **2015**, *16*, 26087–26124. [CrossRef] [PubMed]
128. Chan, S.M.; Cerni, C.; Passey, S.; Seow, H.J.; Bernardo, I.; van der Poel, C.; Dobric, A.; Brassington, K.; Selemidis, S.; Bozinovski, S.; et al. Cigarette smoking exacerbates skeletal muscle injury without compromising its regenerative capacity. *Am. J. Respir. Cell Mol. Biol.* **2020**, *62*, 217–230. [CrossRef]
129. Finkel, T.; Holbrook, N.J. Oxidants, oxidative stress and the biology of ageing. *Nature* **2000**, *408*, 239–247. [CrossRef]
130. Selvaraju, V.; Joshi, M.; Suresh, S.; Sanchez, J.A.; Maulik, N.; Maulik, G. Diabetes, oxidative stress, molecular mechanism, and cardiovascular disease--an overview. *Toxicol. Mech. Methods* **2012**, *22*, 330–335. [CrossRef]
131. Le Brocq, M.; Leslie, S.J.; Milliken, P.; Megson, I.L. Endothelial dysfunction: From molecular mechanisms to measurement, clinical implications, and therapeutic opportunities. *Antioxid. Redox Signal.* **2008**, *10*, 1631–1674. [CrossRef]
132. Falk, E. Pathogenesis of atherosclerosis. *J. Am. Coll. Cardiol.* **2006**, *47* (Suppl. S8), C7–C12. [CrossRef] [PubMed]
133. Megson, I.L.; Whitfield, P.D.; Zabetakis, I. Lipids and cardiovascular disease: Where does dietary intervention sit alongside statin therapy? *Food Funct.* **2016**, *7*, 2603–2614. [CrossRef] [PubMed]
134. Loke, W.M.; Proudfoot, J.M.; Hodgson, J.M.; McKinley, A.J.; Hime, N.; Magat, M.; Stocker, R.; Croft, K.D. Specific dietary polyphenols attenuate atherosclerosis in apolipoprotein E-knockout mice by alleviating inflammation and endothelial dysfunction. *Arterioscler.; Thromb. Vasc. Biol.* **2010**, *30*, 749–757. [CrossRef] [PubMed]
135. Dhalla, N.S.; Temsah, R.M.; Netticadan, T. Role of oxidative stress in cardiovascular diseases. *J. Hypertens.* **2000**, *18*, 655–673. [CrossRef] [PubMed]
136. Sugamura, K.; Keaney, J.F., Jr. Reactive oxygen species in cardiovascular disease. *Free Radical Biol. Med.* **2011**, *51*, 978–992. [CrossRef]
137. Raedschelders, K.; Ansley, D.M.; Chen, D.D. The cellular and molecular origin of reactive oxygen species generation during myocardial ischemia and reperfusion. *Pharmacol. Ther.* **2012**, *133*, 230–255. [CrossRef]
138. Park, W.H.; Kim, S.H. Involvement of reactive oxygen species and glutathione in gallic acid-induced human umbilical vein endothelial cell death. *Oncol. Rep.* **2012**, *28*, 695–700. [CrossRef]
139. Hopkins, P.N. Molecular biology of atherosclerosis. *Physiol. Rev.* **2013**, *93*, 1317–1542. [CrossRef]
140. Cai, H.; Harrison, D.G. Endothelial dysfunction in cardiovascular diseases: The role of oxidant stress. *Circ. Res.* **2000**, *87*, 840–844. [CrossRef]
141. Taniyama, Y.; Griendling, K.K. Reactive oxygen species in the vasculature: Molecular and cellular mechanisms. *Hypertension* **2003**, *42*, 1075–1081. [CrossRef]
142. Paravicini, T.M.; Touyz, R.M. NADPH oxidases, reactive oxygen species, and hypertension: Clinical implications and therapeutic possibilities. *Diabetes Care* **2008**, *31* (Suppl. S2), S170–S180. [CrossRef]
143. Curtin, J.F.; Donovan, M.; Cotter, T.G. Regulation and measurement of oxidative stress in apoptosis. *J. Immunol. Methods* **2002**, *265*, 49–72. [CrossRef]
144. Gerhardt, T.; Ley, K. Monocyte trafficking across the vessel wall. *Cardiovas. Res.* **2015**, *107*, 321–330. [CrossRef] [PubMed]
145. Hansson, G.K.; Libby, P. The immune response in atherosclerosis: A double-edged sword. *Nat. Rev. Immunol.* **2006**, *6*, 508–519. [CrossRef] [PubMed]
146. Martínez-Cayuela, M. Oxygen free radicals and human disease. *Biochimie* **1995**, *77*, 147–161. [CrossRef] [PubMed]
147. Singh, R.B.; Mengi, S.A.; Xu, Y.J.; Arneja, A.S.; Dhalla, N.S. Pathogenesis of atherosclerosis: A multifactorial process. *Exp. Clin. Cardiol.* **2002**, *7*, 40–53.
148. Jennings, L.K. Mechanisms of platelet activation: Need for new strategies to protect against platelet-mediated atherothrombosis. *Thromb. Haemost.* **2009**, *102*, 248–257. [CrossRef]
149. Furie, B.; Furie, B.C. Mechanisms of thrombus formation. *N. Engl. J. Med.* **2008**, *359*, 938–949. [CrossRef]
150. Falk, E.; Shah, P.K.; Fuster, V. Coronary plaque disruption. *Circulation* **1995**, *92*, 657–671. [CrossRef]
151. Fasolo, F.; Di Gregoli, K.; Maegdefessel, L.; Johnson, J.L. Non-coding RNAs in cardiovascular cell biology and atherosclerosis. *Cardiovasc. Res.* **2019**, *115*, 1732–1756. [CrossRef]
152. Olas, B.; Wachowicz, B.; Tomczak, A.; Erler, J.; Stochmal, A.; Oleszek, W. Comparative anti-platelet and antioxidant properties of polyphenol-rich extracts from: Berries of Aronia melanocarpa, seeds of grape and bark of Yucca schidigera in vitro. *Platelets* **2008**, *19*, 70–77. [CrossRef]
153. Jagroop, I.A.; Kakafika, A.I.; Mikhailidis, D.P. Platelets and vascular risk: An option for treatment. *Curr. Pharm. Des.* **2007**, *13*, 1669–1683. [CrossRef]

154. Gawaz, M. Role of platelets in coronary thrombosis and reperfusion of ischemic myocardium. *Cardiovasc. Res.* **2004**, *61*, 498–511. [CrossRef] [PubMed]
155. Davì, G.; Patrono, C. Platelet activation and atherothrombosis. *N. Engl. J. Med.* **2007**, *357*, 2482–2494. [CrossRef] [PubMed]
156. Jagroop, I.A.; Clatworthy, I.; Lewin, J.; Mikhailidis, D.P. Shape change in human platelets: Measurement with a channelyzer and visualisation by electron microscopy. *Platelets* **2000**, *11*, 28–32. [PubMed]
157. van der Pol, A.; van Gilst, W.H.; Voors, A.A.; van der Meer, P. Treating oxidative stress in heart failure: Past, present and future. *Eur. J. Heart Fail.* **2019**, *21*, 425–435. [CrossRef]
158. Rudrapal, M.; Khairnar, S.J.; Khan, J.; Dukhyil, A.B.; Ansari, M.A.; Alomary, M.N.; Alshabrmi, F.M.; Palai, S.; Deb, P.K.; Devi, R. Dietary polyphenols and their role in oxidative stress-induced human diseases: Insights into protective effects, antioxidant potentials and mechanism (s) of action. *Front. Pharmacol.* **2022**, *13*, 283. [CrossRef]
159. Sies, H. Polyphenols and health: Update and perspectives. *Arch. Biochem. Biophys.* **2010**, *501*, 2–5. [CrossRef]
160. Hanif, S.; Shamim, U.; Ullah, M.F.; Azmi, A.S.; Bhat, S.H.; Hadi, S.M. The anthocyanidin delphinidin mobilizes endogenous copper ions from human lymphocytes leading to oxidative degradation of cellular DNA. *Toxicology* **2008**, *249*, 19–25. [CrossRef]
161. Kähkönen, M.P.; Heinonen, M. Antioxidant activity of anthocyanins and their aglycons. *J. Agric. Food Chem.* **2003**, *51*, 628–633. [CrossRef]
162. deGraft-Johnson, J.; Kolodziejczyk, K.; Krol, M.; Nowak, P.; Krol, B.; Nowak, D. Ferric-reducing ability power of selected plant polyphenols and their metabolites: Implications for clinical studies on the antioxidant effects of fruits and vegetable consumption. *Basic Clin. Pharmacol. Toxicol.* **2007**, *100*, 345–352. [CrossRef]
163. Sharma, P.; Jha, A.B.; Dubey, R.S.; Pessarakli, M. Reactive oxygen species, oxidative damage, and antioxidative defense mechanism in plants under stressful conditions. *J. Bot.* **2012**, *2012*, 217037. [CrossRef]
164. Wahid, M.; Ali, A.; Saqib, F.; Aleem, A.; Bibi, S.; Afzal, K.; Ali, A.; Baig, A.; Khan, S.A.; Bin Asad, M.H.H. Pharmacological exploration of traditional plants for the treatment of neurodegenerative disorders. *Phytother. Res.* **2020**, *34*, 3089–3112. [CrossRef] [PubMed]
165. Binsack, R.; Boersma, B.J.; Patel, R.P.; Kirk, M.; White, C.R.; Darley-Usmar, V.; Barnes, S.; Zhou, F.; Parks, D.A. Enhanced antioxidant activity after chlorination of quercetin by hypochlorous acid. *Alcohol. Clin. Exp. Res.* **2001**, *25*, 434–443. [CrossRef] [PubMed]
166. Cotelle, N.; Bernier, J.L.; Catteau, J.P.; Pommery, J.; Wallet, J.C.; Gaydou, E.M. Antioxidant properties of hydroxy-flavones. *Free Radical Biol. Med.* **1996**, *20*, 35–43. [CrossRef]
167. Krinsky, N.I. Mechanism of action of biological antioxidants. Proceedings of the Society for Experimental Biology and Medicine. *Soc. Exp. Biol. Med.* **1992**, *200*, 248–254. [CrossRef]
168. Sies, H. Oxidative stress: From basic research to clinical application. *Am. J. Med.* **1991**, *91*, 31s–38s. [CrossRef]
169. Nijveldt, R.J.; van Nood, E.; van Hoorn, D.E.; Boelens, P.G.; van Norren, K.; van Leeuwen, P.A. Flavonoids: A review of probable mechanisms of action and potential applications. *Am. J. Clin. Nutr.* **2001**, *74*, 418–425. [CrossRef]
170. O'Reilly, J.D.; Mallet, A.I.; McAnlis, G.T.; Young, I.S.; Halliwell, B.; Sanders, T.A.; Wiseman, H. Consumption of flavonoids in onions and black tea: Lack of effect on F2-isoprostanes and autoantibodies to oxidized LDL in healthy humans. *Am. J. Clin. Nutr.* **2001**, *73*, 1040–1044. [CrossRef]
171. Orallo, F.; Alvarez, E.; Camiña, M.; Leiro, J.M.; Gómez, E.; Fernández, P. The possible implication of trans-Resveratrol in the cardioprotective effects of long-term moderate wine consumption. *Mol. Pharmacol.* **2002**, *61*, 294–302. [CrossRef]
172. Mishra, A.; Kumar, S.; Pandey, A.K. Scientific Validation of the Medicinal Efficacy of Tinospora cordifolia. *Sci. World J.* **2013**, *2013*, 292934. [CrossRef]
173. Procházková, D.; Boušová, I.; Wilhelmová, N. Antioxidant and prooxidant properties of flavonoids. *Fitoterapia* **2011**, *82*, 513–523. [CrossRef] [PubMed]
174. Pollard, S.E.; Kuhnle, G.G.; Vauzour, D.; Vafeiadou, K.; Tzounis, X.; Whiteman, M.; Rice-Evans, C.; Spencer, J.P. The reaction of flavonoid metabolites with peroxynitrite. *Biochem. Biophys. Res. Commun.* **2006**, *350*, 960–968. [CrossRef]
175. Turell, L.; Radi, R.; Alvarez, B. The thiol pool in human plasma: The central contribution of albumin to redox processes. *Free Radic. Biol. Med.* **2013**, *65*, 244–253. [CrossRef] [PubMed]
176. Cao, G.; Booth, S.L.; Sadowski, J.A.; Prior, R.L. Increases in human plasma antioxidant capacity after consumption of controlled diets high in fruit and vegetables. *Am. J. Clin. Nutr.* **1998**, *68*, 1081–1087. [CrossRef]
177. Lotito, S.B.; Frei, B. The increase in human plasma antioxidant capacity after apple consumption is due to the metabolic effect of fructose on urate, not apple-derived antioxidant flavonoids. *Free Radic. Biol. Med.* **2004**, *37*, 251–258. [CrossRef] [PubMed]
178. Gao, Y.; Chen, T.; Raj, J.U. Endothelial and Smooth Muscle Cell Interactions in the Pathobiology of Pulmonary Hypertension. *Am. J. Respir. Cell Mol. Biol.* **2016**, *54*, 451–460. [CrossRef]
179. Puzserova, A.; Bernatova, I. Blood pressure regulation in stress: Focus on nitric oxide-dependent mechanisms. *Physiol. Res.* **2016**, *65* (Suppl. S3), S309–S342. [CrossRef]
180. Andriambeloson, E.; Kleschyov, A.L.; Muller, B.; Beretz, A.; Stoclet, J.C.; Andriantsitohaina, R. Nitric oxide production and endothelium-dependent vasorelaxation induced by wine polyphenols in rat aorta. *Br. J. Pharmacol.* **1997**, *120*, 1053–1058. [CrossRef]
181. Duarte, J.; Andriambeloson, E.; Diebolt, M.; Andriantsitohaina, R. Wine polyphenols stimulate superoxide anion production to promote calcium signaling and endothelial-dependent vasodilatation. *Physiol. Res.* **2004**, *53*, 595–602. [CrossRef]

182. Zenebe, W.; Pechánová, O.; Andriantsitohaina, R. Red wine polyphenols induce vasorelaxation by increased nitric oxide bioactivity. *Physiol. Res.* **2003**, *52*, 425–432. [CrossRef] [PubMed]
183. Matsuo, S.; Nakamura, Y.; Takahashi, M.; Ouchi, Y.; Hosoda, K.; Nozawa, M.; Kinoshita, M. Effect of red wine and ethanol on production of nitric oxide in healthy subjects. *Am. J. Cardiol.* **2001**, *87*, 1029–1031. [CrossRef] [PubMed]
184. Ferrara, L.A.; Raimondi, A.S.; d'Episcopo, L.; Guida, L.; Dello Russo, A.; Marotta, T. Olive oil and reduced need for antihypertensive medications. *Arch. Intern. Med.* **2000**, *160*, 837–842. [CrossRef] [PubMed]
185. Diebolt, M.; Bucher, B.; Andriantsitohaina, R. Wine polyphenols decrease blood pressure, improve NO vasodilatation, and induce gene expression. *Hypertension* **2001**, *38*, 159–165. [CrossRef] [PubMed]
186. Andriambeloson, E.; Stoclet, J.C.; Andriantsitohaina, R. Mechanism of endothelial nitric oxide-dependent vasorelaxation induced by wine polyphenols in rat thoracic aorta. *J. Cardiovasc. Pharmacol.* **1999**, *33*, 248–254. [CrossRef]
187. Li, H.F.; Chen, S.A.; Wu, S.N. Evidence for the stimulatory effect of resveratrol on Ca^{2+}-activated K^+ current in vascular endothelial cells. *Cardiovasc. Res.* **2000**, *45*, 1035–1045. [CrossRef]
188. McKenna, E.; Smith, J.S.; Coll, K.E.; Mazack, E.K.; Mayer, E.J.; Antanavage, J.; Wiedmann, R.T.; Johnson, R.G., Jr. Dissociation of phospholamban regulation of cardiac sarcoplasmic reticulum Ca^{2+} ATPase by quercetin. *J. Biol. Chem.* **1996**, *271*, 24517–24525. [CrossRef]
189. Martin, S.; Andriambeloson, E.; Takeda, K.; Andriantsitohaina, R. Red wine polyphenols increase calcium in bovine aortic endothelial cells: A basis to elucidate signalling pathways leading to nitric oxide production. *Br. J. Pharmacol.* **2002**, *135*, 1579–1587. [CrossRef]
190. Ndiaye, M.; Chataigneau, M.; Lobysheva, I.; Chataigneau, T.; Schini-Kerth, V.B. Red wine polyphenol-induced, endothelium-dependent NO-mediated relaxation is due to the redox-sensitive PI3-kinase/Akt-dependent phosphorylation of endothelial NO-synthase in the isolated porcine coronary artery. *FASEB J.* **2005**, *19*, 455–457. [CrossRef]
191. Schramm, D.D.; Wang, J.F.; Holt, R.R.; Ensunsa, J.L.; Gonsalves, J.L.; Lazarus, S.A.; Schmitz, H.H.; German, J.B.; Keen, C.L. Chocolate procyanidins decrease the leukotriene-prostacyclin ratio in humans and human aortic endothelial cells. *Am. J. Clin. Nutr.* **2001**, *73*, 36–40. [CrossRef]
192. Fu, W.; Conklin, B.S.; Lin, P.H.; Lumsden, A.B.; Yao, Q.; Chen, C. Red wine prevents homocysteine-induced endothelial dysfunction in porcine coronary arteries. *J. Surg. Res.* **2003**, *115*, 82–91. [PubMed]
193. Beretz, A.; Anton, R.; Cazenave, J.P. The effects of flavonoids on cyclic nucleotide phosphodiesterases. *Prog. Clin. Biol. Res.* **1986**, *213*, 281–296.
194. Lugnier, C.; Schini, V.B. Characterization of cyclic nucleotide phosphodiesterases from cultured bovine aortic endothelial cells. *Biochem. Pharmacol.* **1990**, *39*, 75–84. [PubMed]
195. Pechánová, O.; Bernátová, I.; Babál, P.; Martínez, M.C.; Kyselá, S.; Stvrtina, S.; Andriantsitohaina, R. Red wine polyphenols prevent cardiovascular alterations in L-NAME-induced hypertension. *J. Hypertens.* **2004**, *22*, 1551–1559.
196. Aviram, M.; Rosenblat, M. Macrophage-mediated oxidation of extracellular low density lipoprotein requires an initial binding of the lipoprotein to its receptor. *J. Lipid Res.* **1994**, *35*, 385–398. [PubMed]
197. Fuhrman, B.; Aviram, M. Flavonoids protect LDL from oxidation and attenuate atherosclerosis. *Curr. Opin. Lipidol.* **2001**, *12*, 41–48.
198. Banach, M.; Markuszewski, L.; Zasłonka, J.; Grzegorczyk, J.; Okoński, P.; Jegier, B. The role of inflammation in the pathogenesis of atherosclerosis. *Przegl. Epidemiol.* **2004**, *58*, 663–670.
199. Cordova, A.C.; Jackson, L.S.; Berke-Schlessel, D.W.; Sumpio, B.E. The cardiovascular protective effect of red wine. *J. Am. Coll. Surg.* **2005**, *200*, 428–439.
200. Stein, J.H.; Keevil, J.G.; Wiebe, D.A.; Aeschlimann, S.; Folts, J.D. Purple grape juice improves endothelial function and reduces the susceptibility of LDL cholesterol to oxidation in patients with coronary artery disease. *Circulation* **1999**, *100*, 1050–1055.
201. Fuhrman, B.; Lavy, A.; Aviram, M. Consumption of red wine with meals reduces the susceptibility of human plasma and low-density lipoprotein to lipid peroxidation. *Am. J. Clin. Nutr.* **1995**, *61*, 549–554.
202. Ciumărnean, L.; Milaciu, M.V.; Runcan, O.; Vesa, Ș.C.; Răchișan, A.L.; Negrean, V.; Perné, M.G.; Donca, V.I.; Alexescu, T.G.; Para, I.; et al. The effects of flavonoids in cardiovascular diseases. *Molecules* **2020**, *25*, 4320. [PubMed]
203. Feng, A.N.; Chen, Y.L.; Chen, Y.T.; Ding, Y.Z.; Lin, S.J. Red wine inhibits monocyte chemotactic protein-1 expression and modestly reduces neointimal hyperplasia after balloon injury in cholesterol-Fed rabbits. *Circulation* **1999**, *100*, 2254–2259. [PubMed]
204. Rolnik, A.; Żuchowski, J.; Stochmal, A.; Olas, B. Quercetin and kaempferol derivatives isolated from aerial parts of Lens culinaris Medik as modulators of blood platelet functions. *Ind. Crops Prod.* **2020**, *152*, 112536.
205. El Haouari, M.; Rosado, J.A. Platelet signalling abnormalities in patients with type 2 diabetes mellitus: A review. *Blood Cells Mol. Dis.* **2008**, *41*, 119–123. [PubMed]
206. Faggio, C.; Sureda, A.; Morabito, S.; Sanches-Silva, A.; Mocan, A.; Nabavi, S.F.; Nabavi, S.M. Flavonoids and platelet aggregation: A brief review. *Eur. J. Pharmacol.* **2017**, *807*, 91–101. [PubMed]
207. Hubbard, G.P.; Stevens, J.M.; Cicmil, M.; Sage, T.; Jordan, P.A.; Williams, C.M.; Lovegrove, J.A.; Gibbins, J.M. Quercetin inhibits collagen-stimulated platelet activation through inhibition of multiple components of the glycoprotein VI signaling pathway. *J. Thromb. Haemost.* **2003**, *1*, 1079–1088.
208. Vanhoutte, P.M. Endothelial dysfunction: The first step toward coronary arteriosclerosis. *Circ. J.* **2009**, *73*, 595–601.

209. de Gaetano, G.; De Curtis, A.; di Castelnuovo, A.; Donati, M.B.; Iacoviello, L.; Rotondo, S. Antithrombotic effect of polyphenols in experimental models: A mechanism of reduced vascular risk by moderate wine consumption. *Ann. N. Y. Acad. Sci.* **2002**, *957*, 174–188.
210. Mladěnka, P.; Zatloukalová, L.; Filipský, T.; Hrdina, R. Cardiovascular effects of flavonoids are not caused only by direct antioxidant activity. *Free Radic. Biol. Med.* **2010**, *49*, 963–975.
211. Demrow, H.S.; Slane, P.R.; Folts, J.D. Administration of wine and grape juice inhibits in vivo platelet activity and thrombosis in stenosed canine coronary arteries. *Circulation* **1995**, *91*, 1182–1188.
212. Hamid, A.A.; Aminuddin, A.; Yunus, M.H.M.; Murthy, J.K.; Hui, C.K.; Ugusman, A. Antioxidative and anti-inflammatory activities of Polygonum minus: A review of literature. *Rev. Cardiovasc. Med.* **2020**, *21*, 275–287. [PubMed]
213. Choy, K.W.; Murugan, D.; Leong, X.F.; Abas, R.; Alias, A.; Mustafa, M.R. Flavonoids as natural anti-inflammatory agents targeting nuclear factor-kappa B (NFκB) signaling in cardiovascular diseases: A mini-review. *Front. Pharmacol.* **2019**, *10*, 1295. [PubMed]
214. Dias, M.C.; Pinto, D.C.; Silva, A.M. Plant flavonoids: Chemical characteristics and biological activity. *Molecules* **2021**, *26*, 5377. [PubMed]
215. Liao, H.; Ye, J.; Gao, L.; Liu, Y. The main bioactive compounds of Scutellaria baicalensis Georgi. for alleviation of inflammatory cytokines: A comprehensive review. *Biomed. Pharmacother.* **2021**, *133*, 110917.
216. Al-Khayri, J.M.; Sahana, G.R.; Nagella, P.; Joseph, B.V.; Alessa, F.M.; Al-Mssallem, M.Q. Flavonoids as potential anti-inflammatory molecules: A review. *Molecules* **2022**, *27*, 2901. [PubMed]
217. Krauth, V.; Bruno, F.; Pace, S.; Jordan, P.M.; Temml, V.; Romano, M.P.; Khan, H.; Schuster, D.; Rossi, A.; Filosa, R.; et al. Highly potent and selective 5-lipoxygenase inhibition by new, simple heteroaryl-substituted catechols for treatment of inflammation. *Biochem. Pharmacol.* **2023**, *208*, 115385.
218. Sychrová, A.; Škovranová, G.; Čulenová, M.; Bittner Fialová, S. Prenylated flavonoids in topical infections and wound healing. *Molecules* **2022**, *27*, 4491.
219. Martinez, J.; Moreno, J.J. Effect of resveratrol, a natural polyphenolic compound, on reactive oxygen species and prostaglandin production. *Biochem. Pharmacol.* **2000**, *59*, 865–870.
220. Pey, A.L.; Megarity, C.F.; Timson, D.J. NAD(P)H quinone oxidoreductase (NQO1): An enzyme which needs just enough mobility, in just the right places. *Biosci. Rep.* **2019**, *39*, BSR20180459.
221. Fukami, T.; Yokoi, T.; Nakajima, M. Non-P450 drug-metabolizing enzymes: Contribution to drug disposition, toxicity, and development. *Annu. Rev. Pharmacol. Toxicol.* **2022**, *62*, 405–425.
222. Sęczyk, Ł.; Świeca, M.; Kapusta, I.; Gawlik-Dziki, U. Protein-Phenolic Interactions as a Factor Affecting the Physicochemical Properties of White Bean Proteins. *Molecules* **2019**, *24*, 408. [CrossRef] [PubMed]
223. Esteves, F.; Rueff, J.; Kranendonk, M. The central role of cytochrome P450 in xenobiotic metabolism—A brief review on a fascinating enzyme family. *J. Xenobiot.* **2021**, *11*, 94–114. [CrossRef] [PubMed]
224. Bhamre Vaibhav, G.; Deore Pranjal, D.; Amrutkar Rakesh, D.; Patil Vinod, R. Polyphenols: The interactions with CYP 450 isoenzymes and effect on pharmacokinetics of drugs. *Curr. Trends Pharm. Pharm. Chem.* **2022**, *4*, 13–23.
225. Mohos, V.; Bencsik, T.; Boda, G.; Fliszár-Nyúl, E.; Lemli, B.; Kunsági-Máté, S.; Poór, M. Interactions of casticin, ipriflavone, and resveratrol with serum albumin and their inhibitory effects on CYP2C9 and CYP3A4 enzymes. *Biomed. Pharmacother.* **2018**, *107*, 777–784. [CrossRef] [PubMed]
226. Lecour, S.; Lamont, K.T. Natural Polyphenols and Cardioprotection. *Mini-Rev. Med. Chem.* **2011**, *11*, 1191–1199.
227. Cory, H.; Passarelli, S.; Szeto, J.; Tamez, M.; Mattei, J. The Role of Polyphenols in Human Health and Food Systems: A Mini-Review. *Front. Nutr.* **2018**, *5*, 87. [CrossRef]
228. Mazzanti, G.; Di Sotto, A.; Vitalone, A. Hepatotoxicity of green tea: An update. *Arch. Toxicol.* **2015**, *89*, 1175–1191. [CrossRef]
229. Bagdas, D.; Çam Etöz, B.; Gul, Z.; Ziyanok, S.; Inan, S.; Turacozen, O.; Gul, N.Y.; Topal, A.; Cinkilic, N.; Taş, S.; et al. In vivo systemic chlorogenic acid therapy under diabetic conditions: Wound healing effects and cytotoxicity/genotoxicity profile. *Food Chem. Toxicol.* **2015**, *81*, 54–61. [CrossRef]
230. Bandele, O.J.; Osheroff, N. Bioflavonoids as poisons of human topoisomerase II alpha and II beta. *Biochemistry* **2007**, *46*, 6097–6108. [CrossRef]
231. B andele, O.J.; Osheroff, N. (-)-Epigallocatechin gallate, a major constituent of green tea, poisons human type II topoisomerases. *Chem. Res. Toxicol.* **2008**, *21*, 936–943. [CrossRef]
232. Bandele, O.J.; Clawson, S.J.; Osheroff, N. Dietary polyphenols as topoisomerase II poisons: B ring and C ring substituents determine the mechanism of enzyme-mediated DNA cleavage enhancement. *Chem. Res. Toxicol.* **2008**, *21*, 1253–1260. [CrossRef] [PubMed]
233. Eghbaliferiz, S.; Iranshahi, M. Prooxidant activity of polyphenols, flavonoids, anthocyanins and carotenoids: Updated review of mechanisms and catalyzing metals. *Phytother. Res.* **2016**, *30*, 1379–1391. [CrossRef] [PubMed]

234. Wang, D.; Wang, Y.; Wan, X.; Yang, C.S.; Zhang, J. Green tea polyphenol (−)-epigallocatechin-3-gallate triggered hepatotoxicity in mice: Responses of major antioxidant enzymes and the Nrf2 rescue pathway. *Toxicol. Appl. Pharmacol.* **2015**, *283*, 65–74. [CrossRef] [PubMed]
235. Zheng, L.F.; Dai, F.; Zhou, B.; Yang, L.; Liu, Z.L. Prooxidant activity of hydroxycinnamic acids on DNA damage in the presence of Cu(II) ions: Mechanism and structure-activity relationship. *Food Chem. Toxicol.* **2008**, *46*, 149–156. [CrossRef] [PubMed]

molecules

MDPI

Review

Health-Promoting Properties: Anti-Inflammatory and Anticancer Properties of *Sambucus nigra* L. Flowers and Fruits

Agnieszka Ewa Stępień [1,*], Julia Trojniak [2] and Jacek Tabarkiewicz [3]

1 Institute of Health Sciences, College of Medical Sciences, University of Rzeszow, 35-959 Rzeszów, Poland
2 Student's Scientific Club Immunology, Institute of Medical Sciences, College of Medical Sciences, University of Rzeszow, 35-959 Rzeszów, Poland; juliatrojniak0@gmail.com
3 Institute of Medical Sciences, College of Medical Sciences, University of Rzeszow, 35-959 Rzeszów, Poland; jtabarkiewicz@ur.edu.pl
* Correspondence: astepien@ur.edu.pl; Tel.: +48-694211973

Abstract: *Sambucus nigra* L. has been used for centuries in traditional medicine thanks to its valuable healing properties. The healing properties result from its high content of biologically active compounds, mainly antioxidants, which contribute to its anti-inflammatory and anticancer properties. In our review, we have presented scientific studies evaluating the anti-inflammatory and anticancer effects of extracts and their components from *S. nigra* L. flowers and fruits. The results of the research show that the effect of antioxidant phytochemicals contained in their composition reduces the level of free radicals and pro-inflammatory cytokines, prevents mutations that increase the risk of cancer development, and inhibits cell proliferation, induction of apoptosis, and changes in intracellular signaling, consequently inhibiting the growth of malignant tumors and the formation of metastases. Flowers and fruits of *S. nigra* L. are a valuable source of nutraceutical and pharmacological substances that can support prevention and anti-inflammatory and oncological therapy without negative side effects for the patient.

Keywords: *Sambucus nigra* L.; anti-inflammatory; anticancer; antioxidants

1. Introduction

In recent years, we have observed a trend in scientific research to return to the assessment of the medicinal properties of plants that are the basis of traditional medicine. Plants are a rich source of biologically active compounds defined as phytochemicals, which are used in the prevention and treatment of many diseases. Bioactive substances present in plants and secondary metabolites are mainly antioxidants with antioxidant properties. Numerous processes in the body generate free radicals, like reactive oxygen or nitrogen species, that are neutralized by endogenous compounds with antioxidative properties. However, a persistently high level of unneutralized free radicals in the body can cause the induction of oxidative stress, which increases the risk of inflammation [1,2].

The inflammation associated with infections, e.g., viral or bacterial infections, plays an important role in the development of carcinogenesis [3]. Exogenous antioxidants present in the diet significantly counteract the development of oxidative stress and decrease the risk of chronic inflammation, leading to cancer [4].

Among medicinal plants, *Sambucus nigra* L. draws special attention as a rich source of many bioactive compounds, such as polyphenols and flavonoids, with antioxidant properties.

Sambucus nigra L. (*S. nigra* L., *S.nigra)*, also called European wild black elder, elderberry, European elderberry, European elder and European black elderberry, belongs to the family *Adoxaceae* Trautv., i.e., *Viburnaceae* Raf. from the order *Dipsacales* and earlier, and also to *Caprifoliaceae* Juss. and *Sambucaceae* Batsch [5–7].

S. nigra L. is found in natural habitats in Asia, North America, Europe, North Africa, New Zealand and Australia as a small deciduous tree or shrub, usually 4 to 12 m tall. *S. nigra* L. grows in the wild, usually in open fields or on forest edges. Its branches are often arching, the bark of which turns green to brownish-gray as it ages and cracks. The leaves are pinnate, arranged opposite each other, and vary in shape from ovate to ovate-lanceolate to ovate-elliptical. They usually measure from 3 to 9 cm in length with a tail (3–4 cm) [8,9] (Figure 1).

Figure 1. The *S. nigra* L. bush (photograph by Julia Trojniak).

The inflorescences of *S. nigra* L. are flattened, apical, and umbellate with five main rays, with milky white flowers with five yellow stamens, and a faint and aromatic fragrance with a slightly bitter taste (Figure 2a) [10,11]. The fruit is a spherical drupe with 3–5 pressed seeds. Initially, the fruits are green, but in the process of ripening, they change color from red to black-purple with a shiny skin and fleshy with a lot of juice [9]. *S. nigra* L. blooms in May and June, while the fruits usually ripen from July/August to October (Figure 2b) [12]. For aesthetic reasons, it is planted in parks and gardens, as well as on plantations, especially to obtain pharmaceutical raw materials.

(**a**) (**b**)

Figure 2. *S. nigra* L. flower (**a**) and fruits (**b**) (photograph by Julia Trojniak).

All raw parts of *S. nigra* L. are poisonous due to the presence of cyanogenic glycosides (CNGs), including the most abundant sambunigrins and prunazines. These compounds, in the presence of β-glucosidase and α-hydroxynitrile lyase enzymes, release highly toxic hydrogen cyanide (prussic acid, HCN) [13]. Elderberry leaves contain more sambunigrin than the flowers and the lowest amount of this compound is in the berries. It was also found that the content of sambunigrin in elderberries varies depending on the height of cultivation. It is necessary for it to undergo thermal treatment, which causes the decomposition of sambunigrin to compounds harmless to the human body [14]. In addition to poisonous compounds, *S. nigra* L. is also rich in nutrients, such as carbohydrates, proteins, fats, fatty acids, organic acids, minerals, vitamins and essential oils [14]. The chemical composition of individual morphological parts of the plant depends on many factors, including environmental conditions, such as soil type, light, temperature, amount and frequency of rainfall, fertilization, cultivation methods and processing and storage conditions, as well as altitude above sea level [15].

For centuries, black elderberry flowers and fruits have been used in the kitchen as an addition to cakes and in the preparation of jams and tinctures due to their taste.

The results of the bioavailability and safety assessment of *S. nigra* L. flowers and fruits showed no toxic effects or side effects on the human body. Flowers and fruits of *S. nigra* L. are recognized by the US Food and Drug Administration (FDA) as a safe food additive, which is indicated by the GRAS (Generally Recognized As Safe) status. The European Medical Agency (EMA) points to the safety of their use but recommends limiting their use by women during pregnancy and lactation as well as in children under 12 years of age [16,17].

S. nigra L., for a long time, has been one of the most frequently used plants in traditional medicine for the treatment of many diseases with diaphoretic, anti-inflammatory, diuretic or antipyretic effects [18,19]. *S. nigra* L. has antioxidant, anti-inflammatory, cytotoxic, anti-allergic, immunomodulating and antiviral, antibacterial, antidepressant, antidiabetic, antiatherosclerotic and hypoglycaemic properties [6,8,19,20]. It also shows anticancer activity, which determines its supporting role in the prevention and treatment of cancers with limited side effects in this form of therapy. All parts of the S. *nigra* L. plant, flowers, leaves, fruits and bark contain numerous bioactive compounds—mainly with antioxidant properties—that determine its health-promoting properties, making them a very valuable pharmaceutical raw material.

Oxidative stress is the result of an imbalance of free radicals-antioxidants in the body as a result of excessive amounts of free radicals (reactive oxygen species (ROS) and nitrogen). Long-term oxidative stress associated with chronic inflammation can lead to the development of many diseases. Reduction of oxidative stress is possible by increasing the level of antioxidants in the body. This indicates the very important role of the presence of vegetables, fruits or medical plants with a high content of polyphenols in the diet because, thanks to the neutralization of free radicals, i.e., reactive oxygen species, they protect cells and tissues against their negative effects, influence the induction of antioxidant enzymes and strengthen the immune response [21].

Flowers (*Sambuci Flos*) and fruits (*Sambuci Fructus*) are the richest in polyphenolic compounds, mainly flavonols, phenolic acids and anthocyanins, which affect their properties see Tables 1 and 2. The results of a previous study showed the high antioxidant capacity of elderberries—an average of 4.31 mmol/Trolox equivalent (TE) [22]. Such high antioxidant activity results from the high content of antioxidants, mainly phenolic compounds, but also minerals, vitamins C and E, and carotenoids that affect these abilities [22].

Table 1. The chemical composition of *Sambuci Flos* [23–26].

The Group of Chemicals	Examples of Substances	Content (%)
Flavonoids	keampferol, quercetin, rutin, astragalin, isoquerlemon, hyperoside, nicotiflorin	3.0%
Phenolic acids and their glycosides	3-O-Caffeoylquinic acid, 4-O-Caffeoylquinic acid, 5-O-Caffeoylquinic acid, 1,5-Di-O-caffeoylquinic acid, 3,5-Di-O-caffeoylquinic acid, 3,4-Di-O-caffeoylquinic acid, 4,5-Di-O-caffeoylquinic acid, 3-O-p-Coumaroylquinic acid, 5-O-p-Coumaroylquinic acid, Chlorogenic acid, P-coumaric acid, Ferulic acid and their glucosides	3.0%
Triterpenes	α- and β-amyrin, ursolic acids, oleanoic acid, benzoic acid	1.0%
Sterols	β-sitosterol, campesterol, stigmasterol, cholesterol	1.0%

This article presents the results of scientific research determining the health-promoting properties of *S. nigra* L. flowers and fruits, including both anti-inflammatory and anticancer properties. The literature review describes current scientific research focusing on important aspects, such as phytochemical analysis, evaluation of the pharmacological aspect of such action of *S. nigra* L. flower and fruit extracts and phytochemicals present in their composition.

2. Anti-Inflammatory Properties of *S. nigra* L.

Inflammation is a natural protective mechanism of the body, preventing damage to its tissues. The mechanism of the inflammatory process is based on the activation of monocytes and/or macrophages present in various tissues that determine its course through the release of inflammatory cytokines such as interleukins (IL) and tumor necrosis factor (TNF-α) and inflammatory mediators such as ROS and nitric oxide (NO), or prostaglandin E2 and cyclooxygenases (COX) and lipoxygenases (LOX) [27]. It was determined that the main role in the development and progression of inflammation and the immune response is played by the activity of regulatory enzymes, such as lipoxygenases and cyclooxygenases. Cyclooxygenase mediates the development and regulation of the inflammatory process through the synthesis of eicosanoids (prostaglandins and prostacyclins) via the arachidonic acid pathway. The increased synthesis of prostaglandins is also affected by the accelerated synthesis of nitric oxide by nitric oxide synthase in the area of inflammation [28].

Currently used therapies are based on drugs acting to suppress the pro-inflammatory response; however, these have limited effectiveness, and importantly, some of them have undesirable health effects for patients. In recent years, the role of preparations based on plant materials, mainly medicinal plants rich in numerous bioactive compounds that do not have side effects for patients, has been focused on in the prevention and therapy of diseases associated with inflammation.

Ho et al. analyzed the effect of *S. nigra* L. fruit and flower extracts and various polyphenols present in these extracts on lipopolysaccharide (LPS)-stimulated RAW 264.7 macrophages. The juice and the acidified methanol extract, both at a concentration of 100 μg/mL, inhibited the secretion of pro-inflammatory factors by macrophages by 30% and 50%, respectively. Among polyphenolic compounds, cyanidins, cyanidin-3-glucoside and cyanidin-3-glucoside sambubioside (100 μM) were reduced by about 60% to 70%, as well as quercetin by 80% (100 μM), and among phenolic acids, chlorogenic acid by about 51.5%. All these compounds are present in elderberry extracts, and their high content in them contributes to the high anti-inflammatory effect of the extracts [29].

The anti-inflammatory properties of quercetin result from its influence on the inhibition of the secretion of pro-inflammatory cytokines such as TNF-α and interleukins (IL-1β, IL-6) and the reduction of cyclooxygenase and lipoxygenase activity [30,31]. In an in vivo study, the administration of quercetin to diabetic rats also reduced the level of prostaglandins and interleukins (IL-1β) [32]. Incubation of quercetin with an LPS-stimulated macrophage line (RAW264.7) resulted in the inhibition of TNF-α generation and nitric oxide synthase activity [33]. It has been shown that it also affects the production of anti-inflammatory cytokines such as IL-10 [34].

In subsequent in vitro studies, the antioxidant and anti-inflammatory effects of *S. nigra* L. fruit extracts were also assessed using LPS-stimulated macrophages (RAW 264.7). An inhibitory effect on the release of pro-inflammatory factors was also observed, confirming the anti-inflammatory effect. Human hepatoma (hepG2) and human colon adenocarcinoma (Caco-2) cells were treated with cytotoxic tert-butyl hydroperoxide, and then *S. nigra* L. extract was added. It inhibited the release of reactive oxygen species as a result of the cytotoxic effect of tert-butyl hydroperoxide in the case of both cell lines, preventing their damage caused by oxidative stress. The high antioxidant and anti-inflammatory activity of *S. nigra* L. fruit extract results from its phytochemical profile, confirming the presence of antioxidants [35].

Santin et al. studied the effect of *S. nigra* L. flower extract on inflammation in vivo and in vitro. They orally administered *S. nigra* L. extract at 30, 100, 300 or 600 mg/kg, respectively, to groups of male mice with carrageenan-induced inflammation. They observed that it affected the inhibition of neutrophil migration and the reduction of the level of the pro-inflammatory cytokines TNF, IL-1β and IL-6. In vitro, neutrophils and macrophages stimulated with lipopolysaccharide were also treated with this extract at concentrations of 1, 10 or 100 μg/mL. They showed a decrease in the levels of reactive nitrogen forms in neutrophils and the pro-inflammatory cytokines TNF, IL-1β and IL-6, and an increase in anti-inflammatory IL-10 and neutralized expression of CD62L and CD18. Researchers point to the role of rutin, the main component of *S. nigra* L. extract, in such an effect. In macrophages, the effect is comparable in reducing NO_2, TNF and IL-6. Extracts from *S. nigra* L. flowers show anti-inflammatory activity by modifying the secretion of pro-inflammatory cytokines by macrophages and neutrophils, which significantly affects the treatment of acute inflammation, confirming the medicinal properties of this plant [36].

The research results indicate that the anti-inflammatory effect of rutin resulting from its antioxidant properties is based on the reduction of the concentration of the pro-inflammatory markers tumor necrosis factor-α, interleukin (IL)-6, cyclooxygenase-2, and IL-1β [37].

In vitro studies have shown that kaempferol, by inhibiting the activity of COX1 and COX2 enzymes, prevents the development of inflammation. This is indicated by the results of in vitro studies where incubation of kaempferol with human hepatocytes decreased COX2 levels [38]. Subsequent studies indicate that kaempferol incubated with macrophage cells (J77, RAW264.7) stimulated with LPS significantly influenced the synthesis of nitric oxide. By inhibiting the synthesis of nitric oxide, it reduced inflammation. In vivo studies have shown that the inclusion of kaempferol in the treatment of patients with diabetic neuropathy reduced the amount of IL-1B, TNF-α and nitric oxide [39,40]. It was observed that kaempferol in diabetic nephropathy had a significant effect on the suppression of inflammation [41].

The results of the above studies show the ability of the extract and its components of elderberry fruit and flowers to protect against the development of oxidative stress and inflammation (Figure 3).

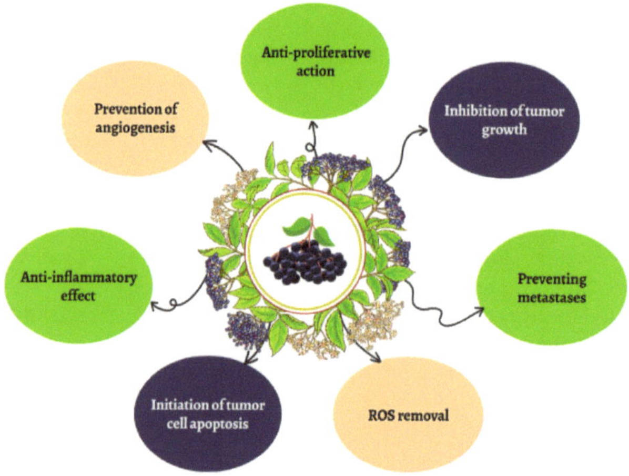

Figure 3. Selected properties of *S.nigra* L.

3. Anticancerogenic Properties of *S. nigra* L.

Worldwide, cancer is the second leading cause of death and more and more new cases of malignancy are being diagnosed, placing a significant burden on healthcare systems to provide full care to cancer patients. Current forms of cancer therapy are often accompanied by numerous negative side effects, which additionally affect the deterioration of the patient's health. For several years, scientists have been focusing on conducting research to develop new, effective methods supporting anticancer therapy as well as reducing side effects.

The causes of cancer development are complex, and both intrinsic and extrinsic factors contribute to this complex process. The consequences of the impact of emerging free radicals, e.g., reactive oxygen species in the human body, are one of the main etiological factors of cancer [3]. Free radicals attacking cell components may contribute to damage to cell structures, and oxidative DNA damage is the cause of mutations in genes or chromosomes, leading to the development of a multi-stage carcinogenesis process [1]. Chronic inflammation as a result of oxidative stress within the cells of a given organ, resulting from the activity of excess free radicals in the human body, plays an important role in the development of carcinogenesis [3]. It is, therefore, important to increase the supply of exogenous antioxidants in the diet that counteract the development of oxidative stress, significantly affecting the deterioration of human health [4].

S. nigra L. contains bioactive substances and secondary metabolites, including antioxidants, that can counteract the negative role of oxidative stress, which increases the risk of cancer development [2]. It exhibits such an anticancer effect due to its influence on, among others, inhibition of cancer cell proliferation and stimulation of immune cells [42].

Flowers and fruits of *S. nigra* L. are a rich source of many biologically active compounds, mainly antioxidants that determine anticancer properties—see Figure 3.

Elderberry flowers are particularly rich in bioactive compounds with antioxidant properties, mainly flavonoids such as kaempferol, quercetin and rutin, phenolic acids and their glycosides, triterpenes and sterols—see Table 1.

In their in vitro study, Periera et al. indicated cytotoxic effects of *S. nigra* L. flower extracts on human bladder carcinoma T24 cells. The UPLC-DAD-MS/MS analysis of the chemical composition of the butane fraction of the extract allowed the identification of nine flavonoids, including rutin, quercetin and kaempferol. The obtained results indicate that flavonoids are probably responsible for the cytotoxic effect on cells of this type of cancer [10].

It has also been shown that the alcoholic extract of elderberry flowers affects the proliferation of a breast cancer cell line (MCF7). The results indicate that incubation of MCF7 cells with different concentrations of this extract significantly affected the level of ERα and PR steroid receptors in these cells. A decrease in ERα and an increase in PR were observed by immunocytochemistry [43].

This indicates the important role of the ingredients present in the *S. nigra* L. flower extract in chemoprevention and/or cancer therapy. Scientists undertook further research to assess the anticancer effect of compounds present in the extract of *Sambucus nigra* L. flowers, especially flavonoids [44].

Scientific research indicates the high antioxidant potential of kaempferol from the group of flavonoids due to the capture of free radicals, mainly reactive oxygen species, which determines its anticancer properties—see Figure 4. Wang et al., in their in vitro study, evaluated the effect of kaempferol on human pancreatic cancer cell lines (MIA PaCa-2 and PANC-1). Thanks to the method based on the CCK-8 tests, the proliferation of cancer cells was analyzed, and the level of ROS and apoptosis of these cells were analyzed by flow cytometry. It was observed that the viability of PANC-1 and MIA PaCa-2 malignant cells decreases in the kaempferol in a dose-dependent manner. However, the results of additional in vivo studies on the effect of kaempferol on mice with pancreatic cancer indicate that tumor weight and volume decreased significantly, depending on the level of ROS. The anticancer properties of kaempferol result from its ability to induce apoptosis of cancer cells dependent on the regulation of ROS levels [45].

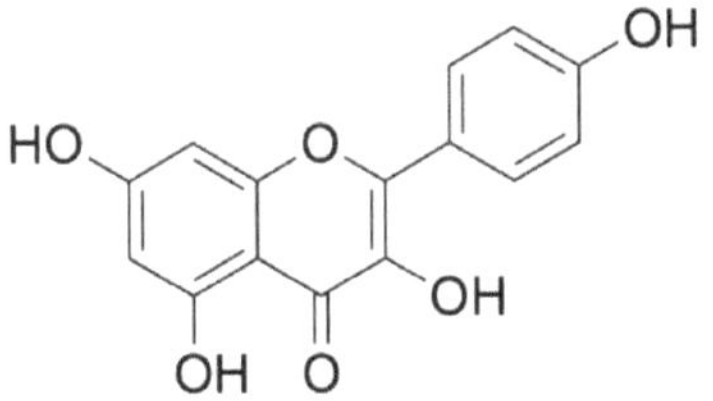

Figure 4. Chemical structure of kaempferol.

Subsequent in vitro studies also confirm the cytotoxic properties of kaempferol. Pham et al., using the MTT test (3-(4,5-dimethylthiazol-2-yl)-2,5-diphenyltetrazolium bromide), analyzed the viability of cancer cells after incubation with kaempferol. They determined its very strong inhibitory effect on the viability of the following human cancer cell lines: ovarian (A2780), lung (H460), skin (A431), pancreas (MIA PaCa-2), prostate (DU145), colon adenocarcinoma (HT29), breast (MCF-7), neuroblastoma (BE2-C) and glioblastoma (U87) [46]. In their in vitro studies, Akram et al. also indicated that kaempferol can arrest the cell cycle of human breast cancer cells (MDA-MB-453) [47]. The results of other studies indicate that the mechanism of action of kaempferol on human breast cancer cells (MDA-MB-231 and BT-474) is mainly the effect on inhibiting the growth of these cells, inducing their apoptosis and inhibiting their migration [48,49]. *Sambuci flos* also contains quercetin, which, thanks to its ability to neutralize free radicals—mainly reactive oxygen species—and to bind to transition metal ions, has chemopreventive properties [50–52]. The analysis of the results of in vitro studies on cancer cell lines indicates the possible anticancer effect of quercetin in relation to various types of cancer, such as ovarian cancer, breast cancer, colorectal cancer and prostate cancer [53–56]. Khan et al. indicate that among the phytochemicals, quercetin may be an alternative therapeutic option with limited side effects in the treatment of ovarian cancer. This emphasizes the interplay between it and miRNA in the regulation of apoptosis in this type of cancer [53]. Khorsandi et al. studied the effect of quercetin on the growth of a human breast cancer cell line (MCF-7) and indicated its high toxicity to cells of this type of cancer. They also determined that the mechanism of MCF-7 cell death as a result of the action of quercetin involves many pathways, mainly necroptosis [54].

In vivo studies in an animal model with colorectal cancer have evaluated the anticancer properties of quercetin. The presence of quercetin in the diet of rats treated with the carcinogenic substance methane nitroxin significantly influenced the development of the disease. Compared to the control and healthy group, it was shown that the presence of quercetin (10 mg/kg of body weight) in the diet reduces cytological and morphological changes in the cells of healthy rats. This indicates the possibility of its use in the prevention and treatment of colon cancer [55].

Quercetin, a bioflavonoid, also has anticancer effects against prostate cancer. Incubation of human prostate carcinoma cells (DU-145 and PC-3, LNCaP) with quercetin significantly improved their viability. A decrease in tumor cell viability was observed compared to the control, with no effect on healthy prostate epithelial cells, depending on the quercetin concentration and incubation time. The authors emphasize that quercetin exerts anticancer effects by modulating the ROS, Akt and NF-κB pathways. They also indicate that it can be used as a chemopreventive option, as well as in combination with chemotherapeutic drugs, to improve the clinical outcomes of patients with prostate cancer [56].

Research has been undertaken to determine the cytotoxic activity of compounds from the group of flavonoids quercetin and rutin. Both compounds have been shown to inhibit the growth of leukemic cells (HL-60) in a dose-dependent manner [57].

Rutin (quercetin-3-rutinoside), from the group of flavonoid glycosides, is present in *Sambucus nigra* L. flowers—see Figure 5. It is characterized by antioxidant properties, inhibition of lipid peroxidation by reducing COX-2 expression, reducing the level of oxidative stress and inducing the activity of nitric oxide synthase (iNOS), as well as exhibiting anti-inflammatory and anticancer properties [58–60].

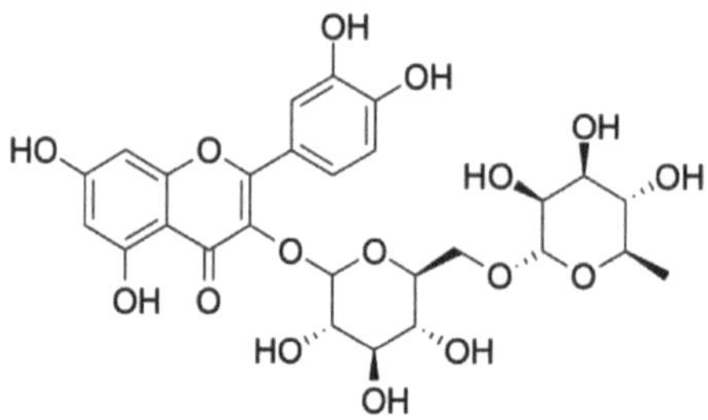

Figure 5. Chemical structure of rutin.

Satari et al. undertook a study to evaluate the effect of the joint action of rutin and the anticancer drug 5-fluorouracil on the human prostate cancer cell line (PC3). In previous studies, they showed that rutin induces apoptosis of prostate cancer cells [58]. They determined cell viability, *p53* gene expression, Bcl-2 signaling protein changes and apoptosis. The results of the studies indicate the synergistic effect of this drug and rutin on the apoptosis of cells of this type of cancer [59]. Subsequent studies confirm the anticancer properties of rutin. It was determined that the incubation of the human neuroblastoma cell line (LAN-5) with rutin had a significant effect on the viability of these cells and induction of G2/M arrest in the cell cycle [60]. Rutin also has a valuable antioxidant, antibacterial, anti-inflammatory, and UV-filtering effect on the skin. Its cytotoxic properties against two human melanoma cell lines (RPMI-7951 and SK-MEL-28) were evaluated. A concentration-dependent decrease in the viability of cancer cell lines was demonstrated, indicating its role in inducing apoptosis of cancer cells by increasing the expression of beta-galactosidase associated with aging (SA-β-gal) [61].

The results of the above studies emphasize the chemotherapeutic role of the natural compound rutin in the prevention and treatment of various types of cancer.

Ursolic and oleanolic acids are common triterpenoid compounds found in *S. nigra* L. flowers (Figure 6). Due to their antioxidant properties, these acids exhibit various pharmacological effects. The potential use of these triterpenoids in anticancer therapies, including treatment, as well as their impact on the immune system, is also indicated [62].

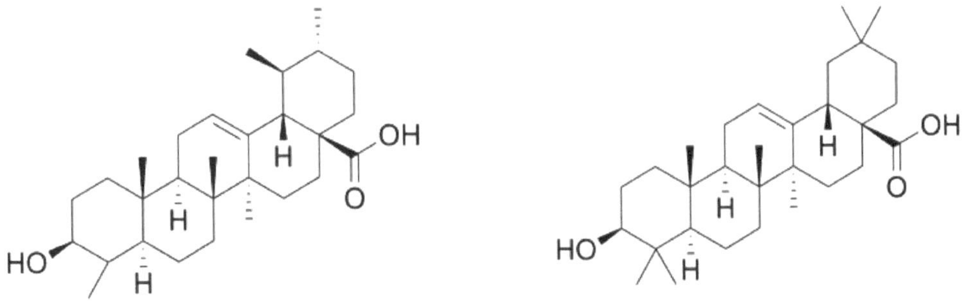

(a) (b)

Figure 6. Chemical structure of triterpene compounds: (**a**) ursolic acid ($C_{30}H_{48}O_3$); and (**b**) oleanolic acid ($C_{30}H_{48}O_3$).

Yang et al., in their research, analyzed the effect of oleanolic and ursolic acids on human liver cancer cell lines (HepG2, Hep3B, Huh7 and HA22T). They observed that both acids significantly reduced the viability of these cells by inducing apoptosis [63].

Studies have shown that ursolic acid may have beneficial effects in various types of cancers, including those caused by inflammation.

The results of a previous study indicate that incubation of ursolic acid with human breast cancer cell lines (MCF-7) inhibits cell proliferation and induces apoptosis, depending on time and concentration [64].

It is important to understand the mechanisms of anticancer activity of ursolic acid at the molecular level. Wang et al., in their in vitro studies, determined the molecular mechanisms of action of ursolic acid on human colon cancer cells (SW480, LoVo). Incubation of these cell lines with ursolic acid significantly inhibited their viability as well as their ability to migrate. The influence of ursolic acid on the signaling pathways MMP9/CDH1, Akt/ERK, COX-2/PGE2 and p300/NF-κB/CREB2 and cytochrome c/caspase pathways of colorectal cancer cells determines its anticancer properties [62].

The results of subsequent in vitro studies indicate the anticancer properties of ursolic acid on breast cancer cells (T47D, MCF-7 and MDA-MB-231). It has been observed that it induces autophagy and apoptosis of these cells through the GSK and Bcl-2/caspase-3 signaling pathways and also prevents the development of inflammation. In contrast, the mechanism of inhibition of breast cancer cell proliferation by ursolic acid is based on the inactivation of the PI3K/AKT signaling pathway. Understanding the mechanism at the molecular level of the effect of this acid on the cells of this type of cancer determines its possible use in breast cancer therapy [65]. Another important element of the anticancer activity of ursolic acid is the effect on the CXCR4/CXCL12 signaling pathway, which is important in the development of metastases. The effect of ursolic acid reduces the expression of CXCR4 by regulating the transcription of mRNA expression and inhibiting the activation of NF-κB [66]. The results of numerous in vitro studies indicate that it activates the growth inhibition of these cell lines due to changes in signaling pathways at the molecular level; in vivo studies also show its cytotoxic effects against these types of cancer. The results of research indicate that it affects numerous pro-inflammatory factors, including cytokines, cell cycle proteins, inflammatory enzymes, growth factors and kinases. It is indicated that by inhibiting the development, growth and metastasis of cancer, it determines its potentially chemopreventive and therapeutic effect [67].

S. nigra L. flowers contain the most phenolic compounds compared to fruits and leaves, which results in the highest antioxidant activity among all parts of this plant [14,18].

The group of phenolic acids present in elderberry flowers includes p-coumaric acid (p-CouA), which is a derivative of cinnamic acid [68] (Figure 7). Scientific research indicates its high biological activity, such as antioxidant properties, the ability to neutralize free radicals, e.g., reactive oxygen species, and anti-inflammatory properties [69]. This affects its anticancer properties, i.e., the ability to induce apoptosis and block the cell cycle, which inhibits their growth [68]. Its cytotoxic effect has been demonstrated against many cancer cell lines. Effects have been observed in vitro on the growth of neuroblastoma (N2a), human lung (A549), colon (HT29-D4) and cancer stem cells, colon cancer (Caco-2), the human endothelial cell line (ECV304), breast cancer (MCF7), and liver cancer cell lines (HPG2) that significantly reduce the viability of these cells [70].

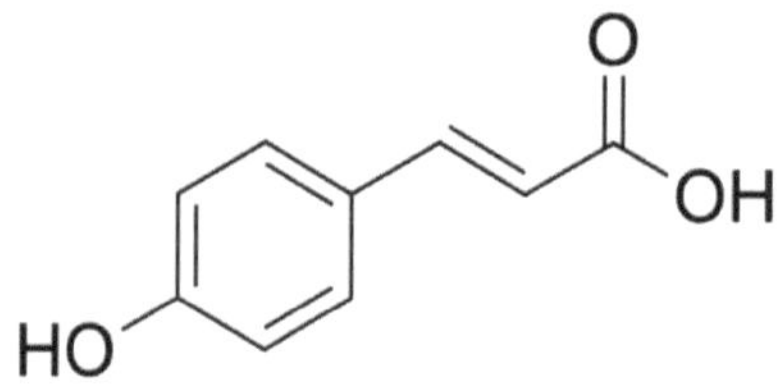

Figure 7. Chemical structure of p-coumaric acid.

The phenolic acid present in the flower of *S. nigra* L. is also chlorogenic acid—see Figure 8. Its antioxidant properties determine its healing effect, including its anticancer effect.

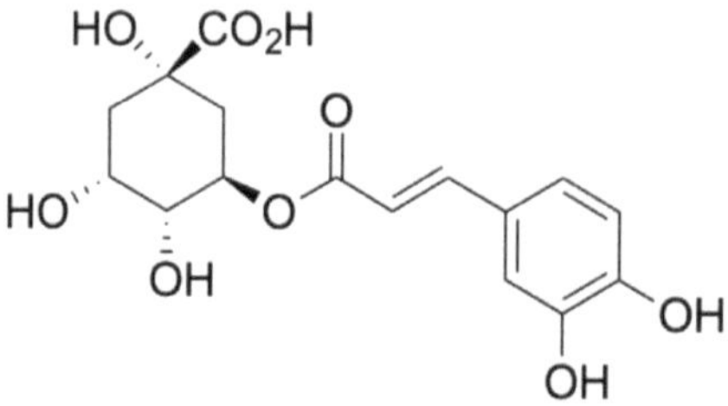

Figure 8. Chemical structure of chlorogenic acid.

The results of research showed that chlorogenic acid affects proliferation and inhibits angiogenesis and the growth of metastases of cancer cells. Its cytotoxic activity against breast cancer cells (MCF-7), colon cancer cells (HT-29), human hepatocarcinoma cells (HepG2), human leukemia cells (U9370) and human lung cancer cells (A549) was demonstrated in a dose- and time-dependent manner [71]. Further studies indicate that chlorogenic acid inhibits the proliferation of human lung cancer cells (A549) by inducing apoptosis [72]. Zeng et al. analyzed the effect of chlorogenic acid on human breast cancer cell lines (MDA-MB-231, MDA-MB-453 and MCF-10A) and the murine breast cancer 4T1 cell line. Incubation of the MDA-MB-231, MDA-MB-453 and 4T1W lines with chlorogenic acid resulted in the inhibition of the growth of these cells. However, its effect on human MCF-10A cells, unlike cancer cells, did not affect their viability and did not inhibit their growth. The results of the in vitro study indicate that chlorogenic acid has the ability to selectively affect the viability and inhibit the proliferation of breast cancer cells [73].

S.nigra L. fruits also contain numerous compounds with high biological activity, among which anthocyanins and polyphenols have the highest activity—see Table 2.

Table 2. Selected bioactive chemical components of *Sambuci Fructus*.

The Group of Chemicals	Examples of Substances		Contents
Carbohydrates	monosaccharides	glucose, fructose	Total: 6.8–11.5%/68.53–104.16 g/kg [14,23] 95% of total
	pectin		7.4% of total
Vitamins	ascorbic acid (vitamin C), riboflavin (vitamin B2), pyridoxine (vitamin B6), niacin (vitamin B3), pantothenic acid (vitamin B5), folic acid (vitamin B9)		Variable
Organic acids	acetic acid, malic acid, shikimic acid, valeric acid, tartaric acid, benzoic acid, ursolic acid, oleanolic acid		1.0–1.3%
Phenolic acids	Chlorogenic acid, neochlorogenic acid, cryptochlorogenic acid		371–432 mg GAE [1]/100 g [74]
Anthocyanins	Cyanidin 3-glucoside—Figure 9a, Cyanidin 3-sambubioside—Figure 9d, Cyanidin 3-sambubioside-5-glucoside—Figure 9c, Cyanidin 3,5-diglucoside—Figure 9e, Cyanidin 3-rutinoside—Figure 9b, Pelargonidin 3-glucoside—Figure 9g, Pelargonidin 3-sambubioside—Figure 9h, Delphinidin 3-rutinoside—Figure 9f		242–283 mg CGE [2]/100 g FW [3]; 272.87 mg/100 g FW; 664–1816 mg CGE/100 g FW; 8.33–101.40 mg CGE/g DW [4]; 1.9–20.2 g CGE/kg; 170–343 mg CGE/100 g; 465.1 mg/100 g FW; 602.9–1265.3 mg CGE/100 g FW; 1374.4 mg CGE/100 g [23,74]
Flavonols	Quercetin 3-O-rutinoside, Quercetin 3-O-galactoside, Quercetin 3-O-vicianoside, Quercetin 3-O-glucoside, Quercetin 3-O-(6″-acetyl)galactoside, Quercetin 3-O-(6″-acetyl)glucoside, Kaempferol 3-O-rutinoside, Kaempferol 3-O-glucoside, Isorhamnetin 3-O-rutinoside, Isorhamnetin 3-O-glucoside, Myricetin 3-O-rutinoside		38.26 mg/100 g FW; 13.6978–20.1836 g/100 g extr [5]; 57.0–102.7 mg QRE [6]/100 g [23]

[1] GAE—gallic acid equivalents; [2] CGE—cyanidin 3-glucoside equivalents; [3] FW—fresh weight; [4] DW—dry weight; [5] extr.—extract; [6] QRE—quercetin 3-rutinoside equivalents.

In vitro studies indicate the antioxidant properties of *Sambucus nigra* fruit extracts. These compounds can also increase the activity of enzymes in the small intestine, liver and lungs, including peroxidase, S-transferase, glutathione reductase and catalase [12,75]. Studies have shown that *S. nigra* L. fruit extracts inhibit both the initiation phase and the development phase of carcinogenesis. This indicates the significant cytotoxic properties of these extracts, which inhibit the proliferation of cancer cells. These data indicate that in addition to the cytotoxic effect on cancer cells, the extracts may have a cytoprotective effect on healthy cells and enhance the immune response, supporting the body's response to cancer [42].

Anthocyanins are the main group of fruit polyphenols, mainly cyanidin-3-glycoside, cyanidin-3-sambubioside and cyanidin-3-diglycoside [15].

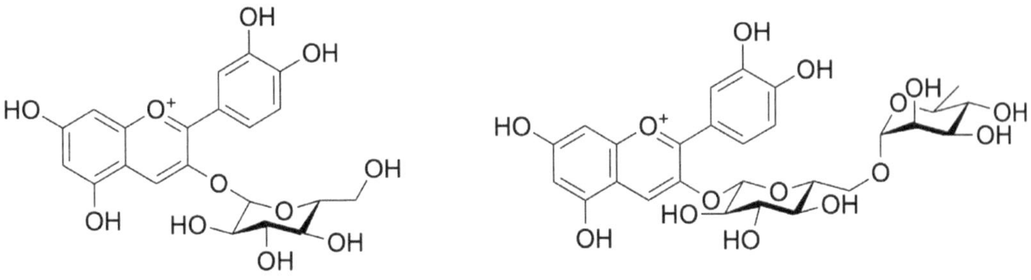

(a) (b)

Figure 9. *Cont.*

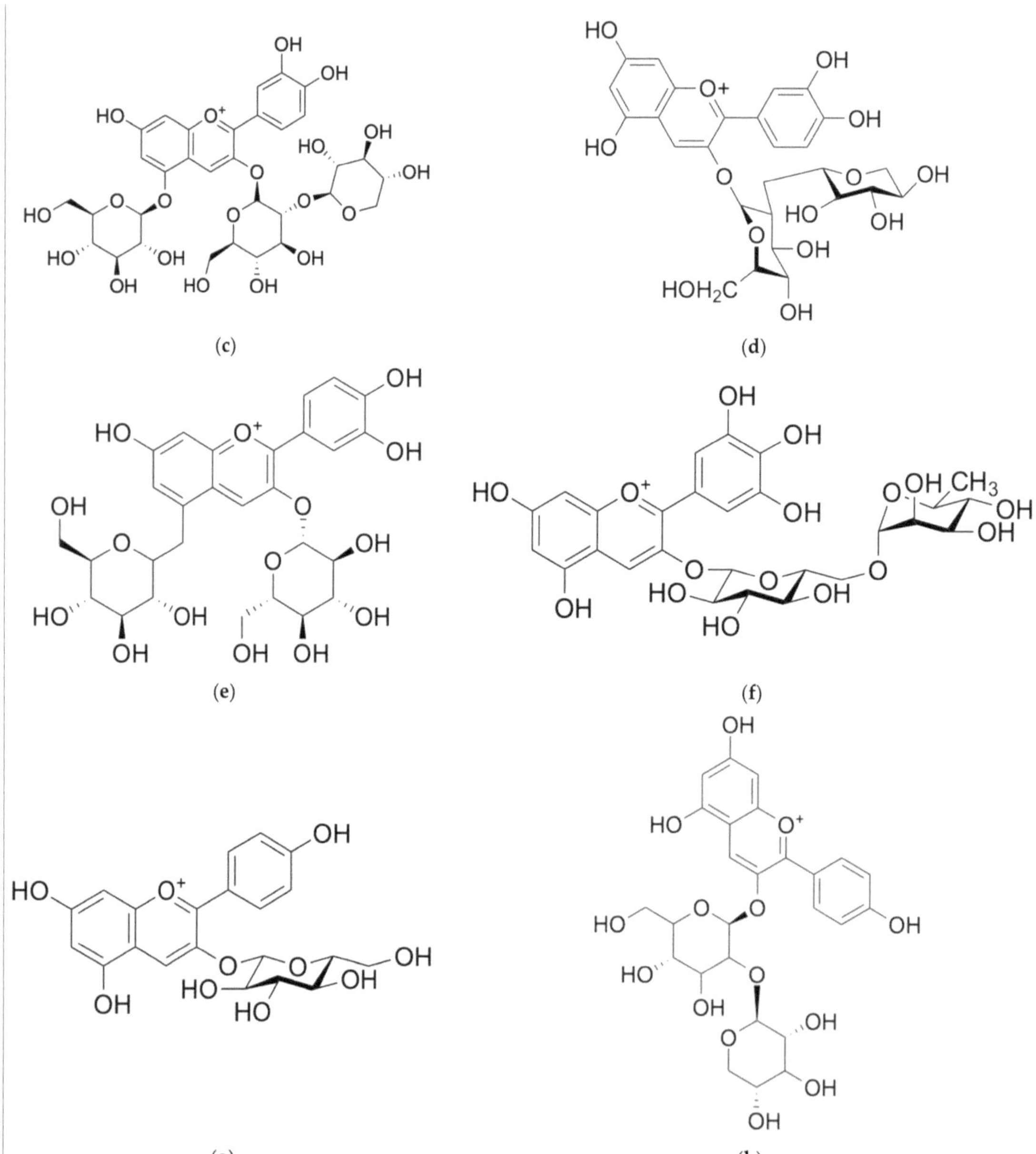

Figure 9. Chemical structure of anthocyanins from *Sambucus nigra* fruits: (**a**) cyanidin 3-glucoside; (**b**) cyanidin 3-rutinoside; (**c**) cyanidin 3-sambubioside-5-glucoside; (**d**) cyanidin 3-sambubioside; (**e**) cyanidin 3,5-diglucoside; (**f**) delphinidin 3-rutinoside; (**g**) pelargonidin 3-glucoside; and (**h**) pelargonidin 3-sambubioside.

The ability of anthocyanins to neutralize free radicals and reduce oxidative stress affects their anti-inflammatory and anticancer properties. Liu et al. studied the in vitro effect of cyanidin-3-glucoside on HER2-positive breast cancer cells. The results indicate that it induces apoptosis of HER2-positive breast cancer cells. In vivo studies showed that it reduced the size and volume of the tumor [76]. In addition, cyanidin-3-glucoside reduces the level of glutathione and phosphorylated MAP kinases, including ERK1/2, p38, JNK1/2 and MKK4, as well as pro-inflammatory cytokines [76]. In subsequent studies, it was determined that cyanidin-3-glucoside weakens breast cancer-induced angiogenesis by inhibiting VEGF, a key cytokine for this process [77]. Research results indicate the photochemopreventive or anticancer abilities of anthocyanins against skin cancer. Studies have been undertaken involving the effect of cyanidin-3-glucoside on the B16-F10 metastatic murine melanoma cell line. It was determined that this anthocyanin isolated from *S. nigra* L. fruit induces apoptosis of melanoma cells and inhibits their proliferation [78].

4. Conclusions

Flowers and fruits of *S. nigra* L. are a natural potential source of valuable nutraceutical and pharmacological substances.

The presence of biologically active substances with antioxidant properties in the flowers and fruits of *S. nigra* L., mainly polyphenols and anthocyanins, determines their health-promoting properties, e.g., anti-inflammatory and anticancer properties.

An important role in these processes is attributed to flavonoids, triterpenoid compounds and anthocyanins, as well as phenolic acids present in extracts from the flowers and fruits of *S. nigra* L. The results of scientific research indicate that these bioactive compounds inhibit the development of inflammation as a result of excessive activity of free radicals, alleviating the symptoms of diseases associated with inflammation by modifying the activity of macrophages and neutrophils and the secretion of pro-inflammatory cytokines, which significantly affects the treatment of acute inflammations, confirming the healing properties of this plant. It has been shown that *S. nigra* L. is a rich source of flavonoids, mainly kaempferol, quercetin and rutin, as well as phenolic acids, and it also inhibits many enzymes and mediators, e.g., prostaglandins and nitric oxide synthase, that cause inflammation.

Research results indicate the cytotoxic properties of extracts from flowers and fruits of *S. nigra* L. and their components—mainly rutin, quercetin, oleanolic and ursolic as well as chlorogenic and p-coumaric acids—against bladder, breast, pancreatic, skin, lung, colon, ovarian, liver, prostate and leukemia cells.

Their influence on cancer cells, metabolic pathways and the regulation of individual mechanisms of proliferation, angiogenesis, growth and induction of apoptosis indicates that *S. nigra* L. extracts can support modern oncological therapy with limited side effects. Both the antioxidant and anti-inflammatory effects of *S. nigra* L. extracts—due to capturing reactive oxygen species and, thus, preventing mutations—and their pro-apoptotic effects have a proven basis for their use in the prevention and supportive treatment of inflammatory and cancer diseases.

Further research on the molecular mechanisms of the pro-inflammatory and anticancer effects of extracts and their individual components of *S. nigra* L. flowers and fruits is necessary to determine the level of the effective dose of such a preparation supporting the prevention and therapy of various inflammatory and cancer diseases. This suggests the need for more human clinical trials. It will also make it possible to determine the level of a safe dose of such a medicine of natural origin that does not pose a threat to health or possible progression of other comorbidities in patients.

It is also important to undertake research to assess the interaction between the phytochemicals present in the flowers and fruits of *S. nigra* L. with other drugs used in anticancer therapy.

Author Contributions: Conceptualization, A.E.S. and J.T. (Jacek Tabarkiewicz); methodology, A.E.S., J.T. (Julia Trojniak) and J.T. (Jacek Tabarkiewicz).; resources, A.E.S. and J.T. (Julia Trojniak); writing—original draft preparation, A.E.S. and J.T. (Julia Trojniak); writing—review and editing, A.E.S., J.T. (Julia Trojniak) and J.T. (Jacek Tabarkiewicz); visualization, A.E.S. and J.T. (Julia Trojniak); supervision, J.T. (Jacek Tabarkiewicz). All authors have read and agreed to the published version of the manuscript.

Funding: This research received no external funding.

Institutional Review Board Statement: Not applicable.

Informed Consent Statement: Not applicable.

Data Availability Statement: Not applicable.

Conflicts of Interest: The authors declare no conflict of interest.

References

1. Schumacker, P.T. Reactive Oxygen Species in Cancer: A Dance with the Devil. *Cancer Cell* **2015**, *27*, 156–157. [CrossRef]
2. Saha, S.K.; Lee, S.B.; Won, J. Correlation between Oxidative Stress, Nutrition, and Cancer Initiation. *Int. J. Mol. Sci.* **2017**, *18*, 1544. [CrossRef] [PubMed]
3. Forni, C.; Rossi, M.; Borromeo, I.; Feriotto, G.; Platamone, G.; Tabolacci, C.; Mischiati, C.; Beninati, S. Flavonoids: A Myth or a Reality for Cancer Therapy? *Molecules* **2021**, *26*, 3583. [CrossRef] [PubMed]
4. Benzie, I.F. Evolution of dietary antioxidants. *Comp. Biochem. Physiol. Part A Mol. Integr. Physiol.* **2003**, *136*, 113–126. [CrossRef]
5. Fazio, A.; Plastina, P.; Meijerink, J.; Witkamp, R.F.; Gabriele, B. Comparative analyses of seeds of wild fruits of *Rubus* and *Sambucus* species from southern Italy: Fatty acid composition of the oil, total phenolic content, antioxidant and anti-inflammatory properties of the methanolic extracts. *Food Chem.* **2013**, *140*, 817–824. [CrossRef] [PubMed]
6. Marțiș, G.S.; Mureșan, V.; Marc, R.M.; Mureșan, C.C.; Pop, C.R.; Buzgău, G.; Mureșan, A.E.; Ungur, R.A.; Muste, S. The Physicochemical and Antioxidant Properties of *Sambucus nigra* L. and *Sambucus nigra Haschberg* during Growth Phases: From Buds to Ripening. *Antioxidants* **2021**, *10*, 1093. [CrossRef] [PubMed]
7. Integrated Taxonomic Information System. Available online: https://www.itis.gov/servlet/SingleRpt/SingleRpt?search_topic=TSN&search_value=35315#null (accessed on 6 October 2021).
8. Krankowski, F.; Tarko, T. Forgotten Fruits as Potential Wine Raw Materials. *Żywność. Nauka. Technol. Jakość/Food Sci. Technol. Qual.* **2022**, *29*, 52–62. [CrossRef]
9. Atkinson, M.D.; Atkinson, E. *Sambucus nigra* L. *J. Ecol.* **2002**, *90*, 895–923. [CrossRef]
10. Pereira, D.I.; Amparo, T.R.; Almeida, T.C.; Costa, F.S.F.; Brandão, G.C.; dos Santos, O.D.H.; da Silva, G.N.; Bianco de Souza, G.H. Cytotoxic Activity of Butanolic Extract from *Sambucus nigra* L. Flowers in Natura and Vehiculated in Micelles in Bladder Cancer Cells and Fibroblasts. *Nat. Prod. Res.* **2020**, *36*, 1100–1104. [CrossRef]
11. Przybylska-Balcerek, A.; Szablewski, T.; Szwajkowska-Michałek, L.; Świerk, D.; Cegielska-Radziejewska, R.; Krejpcio, Z.; Suchowilska, E.; Tomczyk, Ł.; Stuper-Szablewska, K. *Sambucus nigra* Extracts–Natural Antioxidants and Antimicrobial Compounds. *Molecules* **2021**, *26*, 2910. [CrossRef]
12. Liszka, K.; Najgebauer-Lejko, D.; Tabaszewska, M. Owoce Czarnego Bzu (*Sambucus nigra* L.)—Charakterystyka i Możliwości Wykorzystania w Przemyśle Spożywczym. In *Innowacyjne Rozwiązania w Technologii Żywności i Żywieniu Człowieka*; Polskie Towarzystwo Technologów Żywności Oddział Małopolski: Kraków, Poland, 2016; pp. 102–111. (In Polish)
13. Rodríguez Madrera, R.; Suárez Valles, B. Analysis of Cyanogenic Compounds Derived from Mandelonitrile by Ultrasound-Assisted Extraction and High-Performance Liquid Chromatography in *Rosaceae* and *Sambucus* Families. *Molecules* **2021**, *26*, 7563. [CrossRef]
14. Młynarczyk, K.; Walkowiak-Tomczak, D.; Łysiak, G.P. Bioactive Properties of *Sambucus nigra* L. as a Functional Ingredient for Food and Pharmaceutical Industry. *J. Funct. Foods* **2018**, *40*, 377–390. [CrossRef]
15. Waszkiewicz-Robak, B.; Biller, Elżbieta. Właściwości Prozdrowotne Czarnego Bzu/Health benefits of elderberry. *Probl. Hig. Epidemiol.* **2018**, *99*, 217–224. Available online: http://phie.pl/pdf/phe-2018/phe-2018-3-217.pdf (accessed on 20 July 2018).
16. European Medicines Agency. *Assessment Report on Sambucus nigra L., Flos (EMA/HMPC/611504/2016)*; Committee on Herbal Medicinal Products (HMPC): London, UK, 2018. Available online: https://www.ema.europa.eu/en/documents/herbal-report/final-assessment-report-sambucus-nigra-l-flos-revision-1_en.pdf (accessed on 18 July 2023).
17. European Medicines Agency. *Assessment Report on Sambucus nigra L., Fructus (EMA/HMPC/44208/2012)*; Committee on Herbal Medicinal Products (HMPC): London, UK, 2014. Available online: https://www.ema.europa.eu/en/documents/herbal-report/final-assessment-report-sambucus-nigra-l-fructus_en.pdf (accessed on 18 July 2023).
18. Viapiana, A.; Wesołowski, M. The phenolic contents and antioxidant activities of infusions of *Sambucus nigra* L. *Plant Foods Hum. Nutr.* **2017**, *72*, 82–87. [CrossRef] [PubMed]
19. Imenšek, N.; Ivančič, A.; Kraner Šumenjak, T.; Islamčević Rasboršek, M.; Kristl, J. The effect of maturation on chemical composition and harvest of fruits of diverse elderberry interspecific hybrids. *Eur. J. Hortic. Sci.* **2021**, *86*, 223–231. [CrossRef]
20. Duymuş, H.G.; Göger, F.; Başer, K.H.C. In vitro antioxidant properties and anthocyanin compositions of elderberry extracts. *Food Chem.* **2014**, *155*, 112–119. [CrossRef] [PubMed]
21. Osman, A.G.; Avula, B.; Katragunta, K.; Ali, Z.; Chittiboyina, A.G.; Khan, I.A. Elderberry Extracts: Characterization of the Polyphenolic Chemical Composition, Quality Consistency, Safety, Adulteration, and Attenuation of Oxidative Stress- and Inflammation-Induced Health Disorders. *Molecules* **2023**, *28*, 3148. [CrossRef]

22. Salvador, Â.C.; Guilherme, R.J.R.; Silvestre, A.J.D.; Rocha, S.M. Sambucus nigra Berries and Flowers Health Benefits: From Lab Testing to Human Consumption. In *Bioactive Molecules in Food;* Mérillon, J.M., Ramawat, K., Eds.; Reference Series in Phytochemistry; Springer: Cham, Switzerland, 2019; pp. 2261–2295. [CrossRef]
23. Sidor, A.; Gramza-Michałowska, A. Advanced Research on the Antioxidant and Health Benefit of Elderberry (*Sambucus nigra*) in Food—A Review. *J. Funct. Foods* **2015**, *18*, 941–958. [CrossRef]
24. Domínguez, R.; Zhang, L.; Rocchetti, G.; Lucini, L.; Pateiro, M.; Munekata, P.E.S.; Lorenzo José, M. Elderberry (*Sambucus nigra* L.) as potential source of antioxidants. Characterization, optimization of extraction parameters and bioactive properties. *Food Chem.* **2020**, *330*, 127266. [CrossRef]
25. Drugs and Lactation Database (LactMed) Bethesda (MD): National Library of Medicine (US). Elderberry. 2006. Available online: https://www.ncbi.nlm.nih.gov/books/NBK501835/ (accessed on 6 October 2021).
26. Zielińska-Pisklak, M.; Szeleszczuk, Ł.; Młodzianka, A. Bez Czarny (*Sambucus nigra*) Domowy Sposób Nie Tylko Na Grypę i Przeziębienie. *Lek Pol.* **2013**, *23*, 48–54. Available online: www.lekwpolsce.pl (accessed on 21 May 2013). (In Polish)
27. Olejnik, A.; Kowalska, K.; Olkowicz, M.; Rychlik, J.; Juzwa, W.; Myszka, K. Anti-inflammatory effects of gastrointestinal digested *Sambucus nigra* L. fruit extract analysed in co-cultured intestinal epithelial cells and lipopolysaccharide-stimulated macrophages. *J. Funct. Foods* **2015**, *197*, 649–660. [CrossRef]
28. Ricciotti, E.; FitzGerald, G.A. Prostaglandins and inflammation. *Arterioscler. Thromb. Vasc. Biol.* **2011**, *31*, 986–1000. [CrossRef] [PubMed]
29. Ho, G.T.T.; Wangensteen, H.; Barsett, H. Elderberry and Elderflower Extracts, Phenolic Compounds, and Metabolites and Their Effect on Complement, RAW 264.7 Macrophages and Dendritic Cells. *Int. J. Mol. Sci.* **2017**, *18*, 584. [CrossRef] [PubMed]
30. Dower, J.I.; Geleijnse, J.M.; Gijsbers, L.; Schalkwijk, C.; Kromhout, D.; Hollman, P.C. Supplementation of the Pure Flavonoids Epicatechin and Quercetin Affects Some Biomarkers of Endothelial Dysfunction and Inflammation in (Pre)Hypertensive Adults: A Randomized Double-Blind, Placebo-Controlled, Crossover Trial. *J. Nutr.* **2015**, *145*, 1459–1463. [CrossRef] [PubMed]
31. Carullo, G.; Cappello, A.R.; Frattaruolo, L.; Badolato, M.; Armentano, B.; Aiello, F. Quercetin and Derivatives: Useful Tools in Inflammation and Pain Management. *Future Med. Chem.* **2017**, *9*, 79–93. [CrossRef] [PubMed]
32. Ahmed, O.M.; Mohamed, T.; Moustafa, H.; Hamdy, H.; Ahmed, R.R.; Aboud, E. Quercetin and Low Level Laser Therapy Promote Wound Healing Process in Diabetic Rats via Structural Reorganization and Modulatory Effects on Inflammation and Oxidative Stress. *Biomed. Pharmacother.* **2018**, *101*, 58–73. [CrossRef] [PubMed]
33. Endale, M.; Park, S.C.; Kim, S.; Kim, S.H.; Yang, Y.; Cho, J.Y.; Rhee, M.H. Quercetin Disrupts Tyrosine-Phosphorylated Phosphatidylinositol 3-Kinase and Myeloid Differentiation factor-88 Association, and Inhibits MAPK/AP-1 and IKK/NF-κb-Induced Inflammatory Mediators Production in RAW 264.7 Cells. *Immunobiology* **2013**, *218*, 1452–1467. [CrossRef]
34. Seo, M.; Lee, Y.J.; Hwang, J.H.; Kim, K.J.; Lee, B.Y. The Inhibitory Effects of Quercetin on Obesity and Obesity-Induced Inflammation by Regulation of MAPK Signaling. *J. Nutr. Biochem.* **2015**, *26*, 1308–1316. [CrossRef]
35. Ferreira, S.S.; Martins-Gomes, C.; Nunes, F.M.; Silva, A.M. Elderberry (*Sambucus nigra* L.) extracts promote anti-inflammatory and cellular antioxidant activity. *Food Chem. X* **2022**, *15*, 100437. [CrossRef]
36. Santin, J.R.; Benvenutti, L.; Broering, M.F.; Nunes, R.; Goldoni, F.C.; Patel, Y.B.K.; de Souza, J.A.; Kopp, M.A.T.; de Souza, P.; da Silva, R.C.V.; et al. *Sambucus nigra*: A traditional medicine effective in reducing inflammation in mice. *J. Ethnopharm.* **2022**, *283*, 114736. [CrossRef]
37. Muvhulawa, N.; Dludla, P.V.; Ziqubu, K.; Mthembu, S.X.H.; Mthiyane, F.; Nkambule, B.B.; Mazibuko-Mbeje, S.E. Rutin ameliorates inflammation and improves metabolic function: A comprehensive analysis of scientific literature. *Pharmacol. Res.* **2022**, *178*, 106163. [CrossRef] [PubMed]
38. Lee, J.-H.; Kim, G.H. Evaluation of Antioxidant and Inhibitory Activities for Different Subclasses Flavonoids on Enzymes for Rheumatoid Arthritis. *J. Food Sci.* **2010**, *75*, H212–H217. [CrossRef] [PubMed]
39. Rho, H.S.; Ghimeray, A.K.; Yoo, D.S.; Ahn, S.M.; Kwon, S.S.; Lee, K.H.; Cho, D.H.; Cho, J.Y. Kaempferol and Kaempferol Rhamnosides with Depigmenting and Anti-Inflammatory Properties. *Molecules* **2011**, *16*, 3338–3344. [CrossRef] [PubMed]
40. Abo-Salem, O.M. Kaempferol Attenuates the Development of Diabetic Neuropathic Pain in Mice: Possible Anti-Inflammatory and Anti-Oxidant Mechanisms. Maced. *J. Med. Sci.* **2014**, *2*, 424–430. [CrossRef]
41. Sharma, D.; Gondaliya, P.; Tiwari, V.; Kalia, K. Kaempferol attenuates diabetic nephropathy by inhibiting RhoA/Rho-kinase mediated inflammatory signalling. *Biomed. Pharmacother.* **2019**, *109*, 1610–1619. [CrossRef] [PubMed]
42. Banach, M.; Khaidakov, B.; Korewo, D.; Węsierska, M.; Cyplik, W.; Kujawa, J.; Ahrné, L.M.; Kujawski, W. The Chemical and Cytotoxic Properties of *Sambucus nigra* Extracts—A Natural Food Colorant. *Sustainability* **2021**, *13*, 12702. [CrossRef]
43. Schröder, L.; Richter, D.U.; Piechulla, B.; Chrobak, M.; Kuhn, C.; Schulze, S.; Abarzua, S.; Jeschke, U.; Weissenbacher, T. Effects of Phytoestrogen Extracts Isolated from Elder Flower on Hormone Production and Receptor Expression of Trophoblast Tumor Cells JEG-3 and BeWo, as well as MCF7 Breast Cancer Cells. *Nutrients* **2016**, *8*, 616. [CrossRef]
44. Nejabati, H.R.; Roshangar, L. Kaempferol: A potential agent in the prevention of colorectal cancer. *Physiol. Rep.* **2022**, *10*, e15488. [CrossRef]
45. Wang, F.; Wang, L.; Qu, C.; Chen, L.; Geng, Y.; Cheng, C.; Yu, S.; Wang, D.; Yang, L.; Meng, Z.; et al. Kaempferol Induces ROS-Dependent Apoptosis in Pancreatic Cancer Cells via TGM2-Mediated Akt/MTOR Signaling. *BMC Cancer* **2021**, *21*, 396. [CrossRef]

46. Pham, H.N.T.; Sakoff, J.A.; Vuong, Q.V.; Bowyer, M.C.; Scarlett, C.J. Comparative cytotoxic activity between kaempferol and gallic acid against various cancer cell lines. *Data Brief* **2018**, *21*, 1033–1103. [CrossRef]
47. Akram, M.; Iqbal, M.; Daniyal, M.; Khan, A.U. Awareness and current knowledge of breast cancer. *Biol. Res.* **2017**, *50*, 33. [CrossRef] [PubMed]
48. Wang, X.; Yang, Y.; An, Y.; Fang, G. The Mechanism of Anticancer Action and Potential Clinical Use of Kaempferol in the Treatment of Breast Cancer. *Biomed. Pharmacother.* **2019**, *117*, 109086. [CrossRef] [PubMed]
49. Zhu, L.; Xue, L. Kaempferol Suppresses Proliferation and Induces Cell Cycle Arrest, Apoptosis, and DNA Damage in Breast Cancer Cells. *Oncol. Res.* **2019**, *27*, 629–634. [CrossRef] [PubMed]
50. Shafabakhsh, R.; Asemi, Z. Quercetin: A Natural Compound for Ovarian Cancer Treatment. *J. Ovarian Res.* **2019**, *12*, 55. [CrossRef] [PubMed]
51. Reyes-Farias, M.; Carrasco-Pozo, C. The Anti-Cancer Effect of Quercetin: Molecular Implications in Cancer Metabolism. *Int. J. Mol. Sci.* **2019**, *20*, 3177. [CrossRef] [PubMed]
52. Tang, S.M.; Deng, X.T.; Zhou, J.; Li, Q.P.; Ge, X.X.; Miao, L. Pharmacological Basis and New Insights of Quercetin Action in Respect to Its Anti-Ca.ncer Effects. *Biomed. Pharmacother.* **2020**, *121*, 109604. [CrossRef] [PubMed]
53. Khan, K.; Javed, Z.; Sadia, H.; Sharifi-Rad, J.; Cho, W.C.; Luparello, C. Quercetin and MicroRNA Interplay in Apoptosis Regulation in Ovarian Cancer. *Curr. Pharm. Des.* **2020**, *27*, 2328–2336. [CrossRef]
54. Khorsandi, L.; Orazizadeh, M.; Niazvand, F.; Abbaspour, M.R.; Mansouri, E.; Khodadadi, A. Quercetin Induces Apoptosis and Necroptosis in MCF-7 Breast Cancer Cells. *Bratisl. Med. J.* **2017**, *118*, 123–128. [CrossRef]
55. Tezerji, S.; Nazari Robati, F.; Abdolazimi, H.; Fallah, A.; Talaei, B. Quercetin's Effects on Colon Cancer Cells Apoptosis and Proliferation in a Rat Model of Disease. *Clin. Nutr. ESPEN* **2022**, *48*, 441–445. [CrossRef]
56. Ward, A.B.; Mir, H.; Kapur, N.; Gales, D.N.; Carriere, P.P.; Singh, S. Quercetin Inhibits Prostate Cancer by Attenuating Cell Survival and Inhibiting Anti-Apoptotic Pathways. *World J. Surg. Oncol.* **2018**, *16*, 108. [CrossRef]
57. Araújo, K.C.; Costa, E.M.d.M.B.; Pazini, F.; Valadares, M.C.; de Oliveira, V. Bioconversion of quercetin and rutin and the cytotoxicity activities of the transformed products. *Food Chem. Toxicol.* **2013**, *51*, 93–96. [CrossRef] [PubMed]
58. Satari, A.; Amini, S.A.; Raeisi, E.; Lemoigne, Y.; Hiedarian, E. Synergetic Impact of Combined 5-Fluorouracil and Rutin on Apoptosis in PC3 Cancer Cells through the Modulation of P53 Gene Expression. *Adv. Pharm. Bull.* **2019**, *9*, 462–469. [CrossRef] [PubMed]
59. Satari, A.; Ghasemi, S.; Habtemariam, S.; Asgharian, S.; Lorigooini, Z. Rutin: A Flavonoid as an Effective Sensitizer for Anticancer Therapy; Insights into Multifaceted Mechanisms and Applicability for Combination Therapy. *Evid.-Based Complement. Altern. Med.* **2021**, *2021*, 9913179. [CrossRef] [PubMed]
60. Chen, H.; Miao, Q.; Geng, M.; Liu, J.; Hu, Y.; Tian, L.; Pan, J.; Yang, Y. Anti-tumor effect of rutin on human neuroblastoma cell lines through inducing G2/M cell cycle arrest and promoting apoptosis. *Sci. World* **2013**, *2013*, 269165. [CrossRef] [PubMed]
61. Pinzaru, I.; Chioibas, R.; Marcovici, I.; Coricovac, D.; Susan, R.; Predut, D.; Georgescu, D.; Dehelean, C. Rutin Exerts Cytotoxic and Senescence-Inducing Properties in Human Melanoma Cells. *Toxics* **2021**, *9*, 226. [CrossRef] [PubMed]
62. Wang, J.; Liu, L.; Qiu, H.; Zhang, X.; Guo, W.; Chen, W.; Tian, Y.; Fu, L.; Shi, D.; Cheng, J.; et al. Ursolic Acid Simultaneously Targets Multiple Signaling Pathways to Suppress Proliferation and Induce Apoptosis in Colon Cancer Cells. *PLoS ONE* **2013**, *8*, e63872. [CrossRef] [PubMed]
63. Yan, S.; Huang, C.; Wu, S.; Yin, M. Oleanolic Acid and Ursolic Acid Induce Apoptosis in Four Human Liver Cancer Cell Lines. *Toxicol. Vitr.* **2010**, *24*, 842–848. [CrossRef]
64. Wang, J.S.; Ren, T.N.; Xi, T. Ursolic acid induces apoptosis by suppressing the expression of FoxM1 in MCF-7 human breast cancer cells. *Med. Oncol.* **2012**, *29*, 10–15. [CrossRef]
65. Luo, J.; Hu, Y.L.; Wang, H. Ursolic acid inhibits breast cancer growth by inhibiting proliferation, inducing autophagy and apoptosis, and suppressing inflammatory responses via the PI3K/AKT and NF-κB signaling pathways in vitro. *Exp. Ther. Med.* **2017**, *14*, 3623–3631. [CrossRef]
66. Shanmugam, M.K.; Dai, X.; Kumar, A.P.; Tan, B.K.H.; Sethi, G.; Bishayee, A. Ursolic Acid in Cancer Prevention and Treatment: Molecular Targets, Pharmacokinetics and Clinical Studies. *Biochem. Pharmacol.* **2013**, *85*, 1579–1587. [CrossRef]
67. Chan, E.W.C.; Soon, C.Y.; Tan, J.B.L.; Wong, S.K.; Hui, Y.W. Ursolic acid: An overview on its cytotoxic activities against breast and colorectal cancer cells. *J. Integr. Med.* **2019**, *17*, 155–160. [CrossRef]
68. Tehami, W.; Nani, A.; Khan, N.A.; Hichami, A. New Insights into the Anticancer Effects of P-Coumaric Acid: Focus on Colorectal Cancer. *Dose Response* **2023**, *21*, 15593258221150704. [CrossRef] [PubMed]
69. Mozaffari Godarzi, S.; Valizade Gorji, A.; Gholizadeh, B.; Mard, S.A.; Mansouri, E. Antioxidant Effect of P-Coumaric Acid on Interleukin 1-β and Tumor Necrosis Factor-α in Rats with Renal Ischemic Reperfusion. *Nefrol. (Engl. Ed.)* **2020**, *40*, 311–319. [CrossRef]
70. Pei, K.; Ou, J.; Huang, J.; Ou, S. P-Coumaric Acid and Its Conjugates: Dietary Sources, Pharmacokinetic Properties and Biological Activities. *J. Sci. Food Agric.* **2016**, *96*, 2952–2962. [CrossRef] [PubMed]
71. Nwafor, E.O.; Lu, P.; Zhang, Y.; Liu, R.; Peng, H.; Xing, B.; Liu, Y.; Li, Z.; Zhang, K.; Zhang, Y.; et al. Chlorogenic acid: Potential sourceof natural drugs for the therapeutics of fibrosis and cancer. *Transl. Oncol.* **2022**, *15*, 101294. [CrossRef]
72. Wang, L.; Du, H.; Chen, P. Chlorogenic Acid Inhibits the Proliferation of Human Lung Cancer A549 Cell Lines by Targeting Annexin A2 in Vitro and in Vivo. *Biomed. Pharmacother.* **2020**, *131*, 110673. [CrossRef] [PubMed]

73. Zeng, A.; Liang, X.; Zhu, S.; Liu, C.; Wang, S.; Zhang, Q.; Zhao, J.; Song, L. Chlorogenic Acid Induces Apoptosis, Inhibits Metastasis and Improves Antitumor Immunity in Breast Cancer via the NF-KB Signaling Pathway. *Oncol. Rep.* **2021**, *45*, 717–727. [CrossRef]
74. Mustafa, A.; Sezai, E.; Murat, T. Physico-Chemical Characteristics of Some Wild Grown European Elderberry (*Sambucus nigra* L.) Genotypes. *Pharmacogn. Mag.* **2009**, *5*, 320–323. [CrossRef]
75. Pascariu, O.E.; Israel-Roming, F. Bioactive Compounds from Elderberry: Extraction, Health Benefits, and Food Applications. *Processes* **2022**, *10*, 2288. [CrossRef]
76. Liu, W.; Xu, J.; Wu, S.; Liu, Y.; Yu, X.; Chen, J.; Tang, X.; Wang, Z.; Zhu, X.; Li, X. Selective Anti-Proliferation of HER2-Positive Breast Cancer Cells by Anthocyanins Identified by High-Throughput Screening. *PLoS ONE* **2013**, *8*, e81586. [CrossRef]
77. Ma, X.; Ning, S. Cyanidin-3-glucoside attenuates the angiogenesis of breast cancer via inhibiting STAT3/VEGF pathway. *Phytother. Res.* **2019**, *33*, 81–89. [CrossRef]
78. Rugină, D.; Hanganu, D.; Diaconeasa, Z.; Tăbăran, F.; Coman, C.; Leopold, L.; Bunea, A.; Pintea, A. Antiproliferative and Apoptotic Potential of Cyanidin-Based Anthocyanins on Melanoma Cells. *Int. J. Mol. Sci.* **2017**, *18*, 949. [CrossRef]

molecules

MDPI

Review

Hirsutine, an Emerging Natural Product with Promising Therapeutic Benefits: A Systematic Review

Md. Shimul Bhuia [1,†], Polrat Wilairatana [2,*,†], Jannatul Ferdous [3], Raihan Chowdhury [1], Mehedi Hasan Bappi [1], Md Anisur Rahman [4], Mohammad S. Mubarak [5] and Muhammad Torequl Islam [1,*]

1 Department of Pharmacy, Bangabandhu Sheikh Mujibur Rahman Science and Technology University, Gopalganj 8100, Bangladesh; shimulbhuia.pharm@gmail.com (M.S.B.); raihanpharmacy049@gmail.com (R.C.); mehedibappi22@gmail.com (M.H.B.)

2 Department of Clinical Tropical Medicine, Faculty of Tropical Medicine, Mahidol University, Bangkok 10400, Thailand

3 Department of Biotechnology and Genetic Engineering, Bangabandhu Sheikh Mujibur Rahman, Science and Technology University, Gopalganj 8100, Bangladesh; jannat16bge093@gmail.com

4 Department of Pharmacy, Islamic University, Kushtia 7003, Bangladesh; anisurrahmaniupharm@gmail.com

5 Department of Chemistry, The University of Jordan, Amman 11942, Jordan; mmubarak@ju.edu.jo

* Correspondence: polrat.wil@mahidol.ac.th (P.W.); dmt.islam@bsmrstu.edu.bd (M.T.I.)

† These authors contributed equally to this work.

Abstract: Fruits and vegetables are used not only for nutritional purposes but also as therapeutics to treat various diseases and ailments. These food items are prominent sources of phytochemicals that exhibit chemopreventive and therapeutic effects against several diseases. Hirsutine (HSN) is a naturally occurring indole alkaloid found in various Uncaria species and has a multitude of therapeutic benefits. It is found in foodstuffs such as fish, seafood, meat, poultry, dairy, and some grain products among other things. In addition, it is present in fruits and vegetables including corn, cauliflower, mushrooms, potatoes, bamboo shoots, bananas, cantaloupe, and citrus fruits. The primary emphasis of this study is to summarize the pharmacological activities and the underlying mechanisms of HSN against different diseases, as well as the biopharmaceutical features. For this, data were collected (up to date as of 1 July 2023) from various reliable and authentic literature by searching different academic search engines, including PubMed, Springer Link, Scopus, Wiley Online, Web of Science, ScienceDirect, and Google Scholar. Findings indicated that HSN exerts several effects in various preclinical and pharmacological experimental systems. It exhibits anti-inflammatory, antiviral, anti-diabetic, and antioxidant activities with beneficial effects in neurological and cardiovascular diseases. Our findings also indicate that HSN exerts promising anticancer potentials via several molecular mechanisms, including apoptotic cell death, induction of oxidative stress, cytotoxic effect, anti-proliferative effect, genotoxic effect, and inhibition of cancer cell migration and invasion against various cancers such as lung, breast, and antitumor effects in human T-cell leukemia. Taken all together, findings from this study show that HSN can be a promising therapeutic agent to treat various diseases including cancer.

Keywords: hirsutine; pharmacological activities; cancer; diabetes; anti-hypertensive effect; pharmacokinetics

1. Introduction

The term natural products (NPs) refers to substances that originate from natural sources such as plants, animals, and microorganisms. Many of these compounds possess important biological properties [1]. Research findings showed that NPs have provided valuable starting points for the development of numerous highly successful medications that are presently used to treat various illnesses in humans [2]. For centuries, NPs have been utilized as traditional remedies, medicines, potions, and oils without any understanding of the bioactive compounds they contain. People have relied solely on the results of centuries of experimentation to determine their effectiveness [3].

Compared to conventional synthetic molecules, NPs have distinct characteristics that present both benefits and obstacles during the drug discovery process [4]. NPs represent a significant reservoir of orally available medications [5]. In contrast, synthetic drugs are prepared in the laboratory using various methods that have the maximum potential to be toxic or cause side effects in addition to their therapeutic benefits. While herbal medicines may not be as potent as synthetic drugs in certain cases, they are generally considered to be less toxic or cause fewer side effects when compared to synthetic medications [6,7]. In this regard, herbal medications are frequently used in holistic treatments intended to cure a wide variety of diseases as they fulfill the primary criteria for being reliable therapeutics, such as being efficacious and non-toxic. Therefore, numerous chemists are shifting from synthetic chemistry to natural product research to further investigate nature's wonders [8,9].

In the history of drug development from natural sources, Alexander Fleming made the most significant contribution to the field of natural substances at the start of the 20th century through his discovery of penicillin in 1928. Penicillin, derived from *Penicillium chrysogenum*, is a molecule with antibiotic properties and forms the basis of modern anti-lactam antibiotics. The discovery of penicillin led to a concentration of scientific research on isolating NPs from microbial and other sources [10,11]. During this period, various modern pharmaceutical industries focused on developing more antibiotics along with other drugs such as streptomycin, gentamicin, and tetracycline from natural sources on the basis of microbial fermentation technologies [1,12].

Indole alkaloids, which are abundant and a major source of pharmacologically active substances, have had a considerable impact on the preparation of novel anti-cancer medications such as drugs like vincristine and vinblastine [13]. Indole alkaloids are found in various plants such as *Catharanthus roseus* and *Alstonia scholaris*, which exhibited a number of therapeutic properties in addition to their anticancer properties in various preclinical tests. An additional alkaloid, ajmalicine, exhibits anti-arrhythmia and antihypertensive properties, while catharanthine and vindoline have demonstrated diuretic effects, antibacterial activities, and antidiabetic properties derived from plants [14]. Another alkaloid, dragmacidin D, which is an indole-containing alkaloid, can inhibit the development of certain gram-negative and gram-positive microorganisms such as *Bacillus subtilis* and *Escherichia coli*. Furthermore, it has been demonstrated that dragmacidin D can suppress the growth of several opportunistic yeasts like *Candida aeruginosa*, *Candida albicans*, and *Candida neoformans* [15]. Makaluvamine G possesses a moderate ability to inhibit topoisomerase I and can also suppress the activity of DNA and proteins [15].

Hirsutine (HSN) (Figure 1) is the primary indole alkaloid found in Uncaria species, mainly in *Uncaria rhynchophylla*. These plants of the genera are used in local Chinese herbal medicine to treat a variety of symptoms involved with hypertension and cerebrovascular disorders, including spasmolytic, analgesic, and sedative treatments [16]. Published research showed that HSN exhibits a depressant effect on the central nervous system (CNS) in mice, a weak antispasmodic effect on the intestine of mice, and a lowering effect on blood pressure in rats [17]. Pharmacological studies have shown that HSN may have therapeutic benefits in treating certain conditions as it exerts anti-inflammatory (preeclampsia) [18], antiviral (dengue fever and dengue hemorrhagic fever) [19], anticancer activity (breast and lung cancers) [20–24], antitumor [25], anti-diabetic [26], antihypertensive, negative chronotropic, and antiarrhythmic activity [27]. It can also be used to treat and manage myocardial infarction, ischemia-reperfusion (I/R) trauma [28], brain inflammation [29], cerebral ischemia [30], cardiovascular diseases (cardiomyocytes) [31], anti-hypertension [14,32], hypotensive and vasodilatory effects [33], lung cancer [34], thrombolytic effects (thrombocytopenia) [35], and neurogenerative diseases and disorders such as neuronal death [36]. Based on the previous discussion, the aim of the present work was to summarize the literature dealing with the pharmacological effects of HSN and present an overview of the recent developments of its use in the prevention and treatment of different diseases. We hope findings from this can help guide future research and aid in the development of new therapeutic strategies.

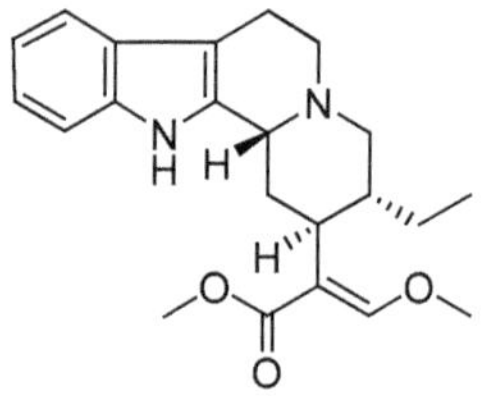

Hirsutine

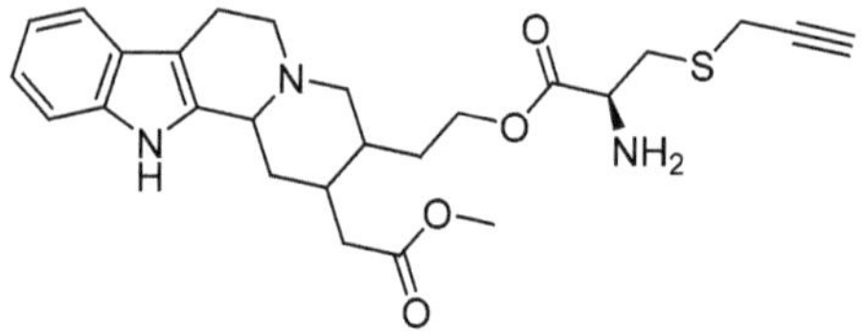

Compound 1

Figure 1. Chemical structures of hirsutine and its derivative (Compound **1**).

2. Results

2.1. Botanical Sources of Hirsutine

Plants are a significant source of medication and are crucial to global health. Medicinal plants play a critical role in protecting from diseases and are integrated into preventive programs, not only for the treatment of illness but as a possible material for preserving good well-being and conditions [37]. Numerous compounds found in plants have been utilized in medicine to aid in the development of novel drugs [38,39]. The phytochemical HSN, which is a type of indole alkaloid, has been obtained from different species of the Uncaria genus. This plant belongs to the Rubiaceae family [31]. In this respect, various species such as *Uncaria rhynchophylla* (Miquel), *U. lancifolia*, *U. hirsuta*, *U. scandens*, *U. homomalla*, *U. sessilifructus*, *U. laevigata*, *U. macrophylla*, *U. yunnanensis*, *U. lanosa*, and *U. rhynchophylloides*, are rich sources of HSN [25,33,40–44]. The species *U. rhynchophylla* has gained attention for its various biological traits such as cardioprotective, antihypertensive, and antiarrhythmic effects [14]. Throughout history, the bark of *U. rhynchophylla* has been used as a remedy for various ailments including convulsions, bleeding, hypertension, autoimmune disorders, and cancer [25]. Several other studies suggested that the plant *U. rhynchophylla* (Miquel) exhibits various pharmaceutical effects such as antihypertensive, anti-inflammatory activity, sedative, and antiarrhythmic actions due to the presence of HSN [18,32]. The other sources of HSN are *U. sinensis* [36], *U. tomentosa* [45], and *Mitragyna hirsute* (https://pubchem.ncbi.nlm.nih.gov/taxonomy/371154, accessed on 30 April 2023). Listed in Table 1 are the botanical sources and the plant portion where HSN is found in significant quantities.

Table 1. Various botanical sources of hirsutine.

Plants	Parts	References
Uncaria rhynchophylla (Miquel)	Bark	[25]
	Dried hooks	[32]
U. hirsuta *U. lancifolia*, *U. scandens*, *U. macrophylla* *U. homomalla*, *U. laevigata*, *U. sessilifructus*, *U. yunnanensis* *U. lanosa*, *U. rhynchophylloides*,	Stems and hooks	[44]

Table 1. *Cont.*

Plants	Parts	References
U. sinensis	Stems and hooks	[36]
Uncaria tomentosa	Leaves and roots	[45]
Mitragyna hirsuta	Leaves and root	https://pubchem.ncbi.nlm.nih.gov/taxonomy/371154, accessed on 30 April 2023

2.2. Physicochemical and Biopharmaceutical Profiles

The role of pharmacokinetics (PKs) in drug discovery is to support the optimization of lead compounds' absorption, distribution, metabolism, and excretion (ADME) properties to develop a clinical candidate that has a concentration-time profile in the body that is sufficient for the sought efficacy and safety profile [46,47]. Bioavailability is referred to as the extent and rate of absorption and availability of the active drug constituent or active moiety from the drug product at the site of drug action [48,49]. The main reasons some drug candidates fail in clinical trials are due to undesirable PK features and unacceptable toxicity [50]. Therefore, it is important to focus on optimizing and characterizing human ADME features and understanding the pharmacokinetic-pharmacodynamic relationship to estimate a clinically relevant dose [51].

HSN ($C_{22}H_{28}N_2O_3$) is a white crystalline powder with a density of 1.20 ± 0.1 g/cm^3 and a molecular mass of 368.5 g/mol. The melting point of the compound is 101 °C, while the boiling point is 531.7 ± 50.0 °C (Predicted). Since HSN is readily soluble in methanol, acid water, chloroform, and lipids, its concentrations rapidly decline as the tissues metabolized the hormone [52], (https://pubchem.ncbi.nlm.nih.gov/compound/3037884, accessed on 4 May 2023; https://www.chembk.com/en/chem/Hirsutine, accessed on 10 July 2023). Research findings showed that the bioavailability of HSN is 4.4% and clearance (CL) is remarkably lower than that of oral administration. In addition, findings indicated that HSN has poor absorption. For oral administration, the volume of distribution (VD), mean residence time (MRT), the highest concentration of a drug in the blood, cerebrospinal fluid, or target organ after a dose is given (C_{max}), and CL of HSN were 70.8 ± 17.8 ng/mL, 3.6 ± 0.6 h, and 21.9 ± 6.6 ng/mL, respectively [52]. In humans, the half-life of HSN in plasma is 3.4 h, and the time it takes to reach its maximum concentration (t_{max}) is 0.50 to 0.83 h [53]. HSN is distributed to the brain, liver, kidney, spleen, heart, and lungs. The liver and kidney tissues attain the highest distribution levels, followed by the lung and spleen. HSN exhibits a modest concentration in brain tissue but was widely distributed, indicating that it might cross the blood-brain barrier (BBB) [52]. Another investigation by Gai et al. (2020) showed that HSN has the capability to cross BBB. These researchers found that HSS inhibits P-glycoprotein mRNA expression in MCF-7/ADR cells [44]. HSN is mostly metabolized in liver tissues before being eliminated in renal tissues, which may be associated with high levels of blood flow and tissue oxygenation in these organs. Furthermore, HSN is metabolized by cytochrome P450s (CYPs) in rat liver microsomes [16]. In rats, HSN undergoes glucuronidation to produce 11-hydroxy metabolites, which are primarily eliminated in bile rather than urine [16].

In our in silico ADME prediction, HSN exhibited better ADME properties to be considered a drug candidate. HSN followed Lipinski's rule of 5, such as H-bond (HB) acceptors (4) and HB donors (1), as well as molar refractivity (MR) 110.39 and TPSA (54.56 Å^2) where all the parameters are within the limit (HB acceptors $\leq$ 10, HB donors $\leq$ 5, MR $\leq$ 140, TPSA $\leq$ 140 Å^2) indicating better physiochemical properties for better ADME as there is no violation of Lipinski's rule of 5 [49] (http://www.swissadme.ch/index.php, accessed on 10 August 2023). ADME prediction also demonstrated that HSN is moderately water soluble, highly absorbable through GIT, capable of permitting BBB, and inhibits P-gp, which is almost the same as reported by different in vivo studies [44,52] (http://www.swissadme.ch/index.php, accessed on 10 August 2023). Different ADME parameters of HSN predicted by SwissADME are shown in Table 2.

Table 2. Different parameters of HSN and their values/status of ADME predicted by SwissADME.

Parameter (s)	Values/Status
Physicochemical properties	
Molecular mass	368.5 g/mol
Number of heavy atoms	27
Number of aromatic heavy atoms	9
Number of rotatable bonds	5
Number H-bond acceptors	4
Number H-bond donors	1
Molar Refractivity	110.39
TPSA	54.56 Å^2
Lipophilicity	
Log Po/w (MLOGP)	2.35
Water Solubility	
Solubility class	Moderately soluble
Pharmacokinetics	
GI absorption	High
BBB permeant	Yes
P-gp substrate	Yes
CYP1A2 inhibitor	No
CYP2C19 inhibitor	No
Drug-likeness	
Lipinski	Yes; 0 violation
Bioavailability Score	0.55

2.3. *Pharmacological Profile of Hirsutine*

2.3.1. Neurobiological Effects

Prevention of Neuroinflammation and Neurotoxicity

Inflammation is a natural immunological reaction that can be provoked by infectious agents, toxic compounds, and damaged cells, among others. These factors can trigger chronic and/or acute inflammatory reactions in the pancreas, heart, liver, lung, kidney, urinary tract, brain, gastrointestinal tract, respiratory tract, and reproductive organs, which may cause tissue injury or disease [54,55]. Immune activation within the CNS is a defining characteristic of ischemia, immune-mediated disorders, neurodegenerative diseases, infections, and trauma, and can frequently cause neuronal injury [56]. Immune and inflammatory responses in CNS are principally mediated by microglia [57]. Activated microglia also contribute to neuronal damage via the secretion of proinflammatory and cytotoxic factors, such as cytokines, reactive oxygen species (ROS), and nitrogen oxides (NO) [58,59]. Microglia can be activated by lipopolysaccharide (LPS), which is a strong immunogenic particle released from gram-negative bacteria and induces inflammation in various experimental animals and cell lines [60]. Mice with SIRS induced by LPS revealed a systemic and local inflammatory response through the discharge of different inflammatory enzymes and cytokines [61].

Published research demonstrated that HSN blocks LPS-related hippocampal cell death and the generation of NO, PGE2, and IL-1β. It also revealed that HSN effectively hinders LPS-mediated NO secretion from cultured rat brain microglia and results in diminished production of PGE2 and intracellular ROS generation, resulting in diminishing neurotoxicity (Table 3) [29]. Another study by Wu et al. (2019) reported that treatment with

Uncaria rhynchophylla alkaloid extract (where HSN presents as an active phytochemical) in LPS-mediated preeclampsia rats inhibited the level of proinflammatory cytokines (IL-6, IL-1β, TNF-α, and IFN-γ), resulting in therapeutic benefits in the complication of pregnant rats by reducing inflammation [18].

Prevention of Neuronal Cell Death

Glutamate is the primary stimulating neurotransmitter in the CNS and is essential in learning, memory development, and metabolism in the brain [62,63]. However, excessive glutamate causes the death of central neurons in various pathological situations such as ischemic-hypoxic injury, epilepsy, and neurodegenerative illness. Various studies reported that ischemia promotes excessive release of glutamate, which stimulates the inotropic glutamate receptors, and overstimulated receptors cause Ca^{2+} influx [64–66]. Ca^{2+} influx triggers cell death by raising the activity of Ca^{2+}-dependent enzymes, including phospholipase C and protein kinase C, which degrade cytoskeletal proteins [67,68]. It is also evident that there is a link between the increase in ROS generation and the inflow of Ca^{2+} into cells [69,70]. In addition, research findings indicated that HSN (3×10^{-4}–10^{-3} M) presents as a bioactive compound in alkaloids extracted from *Uncaria sinensis* is capable of diminishing glutamate-mediated death of neuronal cells in cultured cerebellar granule cells of rats by suppressing Ca^{2+} influx [36].

2.3.2. Cardioprotective Activity

Acute myocardial infarction (AMI) is one of the major causes of disability and mortality in the world [71]. It causes irreparable damage to the cardiac muscle due to an insufficient supply of oxygen (hypoxia) and induction of oxidative stress, which is caused by ischemic reperfusion (I/R) injury [72]. The intracellular changes that occur during I/R, include the accumulation of H^+ and Ca^{2+} and the disruption of the mitochondrial membrane potential, resulting in the production of free radicals or ROS. Accumulation of ROS and consequent activation of pro-inflammatory pathways play a key role in I/R injury. Therefore, oxygen-derived free radicals are a crucial mediator of I/R injury, leading to various forms of oxygen species [73,74], resulting in the apoptosis of cardiac cells [73,75]. I/R injury can be treated by phytochemicals, which can inhibit oxidative stress and has anti-apoptotic and anti-inflammatory effects [76].

An in vivo experiment by Jiang et al. (2023) demonstrated that HSN plays a protective role in inhibiting apoptosis in I/R injury. Pretreatment with HSN (5, 10, and 20 mg/kg) in I/R injured AMI rats revealed reductions in myocardial infarct size, mitochondrial function, and histological injury, and inhibited cardiac cell apoptosis by hindering the AKT/ASK-1/p38 MAPK pathway. Moreover, the results of the study showed that HSN enhances cardiac function and diminishes tissue lactate dehydrogenase (LDH) and ROS content [28]. Another study by Wu et al. (2011) showed that HSN exerts cardioprotective effects due to its antioxidant and anti-apoptotic effects. Additionally, HSN (0.1, 1, and 10 μM) prevented hypoxia-induced myocyte cell death by regulating proapoptotic signaling cascades associated with Bcl-2 family proteins and caspases which inhibit the destruction of hypoxia-induced myocytes. Furthermore, HSN suppressed ROS generation by enhancing the activity of antioxidant enzymes and inhibited lipid peroxidation [31], resulting in the prevention of AMI as the destruction of myocytes is minimized.

2.3.3. Antiviral Activity

The lack of effective treatments for viral diseases is a significant health concern in the current era [77]. Among the different viruses, dengue and flu (influenza) cause major health concerns. Dengue is a viral infection that, in extreme circumstances, can be fatal [78]. The incidence of dengue is on the rise, and the disease is a significant public health concern in tropical regions. Each rainy season is accompanied by a surge of dengue epidemics, where thousands could be affected [79,80]. On the other hand, in humans, influenza viruses produce two distinct types of respiratory illness: the seasonal variety and the pandemic variety [81], especially in infants. However, it is a matter of main concern that resistance to antiviral drugs leads to the development of new drugs.

HSN (10 μM) demonstrated an antiviral effect against all the dengue virus (DENV) serotypes ((DENV-1 (02-20 strain), DENV-2 (16681 strain), DENV-3 (09-59 strain), and DENV-4 (09-48 strain)) by preventing the viral particle assembly, budding, or release phases in the DENV lifecycle, but not the viral translation and replication steps. Using a subgenomic replicon system, it was determined that HSN does not hinder viral genome RNA replication. Additionally, studies suggested that the antiviral activity of HSN may be connected to calcium homeostasis [16]. The antiviral activity of HSN was also evident in a study by Takayama et al. (1997); the results of this study indicated that HSN effectively suppresses the replication of the strains of influenza A (subtype H3N2) with the EC_{50} of 0.4–0.57 μg/mL [82]. Depicted in Figure 2 are the underlying mechanisms of different pharmacological effects of HSN.

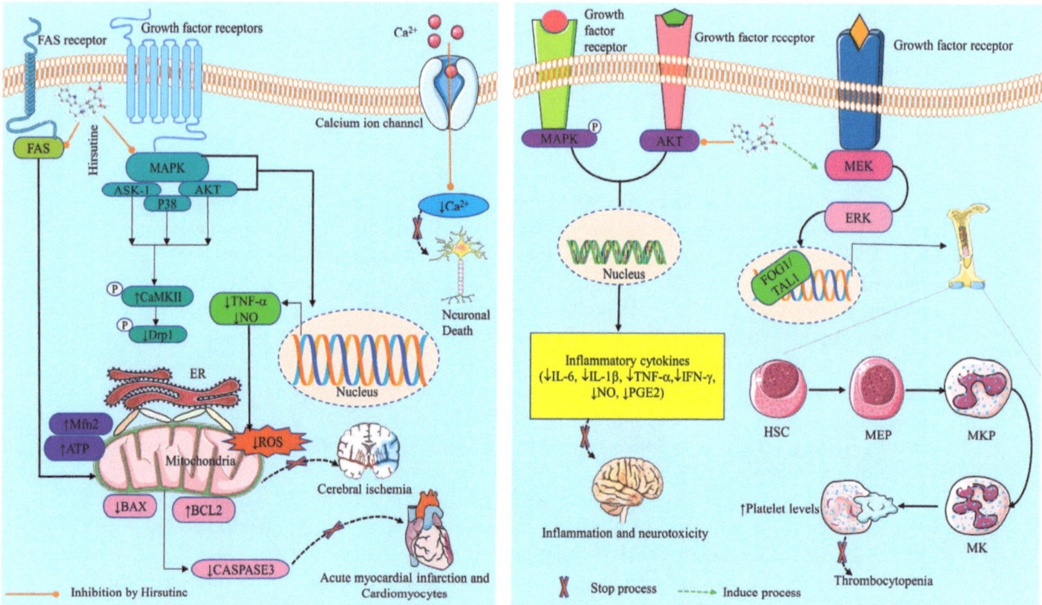

Figure 2. The underlying mechanisms of different pharmacological effects of hirsutine. ↓: Decrease/inhibition/downregulation; ↑: Increase/upregulation/stimulation; HSC: Hematopoietic Stem Cell; MEP: Megakaryocyte-Erythroid Progenitor; MKP: Mitogen-Activated Protein Kinase Phosphatase; NO: Nitric Oxide; PGE2: Prostaglandin E2; IL-6: Interleukin-6; IL-1β: Interleukin-1 beta; TNF-α: Tumor Necrosis Factor-alpha; IFN-γ: Interferon-gamma; ROS: Reactive Oxygen Species.

2.3.4. Anticancer Activity of Hirsutine: Underlying Mechanisms

Induction of Oxidative Stress

An elevated degree of oxidative stress is regarded as a novel target of anticancer therapy. This can be triggered by raising exogenous ROS or blocking the endogenous antioxidant defense system [83,84]. ROS levels can rise as a result of chemotherapy medications [85], which may amplify the cytotoxic effects of the chemotherapeutic agents. ROS are highly reactive molecules that can lead to oxidative damage to cellular components such as proteins, DNA, and lipids [86]. In addition, the elevation in ROS levels can lead to the activation of various signaling pathways, including the p53 pathway, which can induce apoptosis in cancer cells. A recent study found that HSN causes mitochondrial death by causing ATP depletion, the formation of ROS, the loss of mitochondrial membrane potential, and the discharge of cytochrome C (Cyt C) [34]. Additionally, to overcome HSN resistance, it may be helpful to disrupt the ataxia telangiectasia mutated (ATM) pathway, which results in ROS production and a p53-independent DNA damage response in breast cancer cells [21].

Cytotoxicity

Cytotoxicity is the ability of a substance, such as a chemical or drug, to kill or damage cells. Many chemotherapeutic drugs work by exerting cytotoxic effects on rapidly dividing cancer cells, thereby inhibiting their growth and causing them to die [84,87]. In this context, findings revealed that HSN exerts cytotoxic effects against different cell lines in multiple investigations, which raises the possibility of its development as an anticancer drug. According to a study by Meng et al. (2021), HCN showed cytotoxicity against Jurkat clone E6-1 tumor cells [25]. In a different study, it was discovered that HSN exhibits strong cytotoxicity against human breast cancer cell lines, including MCF-7, MDA-MB-231, MCF-10A, BT474, and MDA-MB-453 with concentrations ranging from 6.25−160 M (Table 3) [20–24].

Apoptotic Effect

Chemotherapy drugs can induce apoptosis in cancer cells as a mechanism of their cytotoxicity [88]. Apoptosis is a tightly regulated process of programmed cell death that occurs in response to different stimuli, including chemotherapy drugs [89]. Induction of apoptosis in cancer cells is crucial in cancer treatment, as it allows for the selective elimination of cancer cells without affecting normal ones [90]. The Bcl-2 family protein is involved in the regulation of apoptotic cell death and inhibition of apoptosis [91]. On the other hand, findings revealed that Bax exhibits a potent apoptosis-promoting capacity, resulting in alterations in the membrane potential and structure of mitochondria, and then initiates the caspase-independent apoptotic response via the mitochondrial pathway. Simultaneously, Cyt C is released directly or indirectly from the mitochondria to the cytoplasm to initiate the caspase-dependent apoptotic response via the mitochondrial pathway. Therefore, Bcl-2/Bax plays a crucial role in modulating caspase-dependent and caspase-independent apoptosis induced via the mitochondrial pathway [91,92]. In contrast, the PI3K/AKT signaling pathway is a vital intracellular signal transduction pathway that plays a crucial function in regulating apoptosis and survival [93]. Thus, targeting these pathways is one of the most reliable approaches to anticancer drug design and development.

According to a research report, HSN may cause the human breast cancer MDA-MB-231 cells to undergo programmed cell death by lowering the Bcl-2 to Bax ratio, opening the mitochondrial permeability transition pore (MPTP), secreting Cyt C from the mitochondria, and activating caspases 9 and 3 [22]. Another investigation also found that the cancer cell death mechanism depends on mitochondria. At doses of 10, 25, and 50 μM/L, HSN inhibited Jurkat Clone E6-1 cell death by up-regulating Bcl-2 levels as a preventive compensatory mechanism. The Bax/Bcl-2 ratio may change to influence the apoptotic activity of HSN because HSN therapy markedly diminished Bcl-2 expression and elevated Bax expression [25]. According to another study, HSN suppresses tumor development and promotes apoptosis in an A549 xenograft mouse model via the ROCK1/PTEN/PI3K/Akt/GSK3 signaling pathway [34]. Furthermore, HSN triggers cell apoptosis by mediating DNA damage and inhibiting breast cancer (MCF-7) cell lines [21]. Moreover, HSN causes apoptotic cell death by activating caspases. Additionally, HSN promoted a DNA damage response in MDA-MB-453 cells, as evidenced by an increase in γH2AX expression [22].

Inhibition of Cell Migration and Invasion

Inhibition of cell migration and invasion is an important therapeutic goal in cancer treatment, as the ability of cancer cells to spread and invade surrounding tissues is a primary factor in the progression of the disease [94,95]. While chemotherapy drugs primarily target rapidly dividing cells, some drugs may also possess properties that can inhibit cell migration and invasion [96]. In addition to promoting tumor cell proliferation, inhibiting apoptosis, and attracting angiogenesis, NF-κB activity induces epithelial-mesenchymal transition, thereby facilitating distant metastasis. Under specific conditions, NF-κB activation may also remodel local metabolism and depress the immune system to promote tumor growth. The NF-κB pathway is a prospective therapeutic target because inhibition of NF-κB in myeloid cells or tumor cells typically results in tumor regression [97,98].

An investigation by Lou et al. (2014) demonstrated that HSN (25 μM) inhibits Mouse mammary carcinoma 4T1 cell migration in a dose-dependent manner. Furthermore, the results of the investigation showed that pre-treatment with HSN blocks 4T1 cell haptotaxis toward fibronectin in a Transwell chamber assay. Thus, the findings revealed that HSN pre-treatment significantly suppresses the migration and invasion activity of 4T1 cell lines by suppressing NF-κB signaling pathways which can be a treatment approach for breast cancer [20]. Similarly, a recent investigation by Huang et al. (2017) found that HSN with an IC_{50} value of 62.82 μM/L restricts hypoxia-mediated migration and invasion in human breast cancer MCF-7 cells, most likely through down-regulation of the protein levels of HIF-1α, MMP-9, and Snail and up-regulation of the protein level of E-cadherin [24]. Shown in Figure 3 are the possible anticancer mechanisms of HSN.

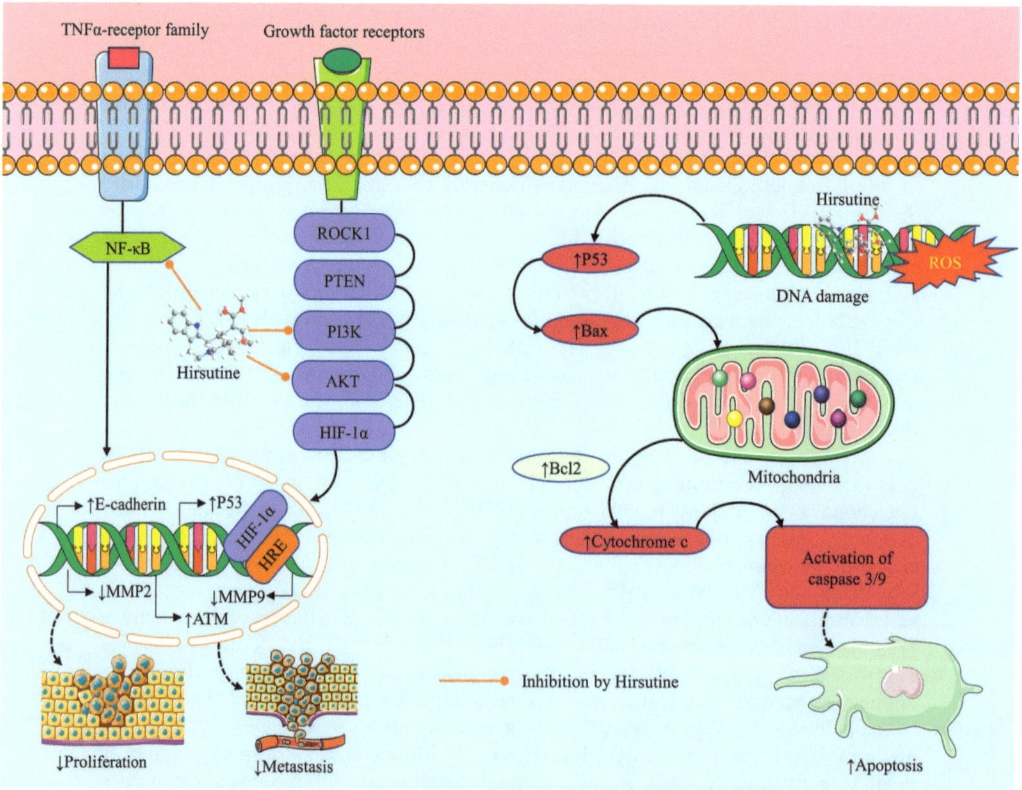

Figure 3. Possible anticancer mechanisms of hirsutine. ↓: Decrease/inhibition/downregulation; ↑: Increase/upregulation/stimulation; ROS: Reactive oxygen species; NF-κB: Nuclear Factor-kappa B; MMP2: Matrix Metalloproteinase 2; MMP9: Matrix Metalloproteinase 9; ATM: Ataxia Telangiectasia Mutated; HIF-1α: Hypoxia-Inducible Factor 1 alpha; AKT: Protein Kinase B; PTEN: Phosphatase and Tensin Homolog; PI3K: Phosphoinositide 3-Kinase; RE: Hypoxia-Response Element.

Anti-Proliferative Effect

By interfering with the normal cell cycle and causing cell death, chemotherapy medications target rapidly dividing cells, including cancer cells. Consequently, these drugs are often referred to as "anti-proliferative" medications, as they can inhibit or ease up the growth and proliferation of cancer cells [99]. In this respect, the cell counting kit-8 (CCK8) assay performed by Meng et al. (2021) demonstrated that HSN, at concentrations of 10, 25, and 50 μM, could remarkably suppress the proliferation of Jurkat clone E6-1 cells over 48 h. Similarly, flow cytometry experiments showed that HSN might cause apoptosis and G0/G1 phase arrest in Jurkat cells by lowering the Bcl-2 expression and, at the same time, enhancing Bax, mRNA, caspase-3, and -9 levels, thus inducing inhibition of cell proliferation in a tumor [25].

Genotoxic Effect

The ability of a chemical compound to damage the genetic material (DNA) of cells is known as genotoxicity. In this regard, numerous chemotherapy drugs may cause genotoxicity [100]. Genotoxicity can result in mutations, chromosomal aberrations, and DNA damage, which can promote cell death, cell transformation, and cancer [101]. Lou et al. (2015) found that HSN may induce genotoxicity. HSN at the concentration of 25, 12.5–50 μM damages the DNA of the HER2-positive/p53-mutated MDA-MB-453 cells by upregulating γH2AX expression and suppressing the NF-κB, HER2, and Akt pathways. HSN can also activate the p38 MAPK pathway in the MDA-MB-453 cells resulting in the induction of genotoxicity which destroys the DNA of breast cancer cells [22]. Another Investigation by Lou et al. (2016) stated that HSN (50 μM) induced the demise of MCF-7 cells and a sustained DNA damage response. This DNA damage was caused by interference with the ATM pathway and the production of ROS, which enhanced the anticancer effect of HSN in breast cancer cells [21].

2.3.5. Effects on Thrombocytopenia

Thrombocytopenia is one of the most common hematological conditions, manifested by an unnaturally low platelet count due to several reasons [102]. Thrombocytopenia is linked to multiple syndromes and diseases and can be an early indicator of hematologic malignancies, thrombotic microangiopathies, infectious diseases, and autoimmune disorders, as well as a common adverse effect of numerous medications [103]. Megakaryocytes (MKs) are a type of functional hematopoietic stem cells; by differentiating and maturing MKs, it is possible to treat thrombocytopenia-related diseases [104]. In the hematopoietic system, transcription factors, cytokines, adhesion factors, and chemokines regulate MKs. The most essential of these mediators for the generation and differentiation of MKs is thrombopoietin, the ligand for the c-MPL receptor. Thrombopoietin is a key regulator of MKs differentiation and can trigger multiple signal transduction pathways, such as JAK2/STAT3/STAT5, MEK-ERK-FOG-1/TAL-1, and PI3K/AKT ((Table 3) [35,105,106]. MEK-ERK-FOG-1/TAL-1 signaling has been linked to MKs' late differentiation and maturation [37]. A study by Kang et al. (2022) showed that HSN not only can promote thrombopoiesis by enhancing MKs differentiation and maturation of K562 and Meg01 cells through activation of MEK-ERK-FOG1/TAL1 signaling but can also lessen the decline of peripheral platelet concentration in mice. In addition, molecular docking simulations confirmed that HSN binds with high affinity to the signaling protein MAP kinase (MEK) [35]. Therefore, HSN can be a promising drug candidate for treating thrombocytopenia.

2.3.6. Metabolic Disease and Disorders

Antihypertensive Effect

Hypertension is a severe health issue and a leading cause of premature death worldwide, with up to one-quarter of men and one-fifth of women, or approximately one billion people, suffering from the condition (https://www.who.int/health-topics/hypertension#tab=tab_1, accessed on 11 June 2023). Globally, hypertension is caused by a combination of factors, including long-term calorie consumption over energy expenditures, chronic supraphysiological ingestion of dietary salt, excessive alcohol use, and psychological stressors. Stroke, myocardial infarction, heart failure, renal insufficiency/failure, retinopathy, dementia, peripheral vascular disease, and premature death are only some of the many severe clinical outcomes associated with elevated BP, especially systolic BP [107]. When it comes to controlling heart activities, intracellular Ca^{2+} is pivotal. Vasoconstriction and an increase in vascular volume via the renin-angiotensin-aldosterone pathway contribute to a rise in vascular resistance and blood pressure as a result of an increase in Ca^{2+} influx into vascular smooth muscle cells [108,109]. Therefore, the regulation of Ca^{2+} influx is an important interventional approach to maintaining hypertension along with other cardiovascular diseases [110].

Various studies have been reported about the antihypertensive effect of HSN, which is due to the inhibition abilities of L-type Ca^{2+} channels [27,32]. HSN (0.1 M to 10 M) altered the action potential waveform and lengthened the cycle of the rabbit SA node. Thus, for the first time, it was demonstrated that HSN exerts immediate inhibitory effects on the cardiac pacemaker. HSN reduced the levels of intracellular Ca^{2+} in isolated vascular smooth muscle cells by inhibiting Ca^{2+} influx via the L-type Ca^{2+} channel, producing a negative chronotropic effect [27]. In this context, Horie et al. (1992) reported that HSN (IC_{50} = 10.51.6 μM) inhibits the discharge of Ca^{2+} from the Ca^{2+} store and promotes Ca^{2+} uptake into the Ca^{2+} store in the smooth muscle of isolated rat aorta, resulting in a decrease of intracellular Ca^{2+} level by obstructing the voltage-dependent Ca^{2+} channel ((Table 3) [32]. A synthetic analog (referred to as compound 1 displayed in Figure 1) of HSN also displayed the antihypertensive effect by the same mechanism (blocking of Ca^{2+} influx through L-type Ca^{2+} channels) as HSN. The compound showed extraordinary activity on the contractile response of thoracic aorta rings from male SD rats in vitro reducing the systolic BP and heart rate [14]. Another study on the aortic arteries of rats by Yano et al. (1991) also demonstrated that HSN at a level of 10^{-6} to 3×10^{-5} M provided a vasodilation effect by inhibiting the trans-membrane Ca^{2+} influx via voltage-dependent Ca^{2+} channels [111]. The antihypertensive mechanism of HSN is depicted in Figure 4.

Anti-Diabetic Effect

Type 2 diabetes mellitus (T2DM) is a metabolic illness with a worldwide incidence that is characterized by high blood sugar levels and insulin resistance (IR) in target tissues and is typically associated with a high risk of multiple complications [112–114]. It is necessary to investigate innovative diabetes treatments. In this respect, reducing hyperglycemia and controlling insulin resistance (IR) are two important steps in the treatment of diabetes [26,115]. An in vivo investigation conducted by Hu et al. (2022) revealed that administration of HSN (5, 10, and 20 mg/kg, p.o.) to HFD-induced diabetic rodents reduces body weight gain, hyperglycemia, and IR. In addition, the study indicated that HSN reverses IR-related hepatic steatosis and enhances *left ventricular* (LV) mass in liver and cardiac examinations [26]. An in vitro study by the same researchers showed that HSN at the concentration of 0.325 μM develops hepatic IR by activating the PI3K/Akt/GSK3β insulin signaling pathway, increasing glycogen synthesis, glucose consumption, and suppressing gluconeogenesis in insulin-resistant HepG2 cells. Similarly, in HGHI H9c2 cells, HSN activated both PI3K/Akt/GSK3β and AMPK/ACC signaling pathways to promote glucose uptake. Furthermore, enhancement of glycolysis was noted in both H9c2 and HGHI HepG2 cells treated by HSN (Table 3). These findings demonstrated the efficacy of HSN in alleviating hepatic and cardiac IR in vivo and in vitro, providing insights into the development of HSN as an IR prevention and treatment for diet-induced diabetes [26]. The anti-diabetic mechanism of HSN is displayed in Figure 4.

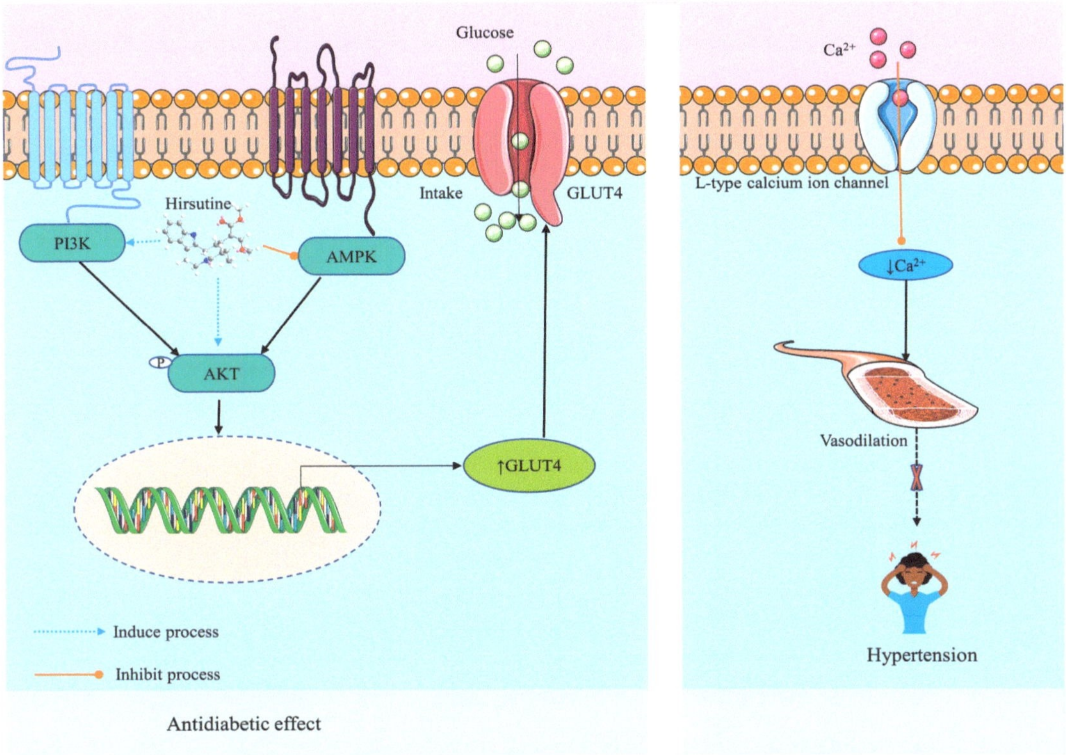

Figure 4. Possible antihypertensive and antidiabetic mechanisms of hirsutine. ↓: Decrease/inhibition/downregulation; ↑: Increase/upregulation/stimulation; AKT: Protein Kinase B; PI3K: Phosphoinositide 3-Kinase; GLUT4: Glucose Transporter 4.

Table 3. Different pharmacological activities of hirsutine and their mechanisms.

Related Disease/Effect	Test Medium/Cell Line/Test System	Compound/ Dose (R/A)/IC_{50}/ Concentration/ Course Interval	Possible Mechanism	Reference
Inflammation	Rats, LPS-induced preeclampsia	35, 70, and 140 mg/kg b.w.	↓ TNF-α, and ↓ IFN-γ, ↓ IL-6, ↓ IL-1β	[18]
	Rat brain microglia (LPS-induced inflammation, 10 μg/mL), in vitro	-	↓ NO, ↓ PGE2 and ↓ IL-1β, ↓ ROS, ↓ phosphorylation of the MAPK, ↓ Akt signaling proteins.	[29]
Thrombocytopenia	The Kunming thrombocytopenia mouse model was established by X-ray irradiation, in vivo	-	↑ MKD/MKM of K562 and Meg01 cells, ↑ platelet levels, ↑ MKD via activation of MEK-ERK-FOG1/TAL1 signaling	[35]
Neuronal death	Rat cerebellar granule cells (glutamate-induced neuronal death), in vitro	10^{-4}–3×10^{-4} M	↓ Ca^{2+} influx	[36]

Table 3. *Cont.*

Related Disease/Effect	Test Medium/Cell Line/Test System	Compound/ Dose (R/A)/IC_{50}/ Concentration/ Course Interval	Possible Mechanism	Reference
Myocardial ischemia-reperfusion	Sprague Dawley Rat Model	5, 10, and 20 mg/kg (p.o.)	↓ Myocardial infarct size, ↑ cardiac function, ↓ LDH, ↓ ROS, ↓ apoptosis, ↑ myocardial ATP, ↑ Mfn2 expression, ↓ p-Drp1, ↑ p-CaMKII, ↓ AKT/ASK-1/p38 MAPK pathway	[28]
Cardiomyocytes cell death	Neonatal rat cardiomyocytes treated with hypoxia	0.1, 1, and 10 μM	↓ Bax, ↓ Fas, ↓ caspase-3. ↑ Bcl-2.	[31]
Hypertension/ negative chronotropic/ antiarrhythmia	In male SD rats, in vitro, vasodilatation induced by the NO/cyclic GMP pathway	$IC_{50} = 1.129 \times 10^{-9} \pm 0.5025$	↓ Ca^{2+} influx, no effect on K^+ channel	[14]
	Male Japanese white rabbits	0.1 to 10 μM	↓ Influx of Ca^{2+} via voltage-dependent Ca^{2+} channels	[27]
	Male Wistar rats	30 μM	↓ Intracellular Ca^{2+} influx	[32]
	Aortic arteries of Wistar male rats, in vitro	10^{-6} to 3×10^{-5} M	↓ Ca^{2+} influx	[112]
	Male Sprague-Dawley rats	3–300 μM, y 60 mM KCl (IC_{50} = 20–30 μM)	↑ Ca^{2+}, ↑ KCl	[33]
Diabetes	Male C57BL/6 J mice, high-fat diet-induced diabetes, in vivo, $n = 9$	5, 10, and 20 mg/kg (p.o)	↓ Ca^{2+}, ↓ glucose tolerance, ↑ glucose uptake, ↑ glycolysis, ↑ phosphatidylinositol 3-kinase (PI3K)/Akt pathways	[26]
	HepG2 and H9c2 cells, high-glucose and high-insulin (HGHI) incubation, in vitro	0.325 μM	↑ p-Akt, ↑ GLUT4 activity, ↓ AMPK.	
Antiviral activity	Human lung carcinoma cells (A549) and baby hamster kidney cells (BHK-21), DENV-1 (02-20 strain), DENV-2 (16681 strain), DENV-3 (09-59 strain), and DENV-4 (09-48 strain)	10 μM	↓ Ca^{2+}, ↓ viral particle assembly, ↓ budding, or release step.	[19]
	Influenza A virus (subtype H3N2), in vitro	EC_{50} = 0.4–0.57 μg/mL	↓ Replication of the strains of Influenza A	[82]
Antitumor	Jurkat clone E6-1 cells, evaluated by CCK8 assay, in vitro	10, 25, and 50 μM for 48 h	↓ Cell proliferation, ↑ pro-apoptotic Bax, cleaved-caspase3, cleaved-caspase9 and Cyt C proteins, ↓ Bcl-2	[25]
Lung cancer	A549 xenograft mouse model, NCI-H1299, and LO2 cells	60–80 μM	↑ Apoptosis, ↑ ROCK1 and PTEN,↓ PI3K/Akt, ↑ caspase-3	[34]

Table 3. *Cont.*

Related Disease/Effect	Test Medium/Cell Line/Test System	Compound/ Dose (R/A)/IC_{50}/ Concentration/ Course Interval	Possible Mechanism	Reference
Breast cancer	MCF-10A, MCF-7 and MDA-MB-231 cells	160 μM/L HSN for 24, 48, and 72 h	↑ Apoptosis, ↓ Bax, ↓ Bcl-2, opening MPTP, releasing Cyt C from mitochondria, and activating caspase 9 and caspase 3.	[23]
	MCF-7	IC_{50} = 62.82 μM/L	Inhibits hypoxia, ↓ migration, and ↓ invasion, ↓ HIF-1α, ↓ snail, ↓ MMP-9, ↑ E-cadherin	[24]
	MCF-7 cell line, ataxia telangiectasia mutated (ATM) pathway	(50 μM) for 24 h	↑ Cell apoptosis by inducing DNA damage, ↑ ATM pathway, ↑ p53-independent DNA damage response, ↑ ROS, ↓ metastasis of breast cancer cells	[21]
	HER2-positive/p53-mutated MDA-MB-453 and BT474 cell lines, in vitro	6.25, 12.5, 25, and 50 μM	↑ Cytotoxicity, ↑ apoptosis, ↑ DNA damage response	[22]
	Mouse mammary carcinoma 4T1 cells, in vitro	(25 μM) for 24 h	↓ NF-κB, ↓ migration and invasion. ↓ MMP-2, MMP-9, and ↓ NF-κB signaling pathways	[20]

LPS: Lipopolysaccharide; ROS: Reactive Oxygen Species; NO: Nitric Oxide; PGE2: Prostaglandin E2; Akt: Ak strain transforming; IL-1β: Interleukin 1β; I_kB: Inhibitor of Nuclear Factor-κB; NF-κB: Nuclear factor kappa B; DENV: Dengue Virus; BHK: Baby Hamster Kidney Cells; HER2: Human Epidermal Growth Factor Receptor 2; BBB: Blood-Brain Barrier; TNF-α: Tumor Necrosis Factor α; IFN-γ: Interferon γ; ATM: Ataxia Telangiectasia Mutated; MMP-2: Matrix metalloproteinase-2; ROCK: Rho-Associated Protein Kinase; PTEN: Phosphatase and Tensin Homolog; Bax: Bcl-2-associated X protein; Bcl-2: B-cell lymphoma 2; PI3K: Phosphoinositide 3-Kinases; MKM: Microdosimetric Kinetic Model; MKD: Mevalonate Kinase Deficiency; NGF: Nerve Growth Factor; CCK: Cell Counting Kit; DRG: Dorsal Root Ganglion; P-gp: P-Glycoprotein; STZ: Streptozotocin; LDH: Lactate Dehydrogenase; Mfn2: Mitofusin2; p-Drp1: Dynamin-Related Protein 1 Phosphorylation; p-CaMKII: Protein Kinase II Phosphorylation; HepG2: The human hepatocellular carcinoma cell line; ADR: Adriamycin-Resistant; MDR: Multidrug Resistance; HFD: High-Fat Diet; HGHI: High-Glucose and High-Insulin; AMPK: AMP-Activated Protein Kinase; MPTP: Mitochondrial Permeability Transition Pore; MAPK: Mitogen-activated protein kinases.

3. Toxicological Profile

Toxicity testing is a crucial component in the identification of potential adverse reactions induced by chemical substances. For example, the manifestation of carcinogenicity, genotoxicity, immunotoxicity, and reproductive and developmental toxicity in humans is commonly found following prolonged exposure to chemicals [116]. The principal aim of toxicology investigations in the drug development process is to assess the safety profile of potential drug candidates [117]. This is achieved using animal models and validated methods [118]. Alkaloids are one of the greatest classes of secondary metabolites found in plants and are present in several economically significant plant families. Due to their toxicity, alkaloids can serve as defense compounds in plants, being effective against pathogens and predators [119]. The observed toxicity of alkaloids has been documented in both animals and humans [120]. The toxicological effects of alkaloids are contingent upon various factors, including the precise quantity administered, duration of exposure, and individual attributes such as sensitivity, location of action, and developmental stage [119].

Research findings showed that HSN exhibits cytotoxicity against different cell lines (NCI-H1299, MCF-10A, MCF-7, MDA-MB-231, and 4T1) at doses ranging from 6.25 to 80 μM in a number of in vitro studies [22,23,34]. However, HSN in vivo was not toxic for normal tissues in lower doses [121]. Acute toxicity analysis of HSN in a mouse model indicated that the LD_{50} of HSN is 110 mg/kg (i.p.) and 35 mg/kg (i.v.) (https://pubchem.ncbi.nlm.nih.gov/compound/3037884#section=Acute-Effects, accessed on 10 August 2023). On the other hand, Compound **1** exerted no toxicity on cell viability in the MTT assay at the normal dosage range. In an in vitro experiment, when VSMCs were treated with compound **1** (Figure 1) (1.6, 8, 40, 200, 1000 μM) for 12 h, only the 1000 μM group showed toxicity on the cells compared with the control group [14]. This indicates that compound **1** will be an ideal and safe medicine candidate.

4. Methodology

4.1. Literature Searching Strategy

The literature search in known databases such as PubMed, Springer Link, Scopus, Wiley Online, Nature, Web of Science, ScienceDirect, and Google Scholar was accomplished using the keyword "hirsutine", then paired with "anti-inflammatory activity", "antioxidant", "oxidative stress", "protective effect", "gastritis", "tumor", "gastroprotective activity", "renoprotective activity", "hepatoprotective activity", "cardioprotective effect", "hepatoprotective activity", "antimicrobial effect", "osteoprotective activity", "antiviral effect", "biological activities", "pharmacological effects", "pharmacological activities", "biological sources", "neurological effect", "pulmoprotective effect", "antidiabetic effect", cerebral ischemia/reperfusion", "chemical features", "pharmacokinetics", "in vivo studies" or "in vitro studies". In our search, no language limitations were imposed. The studies have been assessed in depth, with information regarding the sources, dose, concentration, test system, the proposed mechanism of pharmacological activities, and overall conclusion.

4.2. Inclusion and Exclusion Criteria

Inclusion criteria involved: (1) Studies performed in vitro, ex vivo, or in vivo and in silico with or without employing laboratory animals, including mice, rats, rabbits, and humans, and their derived tissues or cells, (2) Studies with pharmacological activities and botanical sources of HSN, (3) Studies with HSN or its derivatives or preparations, (4) Studies with HSN isolated from natural sources, (5) Studies showing the presence of HSN as a bioactive compound in preclinical studies of plant extracts, (6) Studies indicating that HSN or its derivatives exhibit synergistic effects when combined with other chemical compounds, (7) Studies with or without hypothesized mechanisms of action, (8) Studies carried out on the botanical sources of HSN, and (9) Studies related to the pharmacokinetics of HSN. On the other hand, exclusion criteria included (a) Duplicated data, titles, and/or abstracts which did not meet the inclusion criteria, (b) Papers written in languages other than English, (c) Case reports, letters, editorials, and commentaries, and (d) Studies without full text available.

4.3. In Silico ADME Prediction

In silico ADME of HSN is also predicted through the SwissADME online tool (http://www.swissadme.ch/index.php, accessed on 12 June 2023) to evaluate the PK properties of HSN [122].

4.4. Database Reports

A total of 759 scientific articles were collected as of 1 July 2023, from databases. After that, 96.18% of collected articles were eliminated due to duplication of reports, irrelevant information, lack of sufficient information, and automation systems deeming them unsuitable. Based on the inclusion criteria, we included information in this study from a total of 29 articles on HSN. Among the included articles, 75.86% reported the pharmacological activities of the compound, and 24.14% reported pharmacokinetics (PK), biological sources, and others. In the pharmacological investigation, 27.28% was done in vivo and 72.72% in in vitro test systems. The Preferred Reporting Items for Systematic Reviews and *Meta-Analyses* (PRISMA) analysis of the collected data of HSN is displayed in Figure 5.

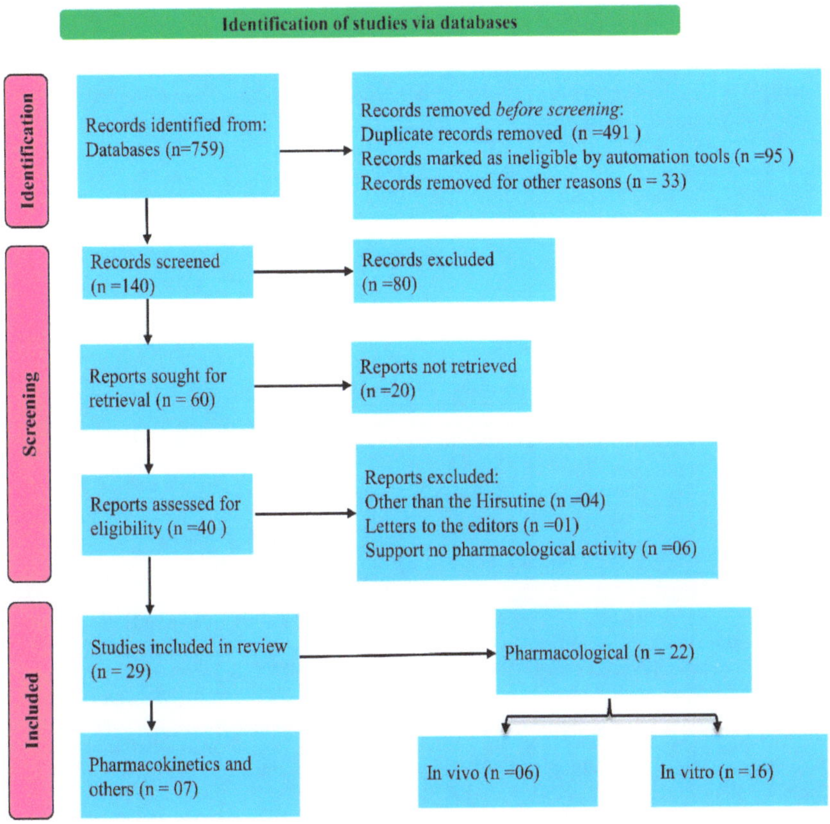

Figure 5. PRISMA analysis of the collected data of hirsutine.

5. Conclusions

At present time, many people depend on intrinsic foodstuffs in the fight against diseases. Thus, people are encouraged to consume natural food products obtained from fruits and vegetables and other food items to treat and manage cardiovascular disorders, cancer, and immune dysfunction, among others. These natural compounds are safer and less expensive than synthetic drugs. The purpose of the present study was to evaluate the therapeutic potential of HSN by investigating the available data from various preclinical studies and the underlying mechanisms behind these effects. Results demonstrated that HSN exhibits various pharmacological activities, including antioxidant, anti-inflammatory, antiviral (against various serotypes of dengue and influenza A viruses), antidiabetic, anticancer, cardioprotective, and stimulating thrombopoiesis via activation of MEK-ERK-FOG1/TAL1 signaling. In addition, it exerts potent activity in preventing various neurogenerative diseases, especially ischemia/reperfusion, neuroinflammation, and neurotoxicity. Furthermore, findings from this study showed that HSN is a potent blocker of Ca^{2+} influx via the L-type Ca^{2+} channels, resulting in reduced hypertension, neuronal cell death, and AMI. In terms of anticancer effects, HSN exhibits potency against breast cancer mainly through the induction of oxidative stress (when it is used in elevated concentration), inhibition of migration and invasion through cytotoxic, apoptotic, genotoxic, and antiproliferative effects. Pharmacokinetics studies demonstrated that HSN is well absorbed and distributed in various organs of the body, but the oral bioavailability of the compound was reported to be low due to the metabolism of the compound in the liver as well as the fact that the compound can cross the BBB. Therefore, the development of an alternative route is needed to increase the bioavailability and efficacy of the lead. However, extensive clinical studies are necessary to establish its efficacy for long-term use in treating human diseases.

Author Contributions: Conceptualization, M.S.B.; Methodology, J.F.; Software, R.C.; Validation, M.T.I. and M.S.B.; Formal analysis, M.S.B., P.W. and M.T.I.; data curation, M.A.R. and J.F.; Writing—original draft, M.S.B., M.H.B. and J.F.; Writing—review and editing, M.S.B., P.W., M.T.I. and M.S.M.; Supervision, M.T.I. and M.S.M.; Project administration, M.T.I. All authors have read and agreed to the published version of the manuscript.

Funding: This research received no external funding.

Institutional Review Board Statement: Not applicable.

Informed Consent Statement: Not applicable.

Data Availability Statement: The processed data are available from the corresponding author upon request.

Conflicts of Interest: The authors declare no conflict of interest.

References

1. Baker, D.D.; Chu, M.; Oza, U.; Rajgarhia, V. The value of natural products to future pharmaceutical discovery. *Nat. Prod. Rep.* **2007**, *24*, 1225–1244. [CrossRef]
2. Cragg, G.M.; Newman, D.J. Drug discovery and development from natural products: The way forward. In Proceedings of the 11th NAPRECA Symposium Book of Proceedings, Antananarivo, Madagascar, 9–12 August 2005; Volume 1, pp. 56–69.
3. Kinghorn, A.D.; Pan, L.; Fletcher, J.N.; Chai, H. The Relevance of Higher Plants in Lead Compound Discovery Programs. *J. Nat. Prod.* **2011**, *74*, 1539–1555. [CrossRef] [PubMed]
4. Atanasov, A.G.; Zotchev, S.B.; Dirsch, V.M.; International Natural Product Sciences Taskforce; Supuran, C.T. Natural products in drug discovery: Advances and opportunities. *Nat. Rev. Drug Discov.* **2021**, *20*, 200–216. [CrossRef] [PubMed]
5. Doak, B.C.; Over, B.; Giordanetto, F.; Kihlberg, J. Oral druggable space beyond the rule of 5: Insights from drugs and clinical candidates. *Chem. Biol.* **2014**, *21*, 1115–1142. [CrossRef] [PubMed]
6. Karimi, A.; Majlesi, M.; Rafieian-Kopaei, M. Herbal versus synthetic drugs; beliefs and facts. *J. Nephropharmacol.* **2015**, *4*, 27–30.
7. Bahmani, M.; Rafieian-Kopaei, M.; Hassanzadazar, H.; Saki, K.; Karamati, S.A.; Delfan, B. A review on most important herbal and synthetic antihelmintic drugs. *Asian Pac. J. Trop. Med.* **2014**, *7*, S29–S33. [CrossRef]
8. Nisar, B.; Sultan, A.; Rubab, S.L. Comparison of Medicinally Important Natural Products versus Synthetic Drugs—A Short Commentary. *Nat. Prod. Chem. Res.* **2018**, *6*, 308. [CrossRef]
9. Bhuia, M.S.; Siam, M.S.H.; Ahamed, M.R.; Roy, U.K.; Hossain, M.I.; Rokonuzzman, M.; Islam, T.; Sharafat, R.; Bappi, M.H.; Mia, M.N.; et al. Toxicity Analysis of Some Frequently Used Food Processing Chemicals Using Allium cepa Biomonitoring System. *Biology* **2023**, *12*, 637. [CrossRef]
10. Mohr, K.I. History of Antibiotics Research. In *How to Overcome the Antibiotic Crisis: Facts, Challenges, Technologies and Future Perspectives*, 1st ed.; Springer: Cham, Switzerland, 2016; pp. 237–272. [CrossRef]
11. Ligon, B.L. Penicillin: Its Discovery and Early Development. *Semin. Pediatr. Infect. Dis.* **2004**, *15*, 52–57. [CrossRef]
12. Bernardini, S.; Tiezzi, A.; Laghezza Masci, V.; Ovidi, E. Natural products for human health: An historical overview of the drug discovery approaches. *Nat. Prod. Res.* **2018**, *32*, 1926–1950. [CrossRef]
13. Xu, D.; Xu, Z. Indole Alkaloids with Potential Anticancer Activity. *Curr. Top. Med. Chem.* **2020**, *20*, 1938–1949. [CrossRef] [PubMed]
14. Zhu, K.; Yang, S.N.; Ma, F.F.; Gu, X.F.; Zhu, Y.C.; Zhu, Y.Z. The Novel Analogue of Hirsutine as an Anti-Hypertension and Vasodilatory Agent Both In Vitro and In Vivo. *PLoS ONE* **2015**, *10*, e0119477. [CrossRef]
15. Dey, P.; Kundu, A.; Kumar, A.; Gupta, M.; Lee, B.M.; Bhakta, T.; Dash, S.; Kim, H.S. Analysis of Alkaloids (Indole Alkaloids, Isoquinoline Alkaloids, Tropane Alkaloids). In *Recent Advances in Natural Products Analysis*; Elsevier Inc.: Amsterdam, The Netherlands, 2020; pp. 505–567. [CrossRef]
16. Nakazawa, T.; Banba, K.I.; Hata, K.; Nihei, Y.; Hoshikawa, A.; Ohsawa, K. Metabolites of Hirsuteine and Hirsutine, the Major Indole Alkaloids of Uncaria Rhynchophylla, in Rats. *Biol. Pharm. Bull.* **2006**, *29*, 1671–1677. [CrossRef]
17. Ozaki, Y. Nihon Yakurigaku Zasshi. *Folia Pharmacol. Jpn.* **1989**, *94*, 17–26. [CrossRef] [PubMed]
18. Wu, L.Z.; Xiao, X.M. Evaluation of the Effects of Uncaria Rhynchophylla Alkaloid Extract on LPS-Induced Preeclampsia Symptoms and Inflammation in a Pregnant Rat Model. *Braz. J. Med. Biol. Res.* **2019**, *52*, e8273. [CrossRef] [PubMed]

19. Hishiki, T.; Kato, F.; Tajima, S.; Toume, K.; Umezaki, M.; Takasaki, T.; Miura, T. Hirsutine, an Indole Alkaloid of Uncaria rhynchophylla, Inhibits Late Step in Dengue Virus Lifecycle. *Front. Microbiol.* **2017**, *8*, 1674. [CrossRef]
20. Lou, C.; Takahashi, K.; Irimura, T.; Saiki, I.; Hayakawa, Y. Identification of Hirsutine as an Anti-Metastatic Phytochemical by Targeting NF-κB Activation. *Int. J. Oncol.* **2014**, *45*, 2085–2091. [CrossRef]
21. Lou, C.; Yokoyama, S.; Abdelhamed, S.; Saiki, I.; Hayakawa, Y. Targeting the Ataxia Telangiectasia Mutated Pathway for Effective Therapy against Hirsutine-Resistant Breast Cancer Cells. *Oncol. Lett.* **2016**, *12*, 295–300. [CrossRef]
22. Lou, C.; Yokoyama, S.; Saiki, I.; Hayakawa, Y. Selective Anticancer Activity of Hirsutine against HER2 Positive Breast Cancer Cells by Inducing DNA Damage. *Oncol. Rep.* **2015**, *33*, 2072–2076. [CrossRef]
23. Huang, Q.W.; Zhai, N.N.; Huang, T.; Li, D.M. Hirsutine induces apoptosis of human breast cancer MDA-MB-231 cells through mitochondrial pathway. *Sheng Li Xue Bao* **2018**, *70*, 40–46.
24. Huang, W.Q.; Chen, S.K. Effect of hirsutine on hypoxia-induced migration and invasion abilities in human breast cancer MCF-7 cells. *Chin. J. Pathophysiol.* **2017**, *33*, 2009–2014.
25. Meng, J.; Su, R.; Wang, L.; Yuan, B.; Li, L. Inhibitory Effect and Mechanism of Action (MOA) of Hirsutine on the Proliferation of T-Cell Leukemia Jurkat Clone E6-1 Cells. *PeerJ* **2021**, *9*, e10692. [CrossRef] [PubMed]
26. Hu, W.; Li, M.; Sun, W.; Li, Q.; Xi, H.; Qiu, Y.; Wang, R.; Ding, Q.; Wang, Z.; Yu, Y.; et al. Hirsutine ameliorates hepatic and cardiac insulin resistance in high-fat diet-induced diabetic mice and in vitro models. *Pharmacol. Res.* **2022**, *177*, 105917. [CrossRef] [PubMed]
27. Masumiya, H.; Saitoh, T.; Tanaka, Y.; Horie, S.; Aimi, N.; Takayama, H.; Tanaka, H.; Shigenobu, K. Effects of Hirsutine and Dihydrocorynantheine on the Action Potentials of Sino-Atrial Node, Atrium and Ventricle. *Life Sci.* **1999**, *65*, 2333–2341. [CrossRef]
28. Jiang, W.; Zhang, Y.; Zhang, W.; Pan, X.; Liu, J.; Chen, Q.; Chen, J. Hirsutine ameliorates myocardial ischemia-reperfusion injury through improving mitochondrial function via CaMKII pathway. *Clin. Exp. Hypertens.* **2023**, *45*, 2192444. [CrossRef] [PubMed]
29. Jung, H.Y.; Nam, K.N.; Woo, B.C.; Kim, K.P.; Kim, S.O.; Lee, E.H. Hirsutine, an indole alkaloid of Uncaria rhynchophylla, inhibits inflammation-mediated neurotoxicity and microglial activation. *Mol. Med. Rep.* **2013**, *7*, 154–158. [CrossRef] [PubMed]
30. Suk, K.; Kim, S.Y.; Leem, K.; Kim, Y.O.; Park, S.Y.; Hur, J.; Baek, J.; Lee, K.J.; Zheng, H.Z.; Kim, H. Neuroprotection by Methanol Extract of Uncaria rhynchophylla against Global Cerebral Ischemia in Rats. *Life Sci.* **2002**, *70*, 2467–2480. [CrossRef]
31. Wu, L.X.; Gu, X.F.; Zhu, Y.C.; Zhu, Y.Z. Protective Effects of Novel Single Compound, Hirsutine on Hypoxic Neonatal Rat Cardiomyocytes. *Eur. J. Pharmacol.* **2011**, *650*, 290–297. [CrossRef]
32. Horie, S.; Yano, S.; Aimi, N.; Sakai, S.; Watanabe, K. Effects of hirsutine, an antihypertensive indole alkaloid from Uncaria rhynchophylla, on intracellular calcium in rat thoracic aorta. *Life Sci.* **1992**, *50*, 491–498. [CrossRef]
33. Zhang, W.B.; Chen, C.X.; Sim, S.M.; Kwan, C.Y. In Vitro Vasodilator Mechanisms of the Indole Alkaloids Rhynchophylline and Isorhynchophylline, Isolated from the Hook of Uncaria Rhynchophylla (Miquel). *Naunyn-Schmiedeberg's Arch. Pharmacol.* **2004**, *369*, 232–238. [CrossRef]
34. Zhang, R.; Li, G.; Zhang, Q.; Tang, Q.; Huang, J.; Hu, C.; Liu, Y.; Wang, Q.; Liu, W.; Gao, N.; et al. Hirsutine Induces mPTP-Dependent Apoptosis through ROCK1/PTEN/PI3K/GSK3β Pathway in Human Lung Cancer Cells. *Cell Death Dis.* **2018**, *9*, 598. [CrossRef] [PubMed]
35. Kang, Y.; Lin, J.; Wang, L.; Shen, X.; Li, J.; Wu, A.; Yue, L.; Wei, L.; Ye, Y.; Yang, J.; et al. Hirsutine, a novel megakaryopoiesis inducer, promotes thrombopoiesis via MEK/ERK/FOG1/TAL1 signaling. *Phytomedicine* **2022**, *102*, 154150. [CrossRef]
36. Shimada, Y.; Goto, H.; Itoh, T.; Sakakibara, I.; Kubo, M.; Sasaki, H.; Terasawa, K. Evaluation of the Protective Effects of Alkaloids Isolated from the Hooks and Stems of Uncaria sinensis on Glutamate-Induced Neuronal Death in Cultured Cerebellar Granule Cells from Rats. *J. Pharm. Pharmacol.* **1999**, *51*, 715–722. [CrossRef] [PubMed]
37. Devi, G.; Sudhakar, K.; Vasupradaa, A.P.; Sravya, V.; Manasa, V.; Yasaswini, E. Medicinal Plants in India and Their Antioxidant Potential—A Review. *Rev. Geintec-Gest. Inov. E Tecnol.* **2021**, *11*, 1397–1405.
38. Chirumbolo, S. Plant Phytochemicals as New Potential Drugs for Immune Disorders and Cancer Therapy: Really a Promising Path? *J. Sci. Food Agric.* **2012**, *92*, 1573–1577. [CrossRef]
39. Uramova, S.; Kubatka, P.; Dankova, Z.; Kapinova, A.; Zolakova, B.; Samec, M.; Zubor, P.; Zulli, A.; Valentova, V.; Kwon, T.K.; et al. Plant Natural Modulators in Breast Cancer Prevention: Status Quo and Future Perspectives Reinforced by Predictive, Preventive, and Personalized Medical Approach. *EPMA J.* **2018**, *9*, 403–419. [CrossRef]
40. Park, H.J. Chemistry and Pharmacological Action of Caffeoylquinic Acid Derivatives and Pharmaceutical Utilization of Chwinamul (Korean Mountainous Vegetable). *Arch. Pharmacal Res.* **2010**, *33*, 1703–1720. [CrossRef]
41. Li, Y.J.; Sun, A.S.; Yu, L.M.; Wu, Q. Inhibition of Rhynchophylline on Cultured Vascular Smooth Muscle Cells Proliferation Induced by Angiotensin II. *Chin. Pharm. J.* **2008**, *143*, 1621–1624.
42. Kim, T.J.; Lee, J.H.; Lee, J.J.; Yu, J.Y.; Hwang, B.Y.; Ye, S.K.; Shujuan, L.; Gao, L.; Pyo, M.Y.; Yun, Y.P. Corynoxeine Isolated from the Hook of Uncaria rhynchophylla Inhibits Rat Aortic Vascular Smooth Muscle Cell Proliferation through the Blocking of Extracellular Signal-Regulated Kinase 1/2 Phosphorylation. *Biol. Pharm. Bull.* **2008**, *31*, 2073–2078. [CrossRef]
43. Gao, J.; Inagaki, Y.; Liu, Y. Research Progress on Flavonoids Isolated from Traditional Chinese Medicine in Treatment of Alzheimer's Disease. *Intractable Rare Dis. Res.* **2013**, *2*, 3–10. [CrossRef]
44. Gai, Y.; Yang, N.; Chen, J. Inhibitory Activity of 8 Alkaloids on P-gp and Their Distribution in Chinese Uncaria Species. *Nat. Prod. Commun.* **2020**, *15*, 1934578X20973506. [CrossRef]
45. Laus, G.; Brössner, D.; Keplinger, K. Alkaloids of Peruvian Uncaria tomentosa. *Phytochemistry* **1997**, *45*, 855–860. [CrossRef]

46. Reichel, A.; Lienau, P. Pharmacokinetics in Drug Discovery: An Exposure-Centred Approach to Optimising and Predicting Drug Efficacy and Safety. *Handb. Exp. Pharmacol.* **2016**, *232*, 235–260. [CrossRef] [PubMed]
47. Bhuia, M.S.; Kamli, H.; Islam, T.; Sonia, F.A.; Kazi, M.A.; Siam, M.S.H.; Rahman, N.; Bappi, M.H.; Mia, M.N.; Hossen, M.M.; et al. Antiemetic Activity of Trans-Ferulic Acid Possibly through Muscarinic Receptors Interaction Pathway: In Vivo and In Silico Study. *Results Chem.* **2023**, *6*, 101014. [CrossRef]
48. Chow, S.C. Bioavailability and Bioequivalence in Drug Development. *Wiley Interdiscip. Rev. Comput. Stat.* **2014**, *6*, 304. [CrossRef]
49. Bhuia, M.S.; Islam, T.; Rokonuzzman, M.; Shamsh Prottay, A.A.; Akter, F.; Hossain, M.I.; Chowdhury, R.; Kazi, M.A.; Khalipha, A.B.R.; Coutinho, H.D.M.; et al. Modulatory Effects of Phytol on the Antiemetic Property of Domperidone, Possibly through the D2 Receptor Interaction Pathway: In Vivo and In Silico Studies. *3 Biotech* **2023**, *13*, 116. [CrossRef]
50. Jia, C.Y.; Li, J.Y.; Hao, G.F.; Yang, G.F. A drug-likeness toolbox facilitates ADMET study in drug discovery. *Drug Discov. Today* **2020**, *25*, 248–258. [CrossRef]
51. Lucas, A.J.; Sproston, J.L.; Barton, P.; Riley, R.J. Estimating Human ADME Properties, Pharmacokinetic Parameters and Likely Clinical Dose in Drug Discovery. *Expert Opin. Drug Discov.* **2019**, *14*, 1313–1327. [CrossRef]
52. Zhou, Q.; Ma, J.; Chen, L. Tissue Distribution of Hirsutine and Hirsuteine in Mice by Ultrahigh-Performance Liquid Chromatography-Mass Spectrometry. *J. Anal. Methods Chem.* **2020**, *2020*, 7204315. [CrossRef]
53. Zhang, Q.; Zhao, J.J.; Xu, J.; Feng, F.; Qu, W. Medicinal Uses, Phytochemistry and Pharmacology of the Genus Uncaria. *J. Ethnopharmacol.* **2015**, *173*, 48–80. [CrossRef]
54. Chen, L.; Deng, H.; Cui, H.; Fang, J.; Zuo, Z.; Deng, J.; Li, Y.; Wang, X.; Zhao, L. Inflammatory responses and inflammation-associated diseases in organs. *Oncotarget* **2017**, *9*, 7204–7218. [CrossRef]
55. Medzhitov, R. Inflammation 2010: New Adventures of an Old Flame. *Cell* **2010**, *140*, 771–776. [CrossRef]
56. Amor, S.; Puentes, F.; Baker, D.; van der Valk, P. Inflammation in neurodegenerative diseases. *Immunology* **2010**, *129*, 154–169. [CrossRef]
57. Hanisch, U.K.; Kettenmann, H. Microglia: Active sensor and versatile effector cells in the normal and pathologic brain. *Nat. Neurosci.* **2007**, *10*, 1387–1394. [CrossRef]
58. Walter, L.; Neumann, H. Role of Microglia in Neuronal Degeneration and Regeneration. *Semin. Immunopathol.* **2009**, *31*, 513–525. [CrossRef] [PubMed]
59. Bhuia, M.S.; Rahaman, M.M.; Islam, T.; Bappi, M.H.; Sikder, M.I.; Hossain, K.N.; Akter, F.; Al Shamsh Prottay, A.; Rokonuzzman, M.; Gürer, E.S.; et al. Neurobiological Effects of Gallic Acid: Current Perspectives. *Chin. Med.* **2023**, *18*, 27. [CrossRef]
60. Skrzypczak-Wiercioch, A.; Sałat, K. Lipopolysaccharide-Induced Model of Neuroinflammation: Mechanisms of Action, Research Application and Future Directions for Its Use. *Molecules* **2022**, *27*, 5481. [CrossRef] [PubMed]
61. Silva, J.F.; Olivon, V.C.; Mestriner, F.L.A.C.; Zanotto, C.Z.; Ferreira, R.G.; Ferreira, N.S.; Silva, C.A.A.; Luiz, J.P.M.; Alves, J.V.; Fazan, R.; et al. Acute Increase in O-GlcNAc Improves Survival in Mice with LPS-Induced Systemic Inflammatory Response Syndrome. *Front. Physiol.* **2020**, *10*, 1614. [CrossRef] [PubMed]
62. Danbolt, N.C. Glutamate uptake. *Prog. Neurobiol.* **2001**, *65*, 1–105. [CrossRef]
63. Nakanishi, S.; Nakajima, Y.; Masu, M.; Ueda, Y.; Nakahara, K.; Watanabe, D.; Yamaguchi, S.; Kawabata, S.; Okada, M. Glutamate Receptors: Brain Function and Signal Transduction. *Brain Res. Brain Res. Rev.* **1998**, *26*, 230–235. [CrossRef]
64. Murrough, J.W.; Abdallah, C.G.; Mathew, S.J. Targeting Glutamate Signalling in Depression: Progress and Prospects. *Nat. Rev. Drug Discov.* **2017**, *16*, 472–486. [CrossRef] [PubMed]
65. Madji Hounoum, B.; Blasco, H.; Coque, E.; Vourc'h, P.; Emond, P.; Corcia, P.; Andres, C.R.; Raoul, C.; Mavel, S. The Metabolic Disturbances of Motoneurons Exposed to Glutamate. *Mol. Neurobiol.* **2018**, *55*, 7669–7676. [CrossRef]
66. Wang, J.; Wang, F.; Mai, D.; Qu, S. Molecular Mechanisms of Glutamate Toxicity in Parkinson's Disease. *Front. Neurosci.* **2020**, *14*, 585584. [CrossRef] [PubMed]
67. Sribnick, E.A.; Del Re, A.M.; Ray, S.K.; Woodward, J.J.; Banik, N.L. Estrogen Attenuates Glutamate-Induced Cell Death by Inhibiting Ca^{2+} Influx through L-Type Voltage-Gated Ca^{2+} Channels. *Brain Res.* **2009**, *1276*, 159–170. [CrossRef] [PubMed]
68. Kristián, T.; Siesjö, B.K. Calcium in Ischemic Cell Death. *Stroke* **1998**, *29*, 705–718. [CrossRef]
69. Görlach, A.; Bertram, K.; Hudecova, S.; Krizanova, O. Calcium and ROS: A mutual interplay. *Redox Biol.* **2015**, *6*, 260–271. [CrossRef]
70. Douda, D.N.; Khan, M.A.; Grasemann, H.; Palaniyar, N. SK3 channel and mitochondrial ROS mediate NADPH oxidase-independent NETosis induced by calcium influx. *Proc. Natl. Acad. Sci. USA* **2015**, *112*, 2817–2822. [CrossRef]
71. Mozaffarian, D.; Benjamin, E.J.; Go, A.S.; Arnett, D.K.; Blaha, M.J.; Cushman, M.; de Ferranti, S.; Després, J.P.; Fullerton, H.J.; Howard, V.J.; et al. Heart Disease and Stroke Statistics—2015 Update: A Report from the American Heart Association. *Circulation* **2015**, *131*, e29–e322. [CrossRef]
72. Neri, M.; Riezzo, I.; Pascale, N.; Pomara, C.; Turillazzi, E. Ischemia/Reperfusion Injury following Acute Myocardial Infarction: A Critical Issue for Clinicians and Forensic Pathologists. *Mediators Inflamm.* **2017**, *2017*, 7018393. [CrossRef]
73. Frank, A.; Bonney, M.; Bonney, S.; Weitzel, L.; Koeppen, M.; Eckle, T. Myocardial ischemia reperfusion injury: From basic science to clinical bedside. *Semin. Cardiothorac. Vasc. Anesth.* **2012**, *16*, 123–132. [CrossRef]
74. Cadenas, S.; Aragones, J.; Landazuri, M.O. Mitochondrial reprogramming through cardiac oxygen sensors in ischaemic heart disease. *Cardiovasc. Res.* **2010**, *88*, 219–228. [CrossRef] [PubMed]

75. Dai, Y.; Zhang, H.; Zhang, J.; Yan, M. Isoquercetin attenuates oxidative stress and neuronal apoptosis after ischemia/reperfusion injury via Nrf2-mediated inhibition of the NOX4/ROS/NF-κB pathway. *Chem. Biol. Interact.* **2018**, *284*, 32–40. [CrossRef] [PubMed]
76. Naito, H.; Nojima, T.; Fujisaki, N.; Tsukahara, K.; Yamamoto, H.; Yamada, T.; Aokage, T.; Yumoto, T.; Osako, T.; Nakao, A. Therapeutic Strategies for Ischemia Reperfusion Injury in Emergency Medicine. *Acute Med. Surg.* **2020**, *7*, e501. [CrossRef] [PubMed]
77. Vasilakis, N.; Cardosa, J.; Hanley, K.A.; Holmes, E.C.; Weaver, S.C. Fever from the Forest: Prospects for the Continued Emergence of Sylvatic Dengue Virus and Its Impact on Public Health. *Nat. Rev. Microbiol.* **2011**, *9*, 532–541. [CrossRef] [PubMed]
78. Rajapakse, S.; Rodrigo, C.; Rajapakse, A. Treatment of Dengue Fever. *Infect. Drug Resist.* **2012**, *5*, 103–112. [CrossRef]
79. Messer, W.B.; Vitarana, U.T.; Sivananthan, K.; Elvtigala, J.; Preethimala, L.D.; Ramesh, R.; Withana, N.; Gubler, D.J.; De Silva, A.M. Epidemiology of Dengue in Sri Lanka before and after the Emergence of Epidemic Dengue Hemorrhagic Fever. *Am. J. Trop. Med. Hyg.* **2002**, *66*, 765–773. [CrossRef]
80. Gupta, E.; Dar, L.; Kapoor, G.; Broor, S. The Changing Epidemiology of Dengue in Delhi, India. *Virol. J.* **2006**, *3*, 92. [CrossRef]
81. Peteranderl, C.; Herold, S.; Schmoldt, C. Human Influenza Virus Infections. *Semin. Respir. Crit. Care Med.* **2016**, *37*, 487–500. [CrossRef]
82. Takayama, H.; Iimura, Y.; Kitajima, M.; Aimi, N.; Konno, K.; Inoue, H.; Fujiwara, M.; Mizuta, T.; Yokota, T.; Shigeta, S.; et al. Discovery of Anti-Influenza A Virus Activity of a Corynanthe-Type Indole Alkaloid, Hirsutine, In Vitro and the Structure-Activity Relationship of Natural and Synthetic Analogs. *Bioorg. Med. Chem. Lett.* **1997**, *7*, 3145–3148. [CrossRef]
83. Van Loenhout, J.; Peeters, M.; Bogaerts, A.; Smits, E.; Deben, C. Oxidative Stress-Inducing Anticancer Therapies: Taking a Closer Look at Their Immunomodulating Effects. *Antioxidants* **2020**, *9*, 1188. [CrossRef]
84. Islam, M.T.; Martorell, M.; González-Contreras, C.; Villagran, M.; Mardones, L.; Tynybekov, B.; Docea, A.O.; Abdull Razis, A.F.; Modu, B.; Calina, D.; et al. An Updated Overview of Anticancer Effects of Alternariol and Its Derivatives: Underlying Molecular Mechanisms. *Front. Pharmacol.* **2023**, *14*, 1099380. [CrossRef] [PubMed]
85. Yang, H.; Villani, R.M.; Wang, H.; Simpson, M.J.; Roberts, M.S.; Tang, M.; Liang, X. The Role of Cellular Reactive Oxygen Species in Cancer Chemotherapy. *J. Exp. Clin. Cancer Res.* **2018**, *37*, 266. [CrossRef] [PubMed]
86. Redza-Dutordoir, M.; Averill-Bates, D.A. Activation of Apoptosis Signalling Pathways by Reactive Oxygen Species. *Biochim. Biophys. Acta (BBA)-Mol. Cell Res.* **2016**, *1863*, 2977–2992. [CrossRef] [PubMed]
87. Bhuia, M.S.; Wilairatana, P.; Chowdhury, R.; Rakib, A.I.; Kamli, H.; Shaikh, A.; Coutinho, H.D.M.; Islam, M.T. Anticancer Potentials of the Lignan Magnolin: A Systematic Review. *Molecules* **2023**, *28*, 3671. [CrossRef] [PubMed]
88. Ozben, T. Oxidative Stress and Apoptosis: Impact on Cancer Therapy. *J. Pharm. Sci.* **2007**, *96*, 2181–2196. [CrossRef]
89. Chaudhry, G.E.; Akim, A.M.; Sung, Y.Y.; Muhammad, T.S.T. Cancer and Apoptosis. *Methods Mol. Biol.* **2022**, *2543*, 191–210. [CrossRef]
90. Carneiro, B.A.; El-Deiry, W.S. Targeting apoptosis in cancer therapy. *Nat. Rev. Clin. Oncol.* **2020**, *17*, 395–417. [CrossRef]
91. Tsujimoto, Y. Role of Bcl-2 Family Proteins in Apoptosis: Apoptosomes or Mitochondria? *Genes Cells* **1998**, *3*, 697–707. [CrossRef]
92. Wang, Q.; Zhang, L.; Yuan, X.; Ou, Y.; Zhu, X.; Cheng, Z.; Zhang, P.; Wu, X.; Meng, Y.; Zhang, L. The Relationship between the Bcl-2/Bax Proteins and the Mitochondria-Mediated Apoptosis Pathway in the Differentiation of Adipose-Derived Stromal Cells into Neurons. *PLoS ONE* **2016**, *11*, e0163327. [CrossRef]
93. Vara, J.Á.F.; Casado, E.; de Castro, J.; Cejas, P.; Belda-Iniesta, C.; González-Barón, M. PI3K/Akt Signalling Pathway and Cancer. *Cancer Treat. Rev.* **2004**, *30*, 193–204. [CrossRef]
94. Kim, E.H.; Song, H.S.; Yoo, S.H.; Yoon, M. Tumor Treating Fields Inhibit Glioblastoma Cell Migration, Invasion and Angiogenesis. *Oncotarget* **2016**, *7*, 65125–65136. [CrossRef] [PubMed]
95. Bhuia, M.S.; Chowdhury, R.; Sonia, F.A.; Kamli, H.; Shaikh, A.; El-Nashar, H.A.S.; El-Shazly, M.; Islam, M.T. Anticancer Potential of the Plant-Derived Saponin Gracillin: A Comprehensive Review of Mechanistic Approaches. *Chem. Biodivers.* **2023**, e202300847. [CrossRef] [PubMed]
96. Appert-Collin, A.; Hubert, P.; Crémel, G.; Bennasroune, A. Role of ErbB Receptors in Cancer Cell Migration and Invasion. *Front. Pharmacol.* **2015**, *6*, 283. [CrossRef] [PubMed]
97. Xia, L.; Tan, S.; Zhou, Y.; Lin, J.; Wang, H.; Oyang, L.; Tian, Y.; Liu, L.; Su, M.; Wang, H.; et al. Role of the NFκB-Signaling Pathway in Cancer. *OncoTargets Ther.* **2018**, *11*, 2063–2073. [CrossRef]
98. Xia, Y.; Shen, S.; Verma, I.M. NF-κB, an Active Player in Human Cancers. *Cancer Immunol. Res.* **2014**, *2*, 823–830. [CrossRef] [PubMed]
99. Cargnello, M.; Tcherkezian, J.; Roux, P.P. The expanding role of mTOR in cancer cell growth and proliferation. *Mutagenesis* **2015**, *30*, 169–176. [CrossRef]
100. Pui, C.H.; Relling, M.V. Can the Genotoxicity of Chemotherapy be Predicted? *Lancet* **2004**, *364*, 917–918. [CrossRef]
101. Fox, J.T.; Sakamuru, S.; Huang, R.; Teneva, N.; Simmons, S.O.; Xia, M.; Tice, R.R.; Austin, C.P.; Myung, K. High-throughput genotoxicity assay identifies antioxidants as inducers of DNA damage response and cell death. *Proc. Natl. Acad. Sci. USA* **2012**, *109*, 5423–5428. [CrossRef]
102. Gauer, R.L.; Braun, M.M. Thrombocytopenia. *Am. Fam. Physician* **2012**, *85*, 612–622.
103. Izak, M.; Bussel, J.B. Management of thrombocytopenia. *F1000Prime Rep.* **2014**, *6*, 45. [CrossRef]

104. Yang, F.; Lai, J.; Deng, J.; Du, J.; Du, X.; Zhang, X.; Wang, Y.; Huang, Q.; Xu, Q.; Yang, G.; et al. The Application of Ethnomedicine in Modulating Megakaryocyte Differentiation and Platelet Counts. *Int. J. Mol. Sci.* **2023**, *24*, 3168. [CrossRef] [PubMed]
105. Wang, L.; Zhang, T.; Liu, S.; Mo, Q.; Jiang, N.; Chen, Q.; Yang, J.; Han, Y.W.; Chen, J.P.; Huang, F.H.; et al. Discovery of a Novel Megakaryopoiesis Enhancer, Ingenol, Promoting Thrombopoiesis through PI3K-Akt Signaling Independent of Thrombopoietin. *Pharmacol. Res.* **2022**, *177*, 106096. [CrossRef] [PubMed]
106. Ng, A.P.; Kauppi, M.; Metcalf, D.; Hyland, C.D.; Josefsson, E.C.; Lebois, M.; Zhang, J.G.; Baldwin, T.M.; Di Rago, L.; Hilton, D.J.; et al. Mpl Expression on Megakaryocytes and Platelets Is Dispensable for Thrombopoiesis but Essential to Prevent Myeloproliferation. *Proc. Natl. Acad. Sci. USA* **2014**, *111*, 5884–5889. [CrossRef] [PubMed]
107. Flack, J.M.; Peters, R.; Shafi, T.; Alrefai, H.; Nasser, S.A.; Crook, E. Prevention of hypertension and its complications: Theoretical basis and guidelines for treatment. *J. Am. Soc. Nephrol.* **2003**, *14*, S92–S98. [CrossRef] [PubMed]
108. Simonetti, G.; Mohaupt, M. Kalzium und Blutdruck [Calcium and Blood Pressure]. *Ther. Umsch. Rev. Ther.* **2007**, *64*, 249–252. [CrossRef] [PubMed]
109. Villa-Etchegoyen, C.; Lombarte, M.; Matamoros, N.; Belizán, J.M.; Cormick, G. Mechanisms Involved in the Relationship between Low Calcium Intake and High Blood Pressure. *Nutrients* **2019**, *11*, 1112. [CrossRef]
110. Oparil, S.; Schmieder, R.E. New Approaches in the Treatment of Hypertension. *Circ. Res.* **2015**, *116*, 1074–1095. [CrossRef]
111. Yano, S.; Horiuchi, H.; Horie, S.; Aimi, N.; Sakai, S.I.; Watanabe, K. Ca^{2+} Channel Blocking Effects of Hirsutine, an Indole Alkaloid from Uncaria Genus, in the Isolated Rat Aorta. *Planta Med.* **1991**, *57*, 403–405. [CrossRef]
112. Kahn, S.E.; Hull, R.L.; Utzschneider, K.M. Mechanisms linking obesity to insulin resistance and type 2 diabetes. *Nature* **2006**, *444*, 840–846. [CrossRef]
113. Taylor, R. Insulin Resistance and Type 2 Diabetes. *Diabetes* **2012**, *61*, 778. [CrossRef]
114. Shahid, M.S.; Ibrahim, M.; Rahman, M.M.; Islam, T.; Bhuia, M.S.; Zaman, S.; Islam, M.T. Phytochemical Group Test and Pharmacological Investigations of Persicaria barbata (L.) H. Hara. *Phytopharmacol. Res. J.* **2023**, *2*, 1–15.
115. Wondmkun, Y.T. Obesity, Insulin Resistance, and Type 2 Diabetes: Associations and Therapeutic Implications. *Diabetes Metab. Syndr. Obes.* **2020**, *13*, 3611–3616. [CrossRef]
116. Guengerich, F.P. Mechanisms of Drug Toxicity and Relevance to Pharmaceutical Development. *Drug Metab. Pharmacokinet.* **2011**, *26*, 3–14. [CrossRef] [PubMed]
117. Parasuraman, S. Toxicological Screening. *J. Pharmacol. Pharmacother.* **2011**, *2*, 74–79. [CrossRef] [PubMed]
118. Dorato, M.A.; Buckley, L.A. Toxicology Testing in Drug Discovery and Development. *Curr. Protoc. Toxicol.* **2007**, *19*, 19.1. [CrossRef]
119. Matsuura, H.N.; Fett-Neto, A.G. Plant Alkaloids: Main Features, Toxicity, and Mechanisms of Action. *Plant Toxins* **2015**, *2*, 1–15.
120. Kuete, V. Health Effects of Alkaloids from African Medicinal Plants. In *Toxicological Survey of African Medicinal Plants*; Elsevier: Amsterdam, The Netherlands, 2014; pp. 611–633.
121. Szabo, I.; Zoratti, M.; Biasutto, L. Targeting Mitochondrial Ion Channels for Cancer Therapy. *Redox Biol.* **2021**, *42*, 101846. [CrossRef] [PubMed]
122. Daina, A.; Michielin, O.; Zoete, V. SwissADME: A Free Web Tool to Evaluate Pharmacokinetics, Drug-Likeness, and Medicinal Chemistry Friendliness of Small Molecules. *Sci. Rep.* **2017**, *7*, 42717. [CrossRef] [PubMed]

molecules

MDPI

Article

Effect of Ultrasonic Irradiation on the Physicochemical and Structural Properties of *Laminaria japonica* Polysaccharides and Their Performance in Biological Activities

Jinhui Wu [1], Huiying Wang [1], Yanfei Liu [1], Baojun Xu [2], Bin Du [1,*] and Yuedong Yang [2,*]

1 Hebei Key Laboratory of Natural Products Activity Components and Function, Hebei Normal University of Science and Technology, Qinhuangdao 066004, China

2 Food Science and Technology Program, Department of Life Sciences, BNU–HKBU United International College, Zhuhai 519087, China

* Correspondence: bindufood@aliyun.com (B.D.); kycyyd@126.com (Y.Y.); Tel.: +86-335-807-7682 (B.D.)

Abstract: Due to the large molecular weight and complex structure of *Laminaria japonica* polysaccharides (LJP), which limit their absorption and utilization by the body, methods to effectively degrade polysaccharides had received more and more attention. In the present research, hot water extraction coupled with three-phase partitioning (TPP) was developed to extract and isolate LJP. Ultrasonic *L. japonica* polysaccharides (ULJP) were obtained by ultrasonic degradation. In addition, their physicochemical characteristics and in vitro biological activities were investigated. Results indicated that ULJP had lower weight-average molecular weight (153 kDa) and looser surface morphology than the LJP. The primary structures of LJP and ULJP were basically unchanged, both contained α-hexopyranoses and were mainly connected by 1,4-glycosidic bonds. Compared with LJP, ULJP had stronger antioxidant activity, α-amylase inhibitory effect and anti-inflammatory effect on RAW264.7 macrophages. The scavenging rate of DPPH free radicals by ULJP is 35.85%. Therefore, ultrasonic degradation could effectively degrade LJP and significantly improve the biological activity of LJP, which provided a theoretical basis for the in-depth utilization and research and development of *L. japonica* in the fields of medicine and food.

Keywords: *Laminaria japonica* polysaccharide; ultrasound; physicochemical characteristics; anti-inflammatory; RAW264.7 macrophages; antioxidant activity

1. Introduction

Named for its scallop shape, *Laminaria japonica* is a widely consumed medicinal and edible brown algae. *L. japonica* is cultured in a large area in the coastal areas of China and the annual output is also increasing year by year, so the cost is very low [1,2]. *L. japonica* has high nutritional value, not only rich in vitamins and minerals, but also containing polysaccharides, crude protein, crude fiber and so on [3–5]. Therefore, *L. japonica* can promote food digestion, beauty, treatment of constipation and other functions [6]. *L. japonica* also has health care functions, such as regulating blood sugar and blood lipids, enhancing human resistance, lowering blood pressure, etc. [7]. In addition, there are still many effects of *L. japonica* that we need to study and develop and the market has great potential for development.

Laminaria japonica polysaccharides (LJP) is one of the main active components of *L. japonica* and is a sulfated polysaccharide that exhibits various biological activities and health benefits, such as anti-inflammatory, antioxidant, hypoglycemic, immunomodulatory and antibacterial effects [8–14]. In recent years, many bioactive polysaccharides have been extracted from dried *L. japonica* samples by physical, chemical and biological methods [15–17]. However, natural polysaccharides have high molecular weight, complex structure, high viscosity, difficult diffusion and are difficult to be absorbed by the body and natural high molecular polysaccharides cannot pass through various barriers of the body and even cell membranes. Changes in polysaccharide molecular weight can lead to changes in polysaccharide structure and biological activity [18]. Therefore, it is necessary to develop a treatment method that can not only maintain the biological activity of LJP, but also reduce its molecular weight, so as to improve the utilization rate of LJP. Moreover, Bhadja et al. [19] investigated six low-molecular-weight seaweed polysaccharides (SPSs) on oxalate-induced damaged human kidney proximal tubular epithelial cells (HK-2). The results showed that these SPSs within 0.1–100 μg/mL did not express cytotoxicity in HK-2 cells and each polysaccharide had a repair effect on oxalate-induced damaged HK-2 cells.

At present, there are various degradation methods for polysaccharides, including acid, oxidative, enzymatic, ultrasonic degradation [20–25] and ultraviolet/hydrogen peroxide method [26]. Among them, the ultrasonic degradation method has the advantages of low cost, high efficiency, simple operation and less side reactions, which is favorable for wide-scale popularization and use. It is speculated that the ultrasonic degradation method can degrade the polysaccharide well and maintain biological activity. However, to the best of our knowledge, there are few studies on reducing the molecular weight of LJP to improve bioactivity based on ultrasonic degradation of LJP.

Therefore, in this study, ULJP were obtained by ultrasonic degradation and the physicochemical properties, structural characteristics and in vitro biological activities of LJP and ULJP were studied and compared. This study provides a theoretical basis for the in-depth utilization and research and development of *L. japonica* in the fields of medicine and food.

2. Results and Discussion

2.1. Physicochemical Properties of LJP and ULJP

The chemical analysis results are shown in Table 1. The molecular weights of LJP and ULJP are one of the important parameters affecting their biological activities, such as anti-inflammatory and antioxidant properties. As shown in Table 1, the Mw, Mn and Mp of ULJP are significantly lower than those of LJP, which indicates that ultrasonic degradation of LJP is effective, The degraded ULJP is a low molecular weight polysaccharide; the larger the polydispersity index, the wider the molecular weight distribution, and the polydispersity index of ULJP is significantly higher than that of LJP, which indicates that ultrasonic degradation effectively degrades high molecular weight LJP into a variety of low molecular weight polysaccharides; too many low molecular weight polysaccharides may cause the polydispersity index of ULJP to be significantly higher than that of LJP [27]. Mao et al. [28] examined the impact of power ultrasound on the molecular properties of a high-molecular weight exopolysaccharide from the Cs-HK1 medicinal fungus. The results showed that the area percentage or the relative proportion of the higher Mw peak in the order of 10^7 decreased, while that of lower Mw of 10^6 decreased with the ultrasound treatment period. In another study, Chen et al. [26] produced the polysaccharides from *Laminaria japonica* by the treatment of ultraviolet/hydrogen peroxide (UV/H2O2) degradation. The molecular weight of LJP decreased from 315 kDa to 20 kDa. The reason for this finding is probably the different degradation method [26].

The monosaccharide composition of LJP and ULJP was analyzed by HPAEC and Table 1 shows the results of monosaccharide composition content of LJP and ULJP. Compared with LJP, the content of fucose, mannose and uronic acid in ULJP decreased significantly, the content of fucose decreased from 13.21% to 0.93%, the content of mannose decreased from 4.97% to 0.56% and the content of uronic acid decreased from 68.42% to 5.37%. Compared with LJP, the galactose content of ULJP increased significantly, from 9.37% to 72.56%, presumably because the fucoidan in LJP was degraded into galactose by ultrasonic. Compared with LJP, the glucose content of ULJP increased significantly, from 0.61% to 16.87%, presumably because the fucoidan in LJP was degraded into glucose ultrasonically [29]. In conclusion, sonication is a practical method to degrade polysaccharides with high efficiency.

Table 1. Chemical constituents, molecular weight parameters and monosaccharide compositions of LJP and ULJP.

Sample	LJP	ULJP
Total sugar (%)	70.10 ± 1.30	70.83 ± 1.21
Protein (%)	1.24% ± 0.02	1.15% ± 0.03
Uronic acid (%)	61.92% ± 1.01	54.95% ± 1.02
Sulfate group (%)	6.79% ± 0.23	7.78% ± 1.03
Mw (kDa)	219.678	153.895
Mn (kDa)	137.254	33.475
Mp (kDa)	181.761	7.84
Polydispersity (Mw/Mn)	1.601	4.597
Fuc	13.21	0.93
Rha	0.23	ND
Ara	0.79	0.77
Gal	9.37	72.56
Glc	0.61	16.87
Xyl	2.23	2.58
Man	4.97	0.56
Fru	ND	ND
Rib	0.17	0.35
Gal-UA	0.20	ND
Gul-UA	5.18	ND
Glc-UA	7.38	4.58
Man-UA	55.66	0.79

2.2. SEM Picture Analysis of LJP and ULJP

As can be seen from Figure 1a,b, the surface morphology of LJP is smooth and flaky; from Figure 1c,d, it can be seen that ULJP is loosely dispersed in fibrous piles, indicating that LJP is degraded into ULJP of various small molecules ultrasonically [30]. This ULJP looks like thin film with a smooth and glittering surface. Furthermore, the SEM scan showed that the exopolysaccharide was made of a homogeneous matrix.

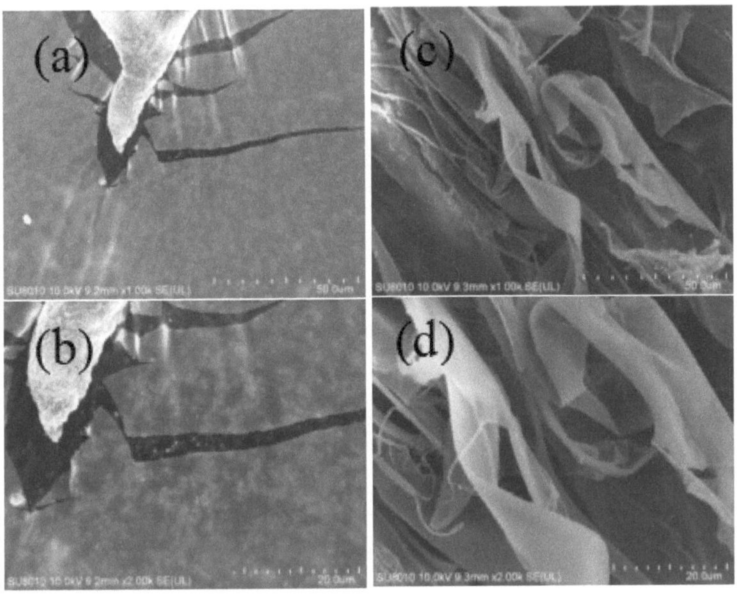

Figure 1. SEM picture of LJP and ULJP. (**a**) ×1.00 k and (**b**) ×2.00 k LJP; (**c**) ×1.00 k and (**d**) ×2.00 k ULJP.

2.3. FT-IR Analysis of LJP and ULJP

Infrared spectroscopy is often used to identify the characteristic functional groups of polysaccharides. Figure 2 shows the infrared spectral results of LJP and ULJP, respectively. The characteristic peaks of polysaccharides are O-H and C-H stretching vibration peaks at 3400 and 2900 cm^{-1} and the characteristic absorption peak of C=O asymmetric stretching vibration is 1600 cm^{-1}, indicating that LJP and ULJP contain uronic acid; 1417 cm^{-1} may be caused by the bending vibration of C-H; the characteristic absorption peak of the stretching vibration of the C-O-C bond of pyranoside is a narrow and strong absorption peak around 1000 cm^{-1}. The stretching vibration peak of O=S=O asymmetry in sulfate is a strong and broad absorption peak at 1250 cm^{-1}, and the characteristic absorption peak of C-O-S stretching vibration and the characteristic peak of sulfate bond are absorption peaks at 820 cm^{-1}, indicating that LJP and ULJP do contain sulfate groups. The functional groups of LJP and ULJP are almost the same, indicating that ultrasonic degradation hardly changes the functional groups of polysaccharides and does not change their primary structure [31]. The important information of the characteristic functional groups in the polysaccharide structure can be qualitatively obtained by the FT-IR map results, but the structural differences between LJP and ULJP cannot be quantitatively analyzed.

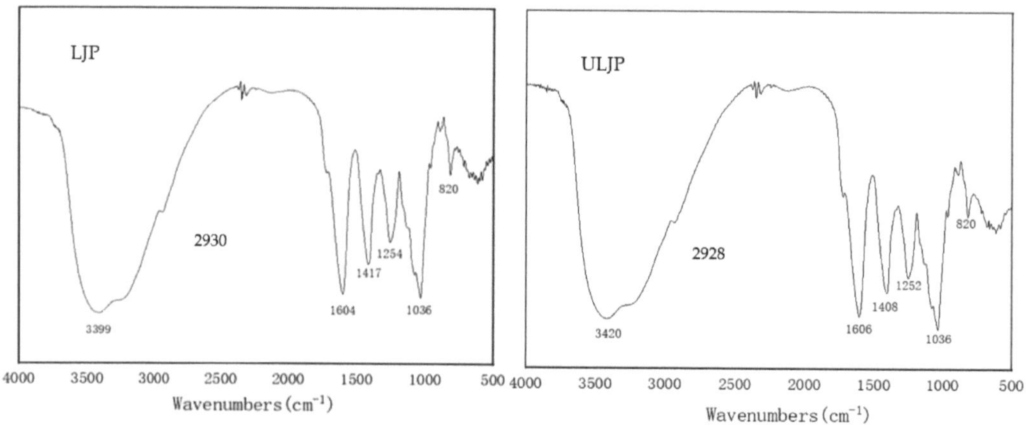

Figure 2. IR spectrum of LJP and ULJP.

2.4. NMR Spectra Analysis of LJP and ULJP

The shielding effect of the hydroxyl group results in the chemical shift of H on the sugar ring carbon, located at 3–4 ppm, which is difficult to resolve. However, the chemical shift of the anomeric hydrogen is in the lower field, between 4.5 and 5.5 ppm, and the monosaccharide species has the same number of proton signals in this region. More than 4.95 ppm is the chemical shift of α-pyranose H and less than 4.95 ppm is the chemical shift of β-pyranose H, so as to determine the sugar ring configuration. In order to further verify the structural information of LJP and ULJP, 1H-NMR spectrum analysis was carried out, as shown in Figure 3, the ^{1}H-NMR spectrum of LJP has a total of 7 peaks, 1H NMR (600 MHz, Deuterium Oxide) δ 5.33 (s, 0H), 4.71 (s, 2H), 3.90 (d, J = 7.9 Hz, 0H), 3.79~3.72 (m, 0H), 3.72~3.61 (m, 0H), 3.61~3.55 (m, 0H), 3.51 (dd, J = 9.4, 3.6 Hz, 0H); as shown in Figure 3, the ^{1}H-NMR spectrum of ULJP has 8 peaks, ^{1}H NMR (600 MHz, Deuterium Oxide) δ 5.33 (s, 3H), 5.28 (s, 1H), 3.95 (s, 2H), 3.88 (dd, J = 14.2, 5.5 Hz, 7H), 3.83~3.77 (m, 11H), 3.77~3.66 (m, 11H), 3.66~3.58 (m, 6H), 3.57 (s, 5H). The ^{1}H-NMR of LJP has an anomeric hydrogen signal, δ 5.33 (s, OH) and the chemical shift is greater than 4.95 ppm, indicating that it is α-pyranose, that is, LJP contains glucose and the chemical shift peak integral area is small, corresponding to the single sugar composition, and the glucose content is low, only 0.61%; the ^{1}H-NMR of LJP also has an anomeric hydrogen signal, δ 5.33 (s, 3H) and the chemical shift is greater than 4.95 ppm, indicating that it is α-pyranose, that is, ULJP contains glucose, the chemical shift peak integral area is larger, at 3H, corresponding to the monosaccharide composition and the glucose content is relatively high, at 16.87% [32].

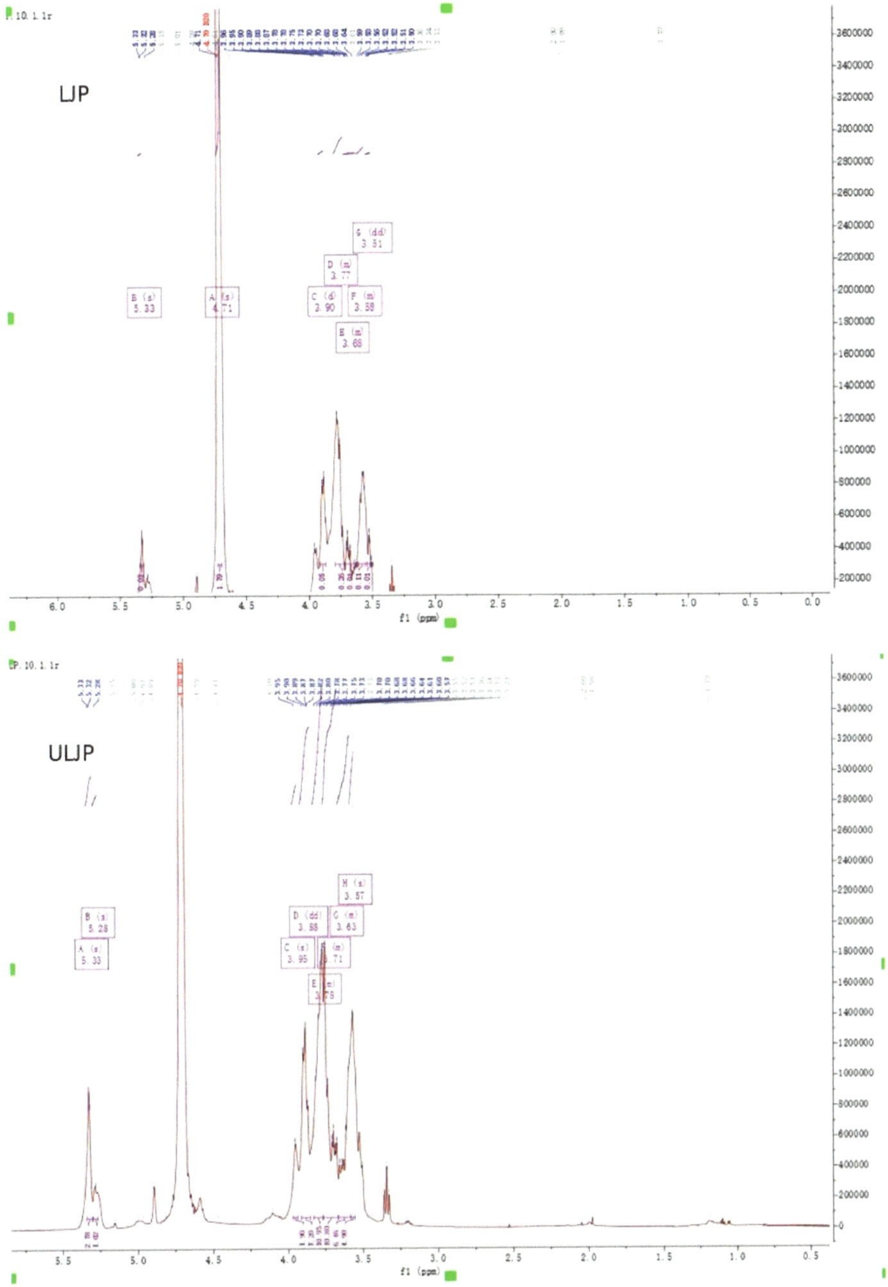

Figure 3. NMR spectra of LJP and ULJP (solvent = D_2O): ^1H NMR spectrum.

The chemical shift range of ^{13}C-NMR is wider than that of ^{1}H-NMR, up to 300 ppm, the anomeric carbon of polysaccharides is between 695 and 110 ppm and the composition of monosaccharides is the same as the number of signals in this range. The conformation of the substituent has a great influence on the chemical shift. For example, the substituent on the anomeric carbon is a vertical bond in the high field. According to this, the sugar ring configuration, D-glucose, α-δ is 97–101 ppm and β-δ is 103 ppm~106 ppm. In order to further verify the structural information of LJP and ULJP, ^{13}C-NMR spectrum analysis was carried out. As shown in Figure 4, the ^{13}C-NMR spectrum of LJP has a total of 8 peaks, ^{13}C NMR (151 MHz, Deuterium Oxide) δ 99.55, 78.86, 73.32, 72.78 (d, J = 24.8 Hz), 71.62 (d, J = 28.9 Hz), 71.17, 69.30, 60.46. As shown in Figure 4, there are 7 peaks in the ^{13}C-NMR spectrum of ULJP, ^{13}C NMR (151 MHz, Deuterium Oxide) δ 100.10, 78.86, 73.34, 72.87, 71.36 (d, J = 50.1 Hz), 69.33, 60.46. The 13C-NMR of LJP has an anomeric carbon signal, δ 99.55, indicating that it consists of a monosaccharide, which is α-glucose; the ^{13}C-NMR of ULJP also has an anomeric carbon signal, δ 100.10, indicating that it consists of a monosaccharide, α-glucose, which is the same as the ^{1}H-NMR result. LJP ^{13}C-NMR 60.46 ppm is the chemical shift of unsubstituted C6; 78.86, 73.32, 72.78, 71.62, 71.17, 69.30 is the signal of non-terminal carbon (C2, C3, C4 and C5), the peak of substituted proton signal being 78.86 ppm. This indicates that LJP is 1,4-glycosidic linkage. ULJP ^{13}C-NMR 60.46 ppm is the chemical shift of unsubstituted C6, 73.34, 72.87, 71.36, 69.33 are non-terminal carbon (C2, C3, C4 and C5) signals, and the peak of substituted proton signal 78.86 ppm indicates that LJP is 1,4-glycosidically linked [33].

Comprehensive analysis of ^{1}H-NMR and ^{13}C-NMR spectral results confirmed that LJP and ULJP contain α-hexo-pyranoses, which are mainly connected by 1,4-glycosidic bonds. Ultrasonic degradation can effectively degrade the fucoidan of macromolecular LJP into small molecule ULJP, such as glucose.

2.5. Antioxidant Activity of LJP and ULJP In Vitro

It can be seen from Figure 5a that the scavenging rate of DPPH free radicals by ULJP and LJP is proportional to the concentration. At 8 mg/mL, the scavenging rate of ULJP is 35.85% and the scavenging rate of LJP is 30.11%. Compared with LJP, the scavenging rate of ULJP is increased by 16.01%. Therefore, the antioxidant activity of ULJP is stronger than that of LJP. This result is similar to that of Du et al. [34]. The corncob polysaccharide was extracted by cellulase enzymatic hydrolysis and its DPPH free radical scavenging rate was significantly stronger than that of the corncob polysaccharide before enzymatic hydrolysis. The molecular weight is reduced and it is easier to contact and react with DPPH free radicals, thereby improving the antioxidant activity of degrading polysaccharides.

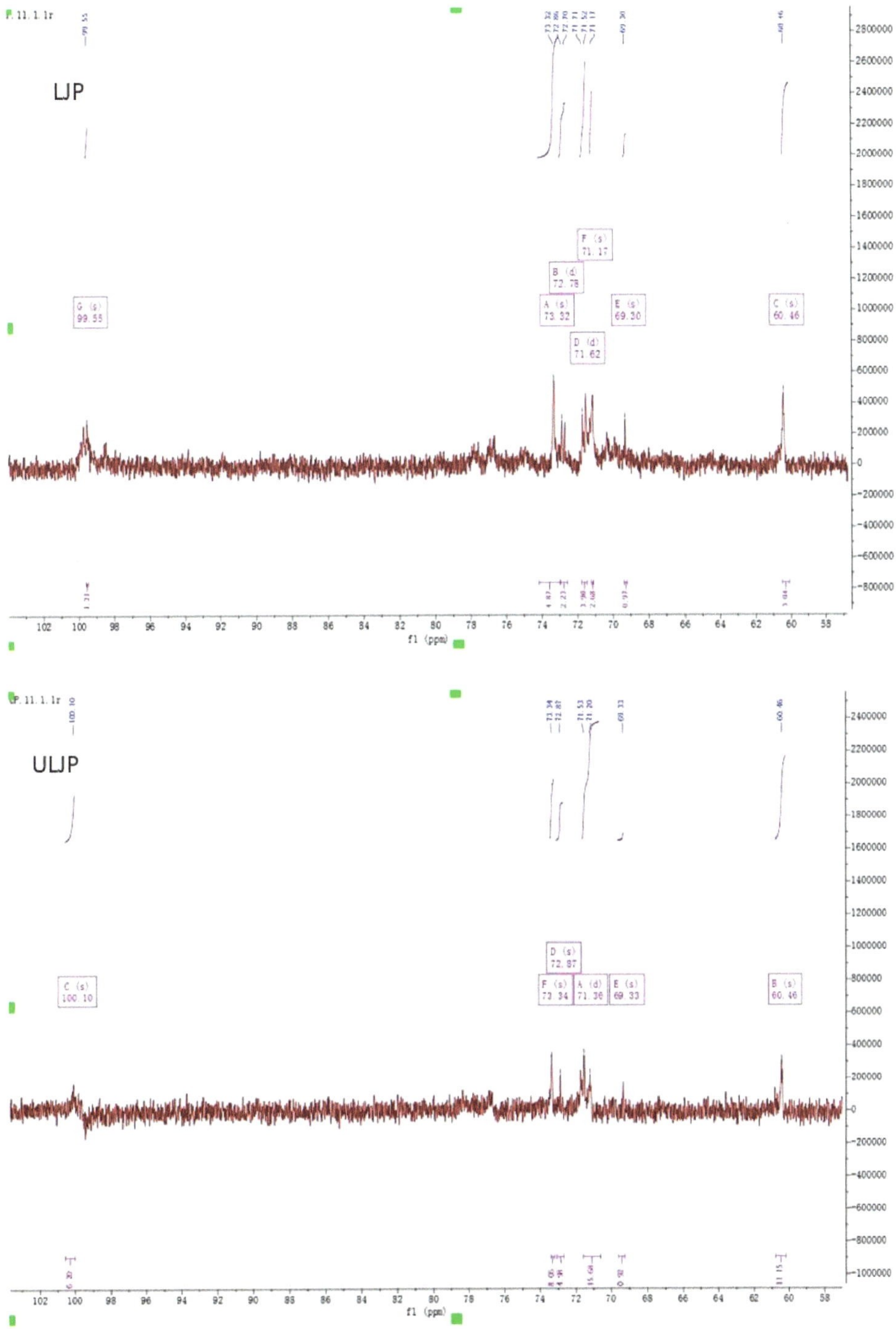

Figure 4. NMR spectra of LJP and ULJP (solvent = D_2O): ^{13}C NMR spectrum.

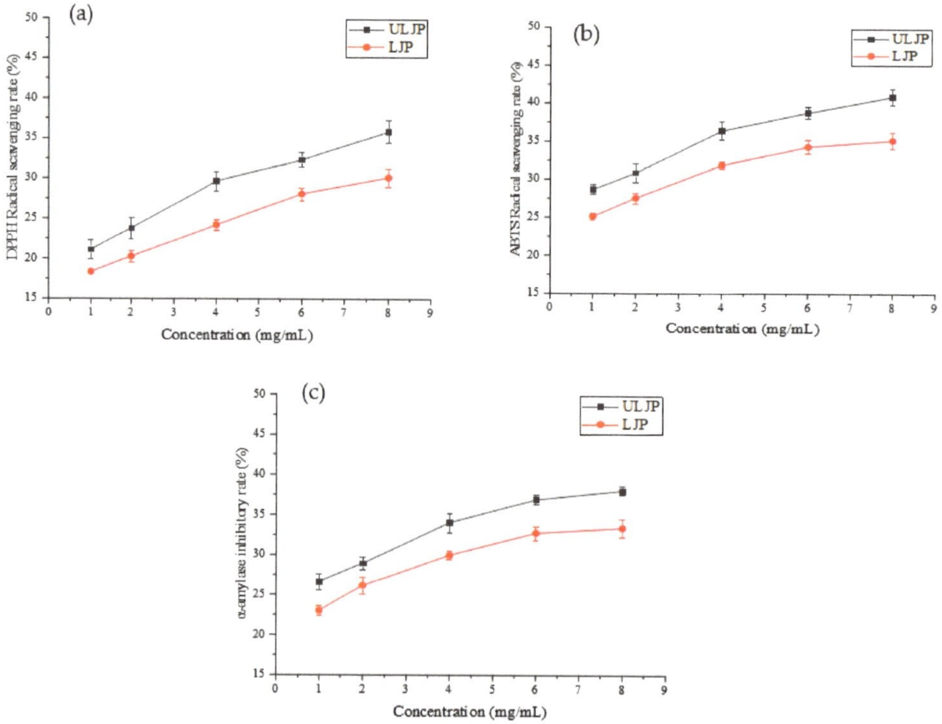

Figure 5. (**a**) Scavenging rate of DPPH free radical by ULJP and LJP. (**b**) Scavenging rate of ABTS free radical by ULJP and LJP. (**c**) Inhibitory rate of α-amylase by ULJP and LJP.

It can be seen from Figure 5b that the scavenging rate of ULJP and LJP to ABTS free radicals is proportional to the concentration. At 8 mg/mL, the scavenging rate of ULJP is 40.97% and the scavenging rate of LJP is 35.22%. Compared with LJP, the scavenging rate of ULJP is increased by 14.03%. Therefore, the antioxidant activity of ULJP is stronger than that of LJP. This is in agreement with many previously published results which showed that, using enzyme and a slight acid degradation method to degrade astragalus polysaccharide, its ABTS free radical scavenging rate is significantly stronger than that of polysaccharide before degradation and its small molecular weight (<3 kDa, 3–10 kDa and >10 kDa) degraded astragalus polysaccharides have good antioxidant capacity in vitro. This is because the degraded small molecular weight polysaccharides have loose structure and are easy to contact and react with ABTS free radicals. Therefore, astragalus polysaccharides degraded by enzymes and slight acid degradation methods also reflect better antioxidant activity [35].

2.6. Hypoglycemic Activity of LJP and ULJP In Vitro

As can be seen from Figure 5c, the inhibition rates of ULJP and LJP on α-amylase increased significantly with the increase of the corresponding concentration. Compared with LJP, the inhibition rate of ULJP on α-amylase increased by 12.27%. Therefore, the inhibition rate of ULJP on α-amylase is greater than that of LJP, that is, the hypoglycemic activity of ULJP is stronger than that of LJP. There were some other reports presenting similar phenomena to those of this work. The H_2O_2-Fe^{2+} methods were used to degrade the pine algae polysaccharide, the solubility of the pine algae polysaccharide after degradation was significantly improved, the particle size was reduced and the inhibitory activity to α-amylase was significant. The hypoglycemic activity of degraded polysaccharides was also significantly enhanced [36]. It has been reported that the hypoglycemic activity of polysaccharide is not only related to the Mw, but also to monosaccharide composition. Yang et al. [37] investigated that a low molecular weight heteropolysaccharide from the fruiting body of *Phellinus pini* exhibitedα-glucosidase inhibition and glucose consumption amelioration in an insulin-resistant HepG2 cell model. The Mw of this polysaccharide is 3.2 kDa.

2.7. Effect of LJP and ULJP on RAW264.7 Cell Viability

The effects of LJP and ULJP on the proliferation activity of RAW264.7 macrophages were detected by CCK-8 method and the results are shown in Figure 6a,b. The proliferation activity of RAW264.7 macrophages was negatively correlated with the concentrations of LJP and ULJP, respectively. When the LJP concentration was 500 μg/mL, the proliferation activity of RAW264.7 macrophages was 92.08%, lower than 100%. At this time, LJP showed an inhibitory effect on cell proliferation; when the concentration of ULJP was 1000 μg/mL, and the proliferation activity of RAW264.7 macrophages was 90.45%, which was lower than 100%. At this time, ULJP showed an inhibitory effect on cell proliferation. Therefore, according to the test results and the actual test, the cytotoxicity of the sample itself should be excluded. For the subsequent activity test of RAW264.7 macrophages, 100, 200, 300 μg/mL can be selected as the test concentrations. This is consistent with the study of Wu [38]; when the concentration of LJP is less than 500 μg/mL, it has no inhibitory effect on the proliferation of RAW264.7 macrophages.

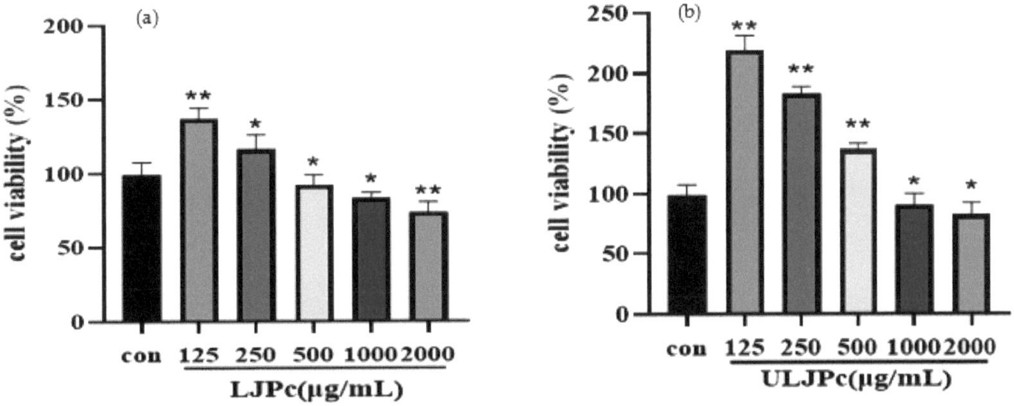

Figure 6. (**a**) Effects of LJP on the proliferation and viability of RAW264.7 macrophages; (**b**) Effects of ULJP on the proliferation and viability of RAW264.7 macrophages. (* $p < 0.05$; ** $p < 0.01$).

2.8. Effect of LJP and ULJP on the RAW264.7 Macrophages

RAW264.7 macrophages usually showed three morphologies: round, slender and obvious branched extensions. The optimal experimental conditions were used for LPS incubation time of 16 h, concentration of 1 μg/mL and incubation time of LJP and ULJP for 2 h to explore the effects of LJP and ULJP on the morphology of RAW264.7 macrophages. As shown in Figure 7, the morphology of RAW264.7 macrophages in the Con control group was round or oval, which was consistent with the normal shape. The morphology changed, the cell volume increased significantly and some antennae were visible around the cells. When LJP and ULJP were added to the DMEM medium that had been added with LPS, compared with the Con control group, the cell body of RAW264.7 cells was gradually enlarged, the antennae were branched and the number of antennae increased significantly. The size of the RAW264.7 macrophages in the LPS group was not as large as that in the LPS group. This morphology could increase the contact area with external substances, which was beneficial to the adherence ability of the RAW264.7 macrophages.

When the concentration of LJP and ULJP increased, the number of antennae of RAW264.7 macrophages decreased and the volume was also relatively reduced, which indicated that both LJP and ULJP had the ability to inhibit the antennae of RAW264.7 macrophages; and at the same concentration of ULJP, the inhibitory effect of the experimental group was better than that of the LJP experimental group.

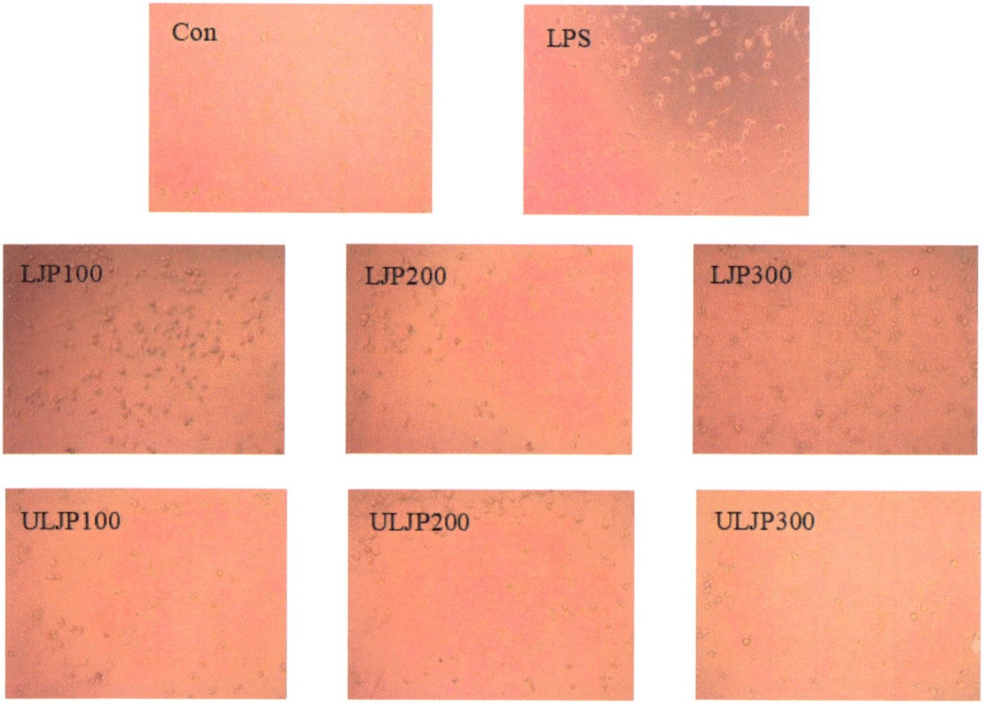

Figure 7. Effects of LJP and ULJP on the morphology of RAW264.7 macrophages.

2.9. *Effect of LJP and ULJP on RAW264.7 Phagocytic Activity*

Activated macrophages help initiate specific defense mechanisms and play a key role in the immune system and phagocytosis is a key indicator of macrophage activity. The phagocytic activity of macrophages was determined by the neutral red uptake assay. From Figure 8a, the phagocytosis of RAW264.7 macrophages by LJP and ULJP decreased in a concentration-dependent manner and, at the same concentration, the inhibitory effect of LJP on the phagocytic activity of RAW264.7 macrophages was lower than that of ULJP. The results showed that both LJP and ULJP could inhibit the phagocytosis of RAW264.7 macrophages and at the same concentration, ULJP had a better inhibitory effect on the phagocytosis of RAW264.7 macrophages than LJP, which was consistent with the morphological characteristics of the cells.

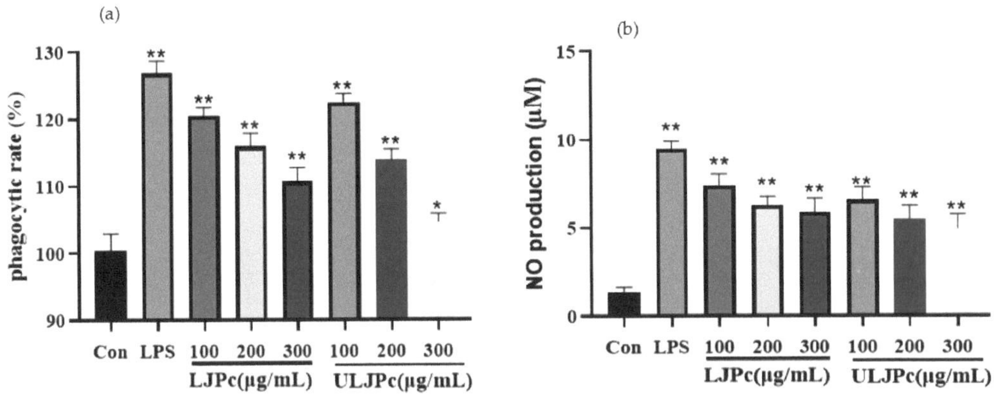

Figure 8. Effects of LJP and ULJP on the morphology of RAW264.7 macrophages. (* $p < 0.05$; ** $p < 0.01$). (**a**) phagocytic rate (%) (**b**) NO production μM.

2.10. Effect of LJP and ULJP on RAW264.7 NO Production

When RAW264.7 macrophages are stimulated, they can alternately be activated to pro-inflammatory subtypes (M1 macrophages) or anti-inflammatory and tissue repair subtypes (M2 macrophages). M1 and M2 cells have opposite effects, M1 macrophages phages have pro-inflammatory effects and release NO and other pro-inflammatory factors, while M2 macrophages have immunosuppressive effects. NO is an indispensable bioactive molecule in the body. As shown in Figure 8b, compared with the LPS group, the NO release of LJP and ULJP was always lower than that of the LPS group, indicating that LJP and ULJP can reduce the release of NO from macrophages. It has an anti-inflammatory effect on macrophages. Adding different concentrations of LJP and ULJP will reduce the release of NO from RAW264.7 macrophages. When the concentration of LJP is 300 μg/mL, the release of NO reaches 5.87 μmol/L ($p < 0.01$), which was 0.74 times that of the control group. When the ULJP concentration was 300 μg/mL, the release of NO reached 4.97 μmol/L ($p < 0.01$), indicating that at the same concentration of ULJP, the anti-inflammatory activity is better than that of LJP.

In this work, LJP and ULJP can reduce the release of NO from cells. LJP and ULJP stimulate RAW macrophages to differentiate into M2 macrophages, thereby reducing NO release. Therefore, it is speculated that LJP can improve non-specific immune function of ULJP. The reduction of NO release in RAW macrophages was significantly lower than that in LJP, indicating that ultrasonic degradation treatment can improve the anti-inflammatory activity of ULJP, which may be due to the smaller molecular weight and the active groups contained in ULJP, such as sulfate groups, are more exposed. It is easier to combine with cells, so it is easier to show strong anti-inflammatory activity.

2.11. Effects of LJP and ULJP on mRNA Expression Levels of Inflammatory Factors in LPS-Induced RAW264.7 Macrophages

The effects of LJP and ULJP on the expression levels of LPS-induced RAW264.7 macrophages are shown in Figures 9 and 10. The results showed that the mRNA expression levels of IL-4, iNOS, IL-1β, IL-6, IL-10, TNF-α and IFN-γ in macrophages were significantly increased with the addition of LPS. The mRNA expression of each inflammatory factor showed a downward trend as a whole. Under the same inflammatory factor and concentration, the inhibitory effect of ULJP was significantly better than that of LJP and the difference was significant compared with the LPS group ($p < 0.01$). That is, ULJP treated by ultrasonic degradation has more excellent anti-inflammatory activity. This result is the same as that of Qiu et al. [39]. Treatment of RAW264.7 macrophages with total flavonoids of gang pine promoted cell proliferation ($p < 0.05$) and inhibited the contents of IL-1β, IL-8 and TGF-β (p

< 0.05), up-regulated the relative expression of IκBα protein ($p < 0.05$), inhibited the mRNA gene expression (mRNA) of NF-κB p65, and the mRNA expression level of NF-κB p65 was concentration-dependent with total flavonoids It is speculated that the inhibition of NF-κB signaling pathway by total flavonoids of pineapple may be one of the mechanisms of its anti-inflammatory activity.

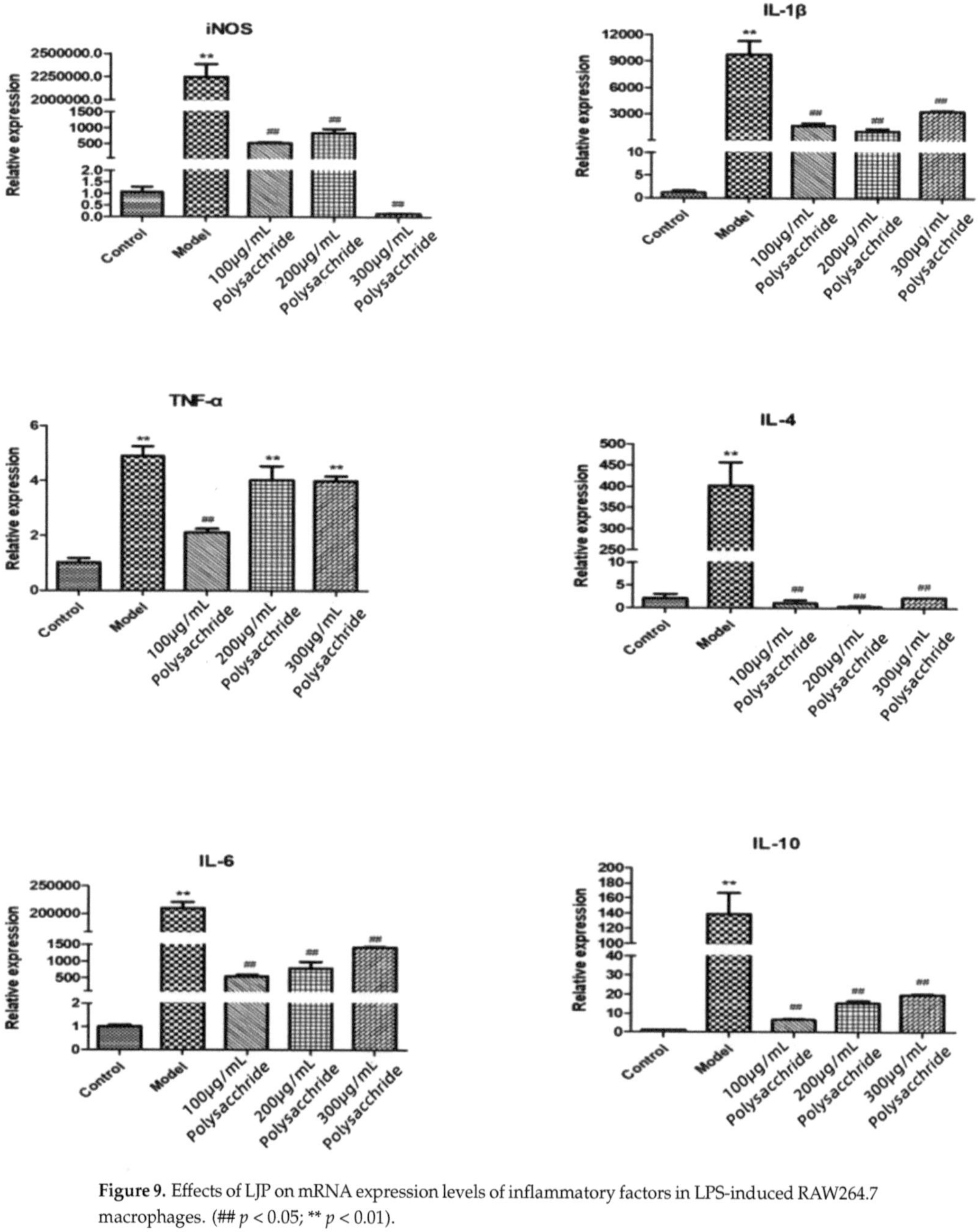

Figure 9. Effects of LJP on mRNA expression levels of inflammatory factors in LPS-induced RAW264.7 macrophages. (## $p < 0.05$; ** $p < 0.01$).

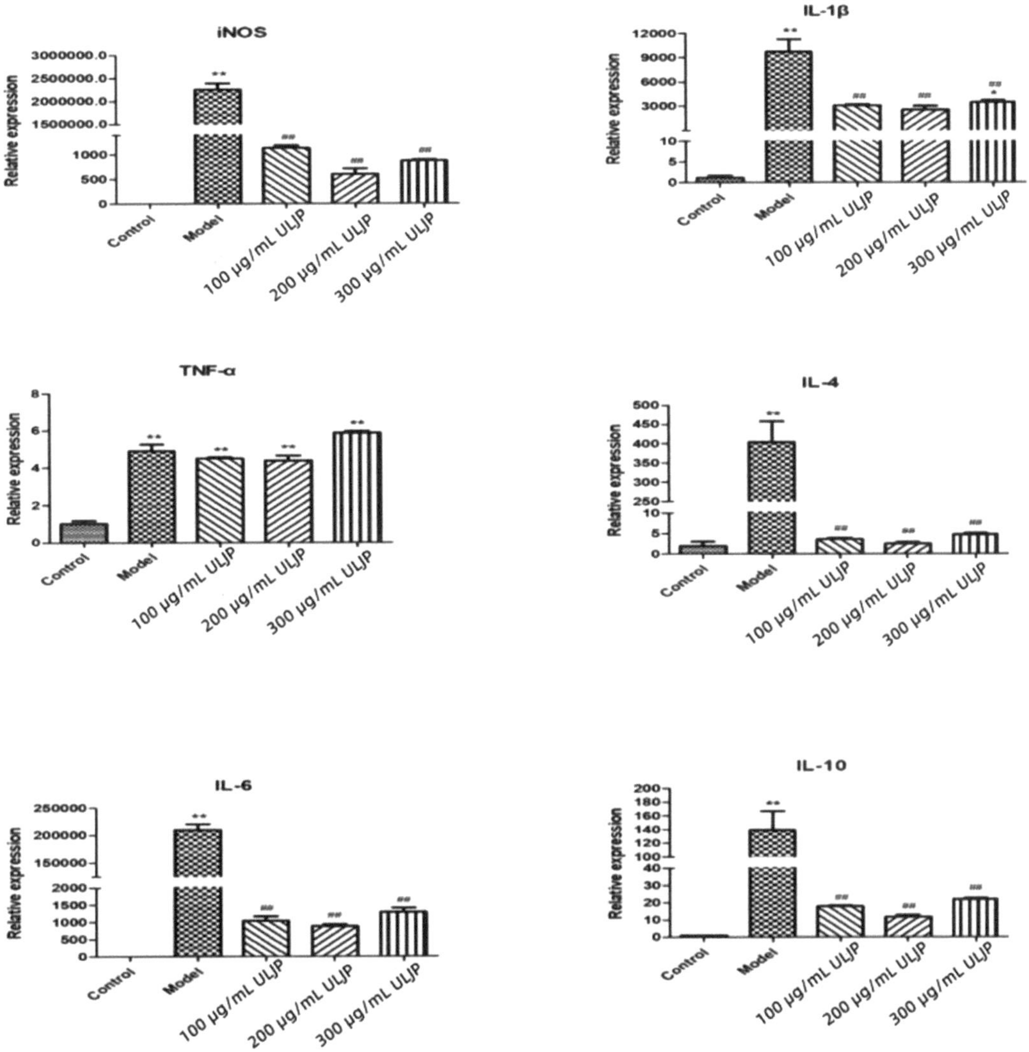

Figure 10. Effects of ULJP on mRNA expression levels of inflammatory factors in LPS-induced RAW264.7 macrophages. (## $p < 0.05$; ** $p < 0.01$; * $p < 0.05$).

3. Materials and Methods

3.1. Materials and Chemicals

Fresh raw *Laminaria japonica* (Place of origin: Qinhuangdao, China), purchased from a local supermarket (Qinhuangdao, Hebei, China) in April 2019, were free from pests and mechanical damage. All chemicals and reagents were of laboratory grade and provided by local authorized suppliers.

3.2. Extraction and Isolation of LJP and ULJP

Hot water extraction coupled with three-phase partitioning (TPP) was developed to extract and isolate LJP from *L. japonica*. According to Wang et al. [40], with some modifications, in short, the optimal extraction conditions for LJP are extraction temperature 70 °C, extraction time 2 h, water-material ratio 50 mL/g, extraction times 2. According to Yan et al. [41], with some modifications, TPP was used to isolate LJP. Briefly, 20 g of *L. Japonica* powder, add 800 mL of distilled water were stirred for 2 h at 70 °C and then subjected to centrifugation (7000× *g*, 5 min) for supernatant collection. Then, 20% (*w*/*w*) $(NH_4)_2SO_4$ was added to the resulting supernatant and stirred until completely dissolved. Subsequently, the mixture was mixed with 1.5 vol oft-butanol (*v*/*v*) and continuously stirred at 35 °C for 30 min. After that, the resultant mixture was centrifuged at 4000× *g* for 15 min, the collected lower aqueous phase was subjected to dialysis (MWCO: 3500 Da) in distilled water for 48 h and then concentrated to half.

3.3. *Ultrasonic Degradation of LJP to Prepare ULJP*

Referring to Wu et al. [42], with a slight modification, the LJP after dialysis and concentration was degraded by a Scientz-IID ultrasonic cell disrupter (Ningbo Scientz Biotechnology Company, Ltd., Ningbo, Zhejiang, China) for 5 h. The frequency is 25 kHz and maximum output power is 950 W. Each ultrasonic wave was 1 h, stopped for 10 min when water was replaced and, after ultrasonic degradation, freeze-dried to obtain ultrasonic *L. japonica* polysaccharide (ULJP).

3.4. *Determination of the Physicochemical Characteristics of LJP and ULJP*

3.4.1. General Analysis

The total sugar, protein, uronic acid and sulfate group contents of polysaccharide were separately determined with the phenol-sulfuric acid, sulfuric acid-carbazole, Bradford and Barium sulfate-turbidimetry methods [43–46].

3.4.2. Determination of Molecular Weight and Polydispersity

According to the method of Chen et al. [47] to determine the molecular weight by gel permeation chromatography (GPC), with a slight modification, the molecular weight of LJP and ULJP was determined by GPC. Polysaccharide was characterized for molecular weight using Agilent 1100 series HPLC system (Agilent Technologies Palo AHO, CA, USA) equipped with a TOSOH TSK-GEL G3000 SW XL column (7.8 mm × 30 cm, 10 μm; TOSOH Corp., Tokyo, Japan) and an refractive index detector. A sample of 20 μL was injected in the system by maintaining a flow rate of 0.5 mL min^{-1} and column temperature of 35 °C. Separation was carried out using 0.05 mol L^{-1} phosphate buffer (pH 6.7) containing 0.05% NaN_3 as mobile phase. The average molecular weight was calculated by the calibration curve obtained using various standard dextrans (738, 5800, 11,220, 21,370, 41,800, 110,000, 118,600, 318,000 and 813,500).

3.4.3. Monosaccharide Composition Analysis

Refer to Li et al. [48] for the determination of monosaccharide composition in polysaccharides by acid hydrolysis and high performance anion exchange chromatography (HPAEC), with minor modifications. LJP (5 ± 0.05 mg) was put into a tube and 0.5 mL of 12 M H_2SO_4 was added in an ice bath with magnetic stirring for 0.5 h to dissolve it completely. After 2 mL of ultrapure water was added, tubes were placed in an oil bath at 105 °C for 4 h. After heating, polysaccharide solution was fixed the volume with a 250 mL volume volumetric flask, 1 mL of sample was taken out and was filtered through 0.22 μm membrane for HPAEC analysis. Based on the use of a pulsed ampere-metric detector (PAD; Dionex ICS 5000 system) on a CarboPac PA-20 anion exchange column (3 × 150 mm; Dionex) on the monosaccharide composition content of LJP and ULJP were analyzed by high performance anion exchange chromatography (HPAEC). Flow rate, 0.5 mL/min; injection volume, 5 μL; solvent system, B: (0.1 M NaOH, 0.2 M NaAc); gradient program, 95:5 *v*/*v* at 0 min, 80:20 *v*/*v* at 30 min, 60:40 *v*/*v* at 30.1 min, 60:40 *v*/*v* at 45 min, 95:5 *v*/*v* at 45.1 min and 95:5 *v*/*v* at 60 min.

3.4.4. Scanning Electron Microscopy Analysis

The morphology of LJP and ULJP was observed by scanning electron microscopy [49]. Samples were fixed on aluminum stub and gold sputtered and examined through KYKY-2800 SEM (KYKY Technology Co., Ltd., Beijing, China) by maintaining an accelerated voltage of 10 kV

3.4.5. Fourier Transform Infrared Analysis

Fourier transform infrared (FT-IR) spectra of freeze-dried LJP and ULJP were captured on a Nexus 670 FT-IR spectrometer in the wavenumber range of 4000–500 cm^{-1} with KBr pellets referenced against air, according to the method by Kibar et al. [50].

3.4.6. NMR Spectra Analysis

According to the method of Takis et al. [51] to measure the sample by nuclear magnetic resonance spectrometer, with some modifications, LJP and ULJP were dissolved in D_2O respectively and Bruker AVANCE III HD 600 MHz NMR was used.

3.5. Antioxidant Activity Evaluation of LJP and ULJP In Vitro

DPPH radical scavenging capacity of samples was evaluated according to the method of Insang et al. [52,53] with slight modifications. Two mL of sample extract and 2.0 mL of 0.1 mmol/L DPPH in ethanol were briefly mixed in a test tube. Before standing in darkness for 30 min at 37 °C, the mixture solution was mixed for 1 min with a vortex. Vitamin C was used as the positive control and ethanol was used as blank. The absorbance was measured at 517 nm with a visible light spectrophotometer. The experiment was carried out in triplicate. Radical scavenging activity was calculated using the following formula:

$$\text{DPPH inhibition} = [1 - (A_1 - A_2)/A_0] \times 100\% \quad (1)$$

where A_0 = absorbance of 2 mL DPPH-ethanol + 2 mL ethanol; A_1 = absorbance of 2 mL DPPH-ethanol + 2 mL sample; A_2 = absorbance of 2 mL ethanol + 2 mL sample.

ABTS radical cation (ABTS+•) was produced by reacting ABTS solution (7.4 mM) with 2.6 mM $K_2S_2O_8$ and allowing the mixture to stand in the dark at room temperature for 12–16 h before use. For the study, the ABTS+• solution was diluted in deionized water or ethanol to an absorbance of 0.7 (±0.02) at 734 nm. An appropriate solvent blank reading was taken. After the addition of 100 μL of sample solutions to 3 mL of ABTS+• solution, the absorbance reading was taken at 30 °C, 10 min after initial mixing. All solution was used on the day of preparation and all determinations were carried out in triplicate [54]. The percentage of inhibition of ABTS+ was calculated using the following formula:

$$\text{ABTS}^{+\bullet}\text{inhibition} = [1 - (A_1 - A_2)/A_0] \times 100\% \quad (2)$$

where A_0 = absorbance of ABTS + dilute solution + water; A_1 = absorbance of ABTS + dilute solution + sample; A_2 = absorbance of PBS + sample.

3.6. In Vitro Hypoglycemic Activity Assays of LJP and ULJP

The hypoglycemic activities of the LJP and ULJP in vitro were determined by α-amylase inhibitory activities following the method described in a previous study Yang et al. [55]. LJP and ULJP were previously dissolved in distilled water at different concentrations (1~8 mg/ mL). The absorbance was determined by a UV-1601 spectrophotometer (Beijing Ruili Analytical Instrument Co., Ltd., Beijin, China) against the reaction mixture without sample was used as the control. The α-amylase inhibitory activity (%) was given by $[1 - (A_{sample} - A_{background})/A_{control}] \times 100\%$.

3.7. Effects of LJP and ULJP on RAW264.7 Macrophages

3.7.1. Cell Culture

RAW264.7 macrophages were cultured in DMEM containing 10% FBS, 100 U/mL penicillin and 100 μg/mL streptomycin at 37 °C in a CO-150 incubator (New Brunswick Scientific, NJ, USA) with an atmosphere of 5% CO_2 [56].

3.7.2. Cell Viability Assay

According to the Rahmawati et al. [57] and other samples, with a slight modification, the blown cells were diluted, counted by a hemocytometer and the cell density was adjusted to 1.5×10^5/mL, with 100 μL/well was seeded in a 96-well plate. To prevent edge effect, a circle of PBS solution should be added around the cells. After culturing for 24 h, the old medium was discarded, 100 μL of medium was added to the control group and 100 μL/well of LJP and ULJP medium solutions were added to the experimental group and 5 duplicate wells were set up. After adding LJP and ULJP for 24 h, add 10 μL/well of CCK-8 solution and incubate in the dark until the OD value is 0.8~1.0.

3.7.3. Effects of LJP and ULJP on the Morphology of RAW264.7 Macrophages

According to published method [58,59] for the effect of the sample on the morphology of RAW264.7 macrophages, with a slight modification, RAW264.7 cells were cultured in 6-well plates with different concentrations of LJP and ULJP (100, 200, 300 μg/mL) and lipopolysaccharide (1 μg/mL) were treated for 24 h and then the morphological changes of the cells were observed.

3.7.4. Effects of LJP and ULJP on Phagocytic Activity of RAW264.7 Macrophages

According to Shin et al. [58] for the effect of the sample on the phagocytic activity of RAW264.7 macrophages, with a slight modification, the neutral red uptake assay was used to measure the phagocytic activity. The cells were seeded in 96-well plates at 1×10^5 cells/well and after culturing for 24 h, the old medium was removed and LJP and ULJP (100, 200, 300 μg/mL) were added respectively, incubated for 1 h and then lipopolysaccharide (1 μg/mL) and incubated for 16 h, with 4 parallel wells in each group. After the incubation, remove the old medium. In order to avoid polysaccharide or LPS residue, wash twice with PBS buffer, add 100 μL of 0.05% neutral red solution to each well and incubate for 1 h, remove the old medium, in order to avoid neutral The residual red solution was washed three times with PBS buffer and 100 μL of lysis buffer (ethanol: glacial acetic acid = 1:1) was added to each well, shaken for 60 min and its OD value was detected at a wavelength of 540 nm and RAW264.7 was calculated according to the OD value.

3.7.5. Effects of LJP and ULJP on NO Release from RAW264.7 Macrophages

The standard curve test of nitrite was prepared in advance, the inflammation model was established according to 2.7.3 and the amount of NO production was measured by Griess kit [59].

3.7.6. Detection of Inflammatory Factor Gene Expression by qPCR

The real-time PCR was conducted to determine iNOS mRNA. Briefly, total RNA was isolated from RAW264.7 cells using a TRIZOL reagent kit (Life technologies, Invitrogen, Carlsbad, CA, USA) according to the manufacturer's instructions. cDNA was synthesized using the SuperScript ® First-Strand synthesis system for RT-PCR (Invitrogen, Carlsbad, CA, USA) under the manufacturer's instruction. Quantitative real-time PCR was performed on the ViiATM 7 Real-Time PCR System (Applied Biosystems, Foster city, CA, USA) with Power SYBR GREEN Master Mix (Applied Biosystems, Foster City, CA, USA). The primer sequences for iNOS and beta-actin are as follows: iNOS, 5′-CACCTTGGAGTTCACCCAGT-3′ and 5′-ACCACTCGTACTTGGGATGC-3′; beta-actin, 5′-GGACAG TGAGGCCA GG ATGG-3′ and 5′-AGTGTGACGTTGACA TCCGTAAAGA-3′.

3.8. Statistical Analysis

All results in this work were expressed as mean ± standard deviation of three replicates. Data in triplicate were analyzed by one-way analysis of variance using SPSS 11.5 software package for Windows (SPSS Inc., Chicago, IL, USA).

4. Conclusions

The results of this study prove that ultrasonic degradation is a simple and efficient method to degrade LJP and obtain ULJP. The two polysaccharides have different physicochemical properties and in vitro biological activities. Compared with LJP, ULJP has a smaller molecular weight, better antioxidant capacity, better inhibitory activity on α-amylase and stronger anti-inflammatory activity on RAW264.7 macrophages. Therefore, the ultrasonic degradation method can easily and efficiently degrade LJP and significantly improve the biological activity of LJP, which provides a theoretical basis for the in-depth utilization and research and development of *L. japonica* in the fields of medicine and food.

Author Contributions: Conceptualization, B.D.; methodology, B.D.; investigation, Y.L.; data curation, H.W.; writing—original draft preparation, J.W.; writing—review and editing, B.X.; supervision, Y.Y.; project administration, B.D.; funding acquisition, B.D. All authors have read and agreed to the published version of the manuscript.

Funding: This research was funded by Hebei province's key research and development program (216Z3201G).

Institutional Review Board Statement: Not applicable.

Informed Consent Statement: Not applicable.

Data Availability Statement: Not applicable.

Acknowledgments: We appreciate Analysis and Testing Center for the technical support.

Conflicts of Interest: The authors declare no conflict of interest.

References

1. Wang, Z.P.; Wang, P.K.; Ma, Y.; Lin, J.X.; Wang, C.L.; Zhao, Y.X.; Zhang, X.Y.; Huang, B.C.; Zhao, S.G.; Gao, L.; et al. *Laminaria japonica* hydrolysate promotes fucoxanthin accumulation in *Phaeodactylum tricornutum*. *Bioresour. Technol.* **2022**, *344*, 126117. [CrossRef] [PubMed]
2. Xu, H.; Zhang, R.; Zhang, J.J. Comparative analysis of fatty acid composition of kelp from different origins. *J. Nuclear Agric.* **2019**, *33*, 759–765.
3. Porrelli, D.; Gruppuso, M.; Vecchies, F. Alginate bone scaffolds coated with a bioactive lactose modified chitosan for human dental pulp stem cells proliferation and differentiation. *Carbohydr. Polym.* **2021**, *273*, 118610. [CrossRef]
4. An, E.K.; Hwang, J.; Kim, S.J. Comparison of the immune activation capacities of fucoidan and laminarin extracted from *Laminaria japonica*. *Int. J. Biol. Macromol.* **2022**, *208*, 230–242. [CrossRef]
5. Chi, S.S.; Wang, G.; Liu, T. Transcriptomic and proteomic analysis of mannitol-metabolism-associated genes in *Saccharina japonica*. *Genom. Proteom. Bioinf.* **2020**, *18*, 415–429. [CrossRef]
6. Bonde, C.S.; Bornancin, L.; Lu, Y. Bio-guided fractionation and molecular networking reveal fatty acids to be principal anti-parasitic compounds in nordic seaweeds. *Front. Pharmacol.* **2021**, *12*, 674520. [CrossRef]
7. Gammone, M.A.; D'Orazio, N. Anti-obesity activity of the marine carotenoid fucoxanthin. *Mar. Drugs* **2015**, *13*, 2196–2214. [CrossRef]
8. Xu, Y.; Zhang, Z.; Feng, H. *Scorias spongiosa* polysaccharides promote the antioxidant and anti-Inflammatory capacity and its effect on intestinal microbiota in mice. *Front. Microbiol.* **2022**, *13*, 865396. [CrossRef]
9. Yu, P.; Yang, S.; Xiao, Z. Structural characterization of sulfated polysaccharide isolated from red algae (*Gelidium crinale*) and antioxidant and anti-inflammatory effects in macrophage cells. *Front. Bioeng. Biotechnol.* **2021**, *9*, 794818.
10. Luan, F.; Zou, J.; Rao, Z.; Ji, Y.; Lei, Z.; Peng, L.; Yang, Y.; He, X.; Zeng, N. Polysaccharides from *Laminaria japonica*: An insight into the current research on structural features and biological properties. *Food Funct.* **2021**, *12*, 4254–4283. [CrossRef]
11. Yuan, L.; Zhong, Z.; Liu, Y. Structures and immunomodulatory activity of one galactose- and arabinose-rich polysaccharide from *Sambucus adnata*. *Int. J. Biol. Macromol.* **2022**, *207*, 730–740. [CrossRef] [PubMed]
12. Swathi, N.; Kumar, A.G.; Parthasarathy, V. EnteromorphaIsolation of species and analyzing its crude extract for the determination of *in vitro* antioxidant and antibacterial activities. *Biomass Convers. Biorefin.* **2022**, *11*, 1–10.

13. Lin, H.; Zhang, J.; Li, S.; Zheng, B.; Hu, J. Polysaccharides isolated from *Laminaria japonica* attenuates gestational diabetes mellitus by regulating the gut microbiota in mice. *Food Front.* **2021**, *2*, 208–217. [CrossRef]
14. Cui, C.; Sun-Waterhouse, D.X.; Zhao, L.X. Polysaccharides from *Laminaria japonica*: Structural characteristics and antioxidant activity. *LWT Food Sci. Technol.* **2016**, *73*, 602–608. [CrossRef]
15. Sun, Y.; Hou, S.; Song, S. Impact of acidic, water and alkaline extraction on structural features, antioxidant activities of *Laminaria japonica* polysaccharides. *Int. J. Biol. Macromol.* **2018**, *112*, 985–995. [CrossRef]
16. Lasunon, P.; Sengkhamparn, N. Effect of ultrasound-assisted, microwave-assisted and ultrasound-microwave-assisted extraction on pectin extraction from industrial tomato waste. *Molecules* **2022**, *27*, 1157. [CrossRef]
17. Ai, J.; Yang, Z.; Liu, J. In Vitro structural characterization and fermentation characteristics of enzymatically extracted black mulberry polysaccharides. *J. Agric. Food. Chem.* **2022**, *70*, 3654–3665. [CrossRef]
18. Yang, M.; Ren, W.; Li, G.; Yang, P.; Chen, R.; He, H. The effect of structure and preparation method on the bioactivity of polysaccharides from plants and fungi. *Food Funct.* **2022**, *13*, 12541–12560. [CrossRef]
19. Bhadja, P.; Tan, C.Y.; Ouyang, J.M.; Yu, K. Repair effect of seaweed polysaccharides with different contents of sulfate group and molecular weights on damaged HK-2 cells. *Polymers* **2016**, *8*, 188. [CrossRef]
20. Zha, S.H.; Zhao, Q.S.; Ouyang, B.; Mo, J.; Chen, J.L. Molecular weight controllable degradation of *Laminaria japonica* polysaccharides and its antioxidant properties. *J. Ocean Univ. China* **2016**, *15*, 637–642. [CrossRef]
21. Xu, Z.; Zuo, Z.Q.; Gaowa, B.; Gu, Y.Y.; Hui, C.; Shen, Y.L.; Xu, H.P. The antithrombotic effects of low molecular weight fragment from enzymatically modified of *Laminaria Japonica* polysaccharide. *Med. Sci. Monit.* **2020**, *26*, e920221. [CrossRef] [PubMed]
22. Chen, Y.; Huang, W.; Chen, Y.; Wu, M.; Jia, R.; You, L. Influence of molecular weight of polysaccharides from *Laminaria japonica* to LJP-based hydrogels: Anti-inflammatory activity in the wound healing process. *Molecules* **2022**, *27*, 6915. [CrossRef] [PubMed]
23. Wu, Y.M. *Degradation of Laminarin LJP4, Analysis and Identification of Its Products and Study on Anticoagulant Activit*; Northwestern University: Xi'an, China, 2016.
24. Mao, Y.H.; Song, A.X.; Li, L.Q.; Yang, Y.; Yao, Z.P.; Wu, J.Y. A high-molecular weight exopolysaccharide from the Cs-HK1 fungus: Ultrasonic degradation, characterization and in vitro fecal fermentation. *Carbohydr. Polym.* **2020**, *246*, 116636. [CrossRef] [PubMed]
25. Chen, Q.; Kou, L.; Wang, F. Size-dependent whitening activity of enzyme-degraded fucoidan from *Laminaria japonica. Carbohydr. Polym.* **2019**, *225*, 115211. [CrossRef] [PubMed]
26. Feng, J.X. *Study on Chemical Structure, Chain Conformation and Anti-Inflammatory Activity of Polysaccharide Degraded by Ultrasoun*; Hebei Normal University of Science and Technology: Qinhuangdao, China, 2021.
27. Larsen, L.R.; van der Weem, J.; Caspers-Weiffenbach, R. Effects of ultrasound on the enzymatic degradation of pectin. *Ultrason. Sonochem.* **2021**, *72*, 105465. [CrossRef] [PubMed]
28. Gao, J. *Structural Characterization of Laminarin and Its Effects on Dyslipidemia-Related Intestinal Flora*; South China University of Technology: Guangzhou, China, 2019.
29. Lee, I.S.; Ko, S.J.; Lee, Y.N.; Lee, G.; Rahman, H.; Kim, B. The effect of *Laminaria japonica* on metabolic syndrome: A systematic review of its efficacy and mechanism of action. *Nutrients* **2022**, *14*, 3046. [CrossRef]
30. Yu, L. *Study on the Structure and Biological Activity of Functional Components in Wakame*; Shanghai Normal University: Shanghai, China, 2021.
31. He, Y.F.; Qian, J.Y.; Li, H. GC-MS combined with two-dimensional nuclear magnetic resonance to analyze the structure of Clam polysaccharide Fr.2A. *Mar. Sci.* **2017**, *41*, 50–57.
32. Li, H.Y.; Yi, Y.L.; Guo, S.; Zhang, F.; Yan, H.; Zhan, Z.L.; Zhu, Y.; Duan, J.A. Isolation, structural characterization and bioactivities of polysaccharides from *Laminaria japonica*: A review. *Food Chem.* **2022**, *370*, 131010. [CrossRef]
33. Yang, K.; Jin, Y.; Cai, M.; He, P.; Tian, B.; Guan, R.; Yu, G.; Sun, P. Separation, characterization and hypoglycemic activity in vitro evaluation of a low molecular weight heteropolysaccharide from the fruiting body of *Phellinus pini. Food Funct.* **2021**, *12*, 3493–3503. [CrossRef]
34. Du, J.; An, X.P.; Liu, N. Effects of enzymatic hydrolysis on the structure and antioxidant activity of corncob polysaccharides in vitro. *Feed Ind.* **2021**, *42*, 45–50.
35. Pan, W.F. *Preparation of Astragalus Polysaccharide Degradation Products and Their Anti-Aging Effects*; Guangdong Pharmaceutical University: Guangzhou, China, 2020.
36. Yan, S.L.; Pan, C.; Yang, X.Q. Degradation, structural characterization and determination of hypoglycemic activity of *Pinus alba* polysaccharide. *Food Ferment. Ind.* **2021**, *47*, 119–126.
37. Wu, Z.Y. *Extraction of Kelp Polysaccharide and its Effect on Macrophage Inflammatory Response and Alleviation of CCl4-Induced Liver Injury in Mice*; Yantai University: Yantai, China, 2021.
38. Qiu, H.C.; Lai, K.D.; Liang, B. Effects of total flavonoids of pine on lipopolysaccharide-induced inflammation in RAW264.7 cells. *Guangxi Sci.* **2021**, *28*, 396–401.
39. Wang, C.; Zhang, Y.; Xue, H. Extraction kinetic model of polysaccharide from *Codonopsis pilosula* and the application of polysaccharide in wound healing. *Biomed. Mater.* **2022**, *17*, 025012. [CrossRef] [PubMed]
40. Yan, J.K.; Wang, C.; Yu, Y.B. Physicochemical characteristics and *in vitro* biological activities of polysaccharides derived from raw garlic (*Allium sativum* L.) bulbs via three-phase partitioning combined with gradient ethanol precipitation method. *Food Chem.* **2020**, *339*, 128081. [CrossRef] [PubMed]

41. Wu, X.; Liu, Z.; Liu, Y. Immunostimulatory effects of polysaccharides from *Spirulina platensis in vivo* and *vitro* and their activation mechanism on RAW246.7 macrophages. *Mar. Drugs* **2020**, *18*, 538. [CrossRef] [PubMed]
42. Yi, Y.; Hua, H.; Sun, X. Rapid determination of polysaccharides and antioxidant activity of *Poria cocos* using near-infrared spectroscopy combined with chemometrics. *Spectrochim. Acta A Mol. Biomol. Spectrosc.* **2020**, *240*, 118623. [CrossRef]
43. Li, J.; Zhao, Y.; Jiang, X. Quantitative analysis of protein in thermosensitive hydroxypropyl chitin for biomedical applications. *Anal. Biochem.* **2020**, *599*, 113745. [CrossRef]
44. Su, D.L.; Li, P.J.; Quek, S.Y. Efficient extraction and characterization of pectin from orange peel by a combined surfactant and microwave assisted process. *Food Chem.* **2019**, *286*, 1–7. [CrossRef]
45. Li, Q.M.; Zha, X.Q.; Zhang, W.N.; Liu, J.; Pan, L.H.; Luo, J.P. *Laminaria japonica* polysaccharide prevents high-fat-diet-induced insulin resistance in mice via regulating gut microbiota. *Food Funct.* **2021**, *12*, 5260–5273. [CrossRef]
46. Chen, Y.; Wang, T.; Zhang, X. Structural and immunological studies on the polysaccharide from spores of a medicinal entomogenous fungus *Paecilomyces cicadae*. *Carbohydr. Polym.* **2021**, *254*, 117462. [CrossRef]
47. Li, Z.; Xiao, W.; Xie, J. Isolation, characterization and antioxidant activity of yam polysaccharides. *Food* **2022**, *11*, 800. [CrossRef] [PubMed]
48. Gao, J.; Liu, C.; Shi, J. The regulation of sodium alginate on the stability of ovalbumin-pectin complexes for VD encapsulation and *in vitro* simulated gastrointestinal digestion study. *Food Res. Int.* **2021**, *140*, 110011. [CrossRef] [PubMed]
49. Kibar, H.; Arslan, Y.E.; Ceylan, A. Weissella cibaria EIR/P2-derived exopolysaccharide: A novel alternative to conventional biomaterials targeting periodontal regeneration. *Int. J. Biol. Macromol.* **2020**, *165*, 2900–2908. [CrossRef] [PubMed]
50. Takis, P.G.; Jiménez, B.; Sands, C.J. SMolESY: An efficient and quantitative alternative to on-instrument macromolecular H-NMR signal suppression. *Chem. Sci.* **2020**, *11*, 6000–6011. [CrossRef] [PubMed]
51. Insang, S.; Kijpatanasilp, I.; Jafari, S. Ultrasound-assisted extraction of functional compound from mulberry (*Morus alba* L.) leaf using response surface methodology and effect of microencapsulation by spray drying on quality of optimized extract. *Ultrason. Sonochem.* **2022**, *82*, 105806. [CrossRef]
52. Frühbauerová, M.; Červenka, L.; Hájek, T. Bioaccessibility of phenolics from carob (*Ceratonia siliqua* L.) pod powder prepared by cryogenic and vibratory grinding. *Food Chem.* **2022**, *377*, 131968. [CrossRef]
53. Yang, H.R.; Chen, L.H.; Zeng, Y.J. Structure, antioxidant activity and *in vitro* hypoglycemic activity of a polysaccharide purified from *Tricholoma matsutake*. *Food* **2021**, *10*, 2184. [CrossRef]
54. Yu, H.T.; Chen, Y.F.; Sun, M.C. A novel polymeric nanohybrid antimicrobial engineered by antimicrobial peptide MccJ25 and chitosan nanoparticles exerts strong antibacterial and anti-inflammatory activities. *Front. Immunol.* **2021**, *12*, 811381.
55. Rahmawati, L.; Park, S.H.; Kim, D.S. *Prasiola japonica* anti-inflammatory activities of the ethanol extract of, an edible freshwater green algae, and its various solvent fractions in LPS-induced macrophages and carrageenan-induced paw edema via the AP-1 pathway. *Molecules* **2021**, *27*, 194. [CrossRef]
56. Kong, X.P.; Chen, Z.H.; Xia, Y.J. Optimization of RAW264.7 cell inflammation model and study on anti-inflammatory activity of isoquinoline alkaloids. *Mod. Med. Clin.* **2020**, *35*, 2293–2299.
57. Zhang, N.; Ma, H.; Zhang, Z.F. Characterization and immunomodulatory effect of an alkali-extracted galactomannan from *Morchella esculenta*. *Carbohydr. Polym.* **2022**, *278*, 118960. [CrossRef] [PubMed]
58. Shin, S.W.; Hwang, Y.S. Anti-priodontitis efect of ehanol etracts of seds. *Nutrients* **2021**, *14*, 136. [CrossRef] [PubMed]
59. Hou, M.D.; Gao, J.; Liu, Z.Q. Antioxidant and immunomodulatory activities In Vitro of a neutral polysaccharide from ginger (*Zingiber Officinale*). *Starch Stärke* **2021**, *73*, 2100048.

molecules

MDPI

Article

Sodium Butyrate Attenuates AGEs-Induced Oxidative Stress and Inflammation by Inhibiting Autophagy and Affecting Cellular Metabolism in THP-1 Cells

Man Yan [1,†], Xiang Li [2,†], Chang Sun [1], Jiajun Tan [1], Yuanyuan Liu [3], Mengqi Li [1], Zishang Qi [1], Jiayuan He [4], Dongxu Wang [5,*] and Liang Wu [1,*]

1 Department of Laboratory Medicine, School of Medicine, Jiangsu University, Zhenjiang 212013, China
2 Medical Laboratory Department, Huai'an Second People's Hospital, Huai'an 223022, China
3 Department of Endocrinology, The Affiliated Huai'an No. 1 People's Hospital of Nanjing Medical University, Huai'an 223300, China
4 Zhenjiang Center for Disease Control and Prevention, Zhenjiang 212002, China
5 School of Grain Science and Technology, Jiangsu University of Science and Technology, Zhengjiang 212100, China
* Correspondence: wdx@just.edu.cn (D.W.); wlujs@ujs.edu.cn (L.W.)
† These authors contributed equally to this work.

Abstract: In recent years, sodium butyrate has gained increased attention for its numerous beneficial properties. However, whether sodium butyrate could alleviate inflammatory damage by macrophage activation and its underlying mechanism remains unclear. The present study used an advanced glycosylation products- (AGEs-) induced inflammatory damage model to study whether sodium butyrate could alleviate oxidative stress, inflammation, and metabolic dysfunction of human monocyte-macrophage originated THP-1 cells in a PI3K-dependent autophagy pathway. The results indicated that sodium butyrate alleviated the AGEs-induced oxidative stress, decreased the level of reactive oxygen species (ROS), increased malondialdehyde (MDA) and mRNA expression of pro-inflammatory cytokines of interleukin (IL)-1β and tumor necrosis factor (TNF)-α, and increased the content of superoxide dismutase (SOD). Sodium butyrate reduced the protein expression of the NLR family, pyrin domain-containing protein 3 (NLRP3) and Caspase-1, and decreased the nucleus expression of nuclear factor-kappaB (NF-κB). Sodium butyrate decreased the expression of light-chain-associated protein B (LC3B) and Beclin-1, and inhibited autophagy. Moreover, sodium butyrate inhibited the activation of the PI3K/Akt pathway in AGEs-induced THP-1 cells. In addition, the metabolomics analysis showed that sodium butyrate could affect the production of phosphatidylcholine, L-glutamic acid, UDP-N-acetylmuraminate, biotinyl-5'-AMP, and other metabolites. In summary, these results revealed that sodium butyrate inhibited autophagy and NLRP3 inflammasome activation by blocking the PI3K/Akt/NF-κB pathway, thereby alleviating oxidative stress, inflammation, and metabolic disorder induced by AGEs.

Keywords: diabetic nephropathy; sodium butyrate; advanced glycation end products; inflammatory damage; cellular metabolism

1. Introduction

The excessive activation of macrophages and high oxidative stress in diabetic patients is the important cause of diabetic kidney injury [1,2]; about 25–40% of diabetic patients develop kidney damage that eventually leads to end-stage renal disease, which is one of the main causes of death in patients with diabetes [3]. Due to abnormal glucose metabolism, advanced glycation end products (AGEs) and other glucose toxic products can be produced in diabetic patients through polyol and hexosamine pathways [4,5]. The AGEs can bind to AGEs receptors (receptor for advanced glycation end products, RAGEs) on the surface of macrophages to induce excessive inflammation and oxidative stress [6]. It can promote the migration, activation, and aggregation of macrophages to the kidney, leading to the recruiting of a large number of macrophages in the kidney and eventually leading to irreversible kidney damage [7,8]. Therefore, prevention and management of AGEs-induced macrophage inflammation and oxidative stress are the key to preventing diabetic kidney damage.

The clinical studies have also shown that the increasing dietary fiber intake can effectively improve symptoms and prevent complications in type 2 diabetes (T2DM) [9,10]. Short-chain fatty acids (SCFAs) are produced by the fermentation of indigestible dietary fiber by some anaerobic bacteria in the colon, mainly composed of acetic acid, propionic acid, and butyric acid, and other organic acids with a carbon atom number less than six [11]. A central function of inflammation control is played by microbiota-derived metabolites, specifically SCFAs such as acetate, propionate, and butyrate. As an SCFA, butyrate mainly originates in the gut when dietary fiber ferments; however, it can reach the bloodstream and plays a role in inflammation- and immunity-associated diseases such as inflammatory bowel disease, asthma, and arthritis [12–14]. There has been evidence that sodium butyrate (NaB) may be effective in the prevention and treatment of diabetic kidney damage in vivo and in vitro in previous studies [15]. However, the mechanism of NaB ameliorating diabetic kidney damage is unclear; it is speculated that the autophagy and related cell metabolism are a possible signaling pathway [16].

There is a crucial and rudimentary biological process in chronic kidney diseases called autophagy that plays an important role in both physiological and pathological processes [17]. Physiologically, it is the process of degrading proteins and organelles mediated by lysosomes, and regulates cell metabolism and survival. Many studies have shown that a number of signal transduction pathways are involved in the regulation of autophagy, i.e., PI3K/AKT, and the signaling crosstalk between PI3K/AKT and the NOD-like receptor pyrin domain-containing protein 3 (NLRP3) that controls mTOR activity, an important member of the NOD-like receptor (NLR) family, in the cytosol and activates sterile inflammation that was shown to be up-regulated in diabetic patients with kidney injury [18–20]. The activation of autophagy may reduce hyperuricemia-induced inflammasome activity by targeting ubiquitylated inflammasomes for degradation and decreasing reactive oxygen species (ROS) production and downstream inflammatory responses [21–23]. It is well known that NaB is an important energy substance for intestinal epithelial cells and has the functions of regulating energy metabolism by modulating the gut microbiota. Zhang et al. [24] reported that feeding NaB to insulin-resistant db-/db- mice can improve liver glycogen metabolism. However, whether NaB can alleviate oxidative stress and inflammation through other aspects of cell metabolism remains unclear.

In this study, we investigated the effects of NaB on AGEs-induced inflammatory damage in the human monocyte-macrophage originated THP-1 cells, and autophagy signal pathway and cell metabolism were detected. Our findings elucidated the potential mechanisms of NaB protective function in THP-1 cell inflammation and gave some clues for the potential therapy for diabetic nephropathy.

2. Results

2.1. NaB Inhibits Cellular Inflammation Induced by AGEs

To determine the ability of AGEs to induce inflammation, THP-1 cells were activated by high, medium, and low concentrations of AGEs (100, 200, and 400 μg/mL), and the expression of inflammatory cytokines was detected by qRT-PCR. The results showed that, compared with the NC group, the mRNA expressions of IL-1β, TNF-α, NLRP3, and Caspase-1 of THP-1 cells were significantly increased when treated with the concentrations of 200 and 400 μg/mL AGEs (Figure 1A). In a subsequent experiment, we used 400 μg/mL AGEs to activate THP-1 cells and 400 μmol/L NaB to inhibit the inflammatory response of THP-1 cells. Compared with the AGEs group, when treated with NaB (400 μmol/L), the mRNA expressions of IL-1β, TNF-α, NLRP3 and Caspase-1 were significantly decreased in the AGEs+NaB group (Figure 1B).

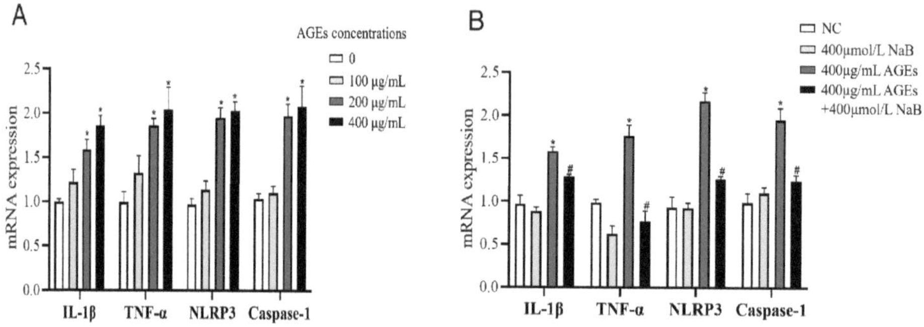

Figure 1. The mRNA expression of inflammatory cytokines of THP-1 cells (n = 3/group). (**A**): THP-1 cells were treated with three different concentrations of AGE. (**B**): THP-1 cell were pre-treated with 400 μmol/L of NaB for 24 h, and then treated with 400 μg/mL of AGEs to induce inflammation. NC: normal control; L-AGEs: 100 μg/mL AGEs; M-AGEs: 200 μg/mL AGEs; H-AGEs: 400 μg/mL AGEs; NaB: 400 μmol/L NaB; AGEs: 400 μg/mL AGEs; AGEs+NaB: pre-treated with 400 μmol/L of NaB for 24 h, and then treated with 400 μg/mL of AGEs. *: compared with the NC group, $p < 0.05$; #: compared with the AGEs group, $p < 0.05$.

2.2. NaB Inhibits Cellular Oxidative Stress Induced by AGEs

Oxidative stress plays an important role in the occurrence and development of inflammation. In this study, we determined the concentrations of ROS, MDA, and SOD in THP-1 cells. Compared with the NC group, the concentrations of ROS and MDA in the AGEs group were significantly increased, while the concentration of SOD was significantly decreased ($p < 0.05$) (Figure 2). Compared with the AGEs group, NaB can significantly increase the expression of SOD and decrease the expressions of ROS and MDA ($p < 0.05$) (Figure 2).

2.3. NaB Inhibits the Activation of Autophagy Pathway Induced by AGEs

Compared with the AGEs group, after treatment with NaB, the expression of NF-κB p65 in the nucleus was significantly decreased, and the expression of NF-κB p65 in the cytoplasm was significantly increased ($p < 0.05$) (Figure 3A). After treatment with NaB, compared with the AGEs group, the expressions of NLRP3 and Caspase-1 were significantly decreased ($p < 0.05$) (Figure 3B). It is known that beclin-1 and light-chain-associated protein B (LC3B) are key proteins in autophagy, and autophagy plays an important role in regulating inflammation. Compared with the AGEs group, the expressions of Beclin-1 and LC3B in the AGEs+NaB group were significantly decreased ($p < 0.05$) (Figure 3C). The PI3K/Akt pathway is an upstream pathway of NLRP3 and NF-κB and is the important regulatory signal of autophagy. In this study, with the pre-treatment of NaB, the expressions of p-PI3K and p-Akt were significantly inhibited in AGEs-treated cells ($p < 0.05$) (Figure 3D).

2.4. NaB Affects the Constitution of Cellular Metabolites

A total of 173 metabolites were detected by comprehensive analysis of the metabolic data of THP-1 cell samples from each group. The principal component analysis (PCA) of these metabolites showed a good clustering among the groups. The NC and AGEs groups could be clearly distinguished, which indicated that the cell metabolites in those two groups above changed significantly when treated with AGEs (Figure 4A). The OPLS-DA model was further used to evaluate the contribution of differential metabolites in THP-1 after AGEs and NaB treatment. In the OPLS-DA score plot, the abscissa was predictive principal component analysis, and the ordinate was orthogonal principal component analysis. The samples of each groups were clearly separated and well-clustered. In this score plot, R2X (the model's interpretation of X variable) was 0.839, R2Y (the model's interpretation of Y variable) was 0.989, and Q2 (the model's predictability) was 0.961. The results showed that there were significant differences in metabolite information among the three groups of samples (Figure 4B).

Next, the differential metabolites among each group were evaluated using variable influence on the project score (VIP > 1) of OPLS-DA and student's *t*-test analysis (with $p < 0.05$). The variation trends of the above differential metabolites are shown in Table 1. Compared with the NC group, the down-regulated metabolites were L-glutamic acid, ceramide (d18:1/12:0), elaidic acid, and phosphatidylcholine; the up-regulated products were biotinyl-5′-AMP, oxidized glutathione, Prostaglandin F1a (PGF1a), N-methyltryptamine, androsterone, palmitic acid, and 13S-hydroxyoctadecadienoic acid. Compared with the AGEs group, undecanoic acid, PGF1-α, phosphatidylcholine, and L-glutamic acid were significantly up-regulated, and UDP-N-acetylmuraminate, biotinyl-5′-AMP, sphingomyelin (SM) (d18:1/16:0), and thromboxane B2 (TXB2) were significantly down-regulated.

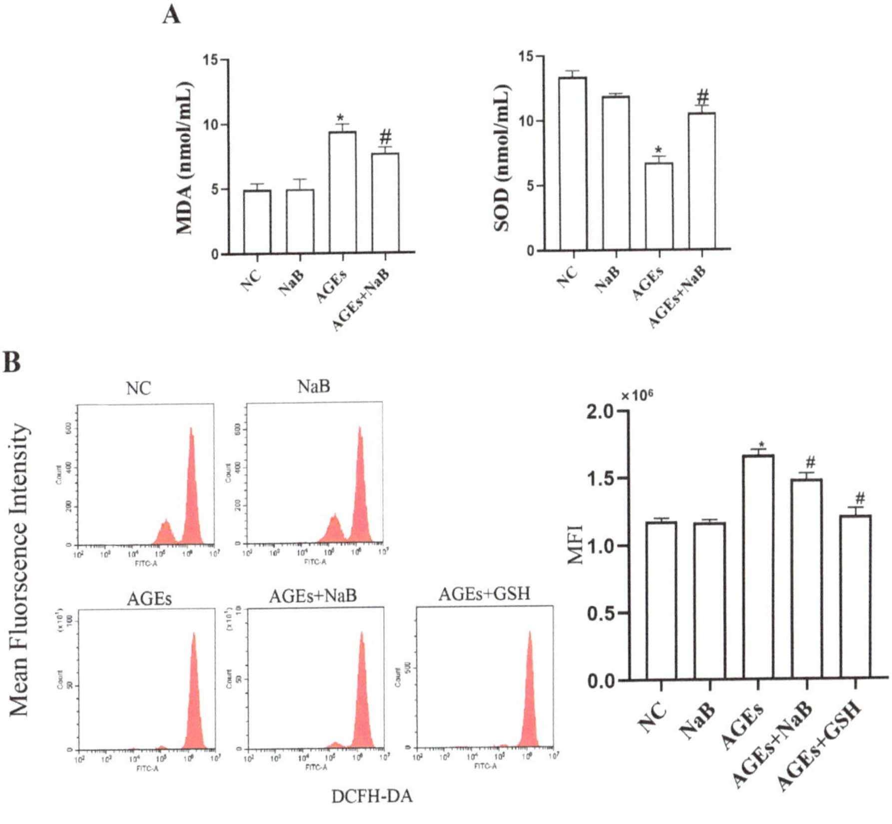

Figure 2. The oxidative stress-related ROS, MDA, and SOD level in THP-1 cells (n = 3/group). (**A**): Intracellular MDA level and intracellular SOD level. The MDA level was determined by thibabituric acid (TBA) method. The SOD level was determined by xanthine oxidase method (hydroxylamine method). (**B**): Intracellular ROS level. The intracellular ROS level was determined by DCFH-DA fluorescent probe kit. NC: normal control; NaB: 400 μmol/L NaB; AGEs: 400 μg/mL AGEs; AGEs+NaB: pre-treated with 400 μmol/L of NaB for 24 h, and then treatment with 400 μg/mL of AGEs. *: compared with the NC group, $p < 0.05$; #: compared with the AGEs group, $p < 0.05$.

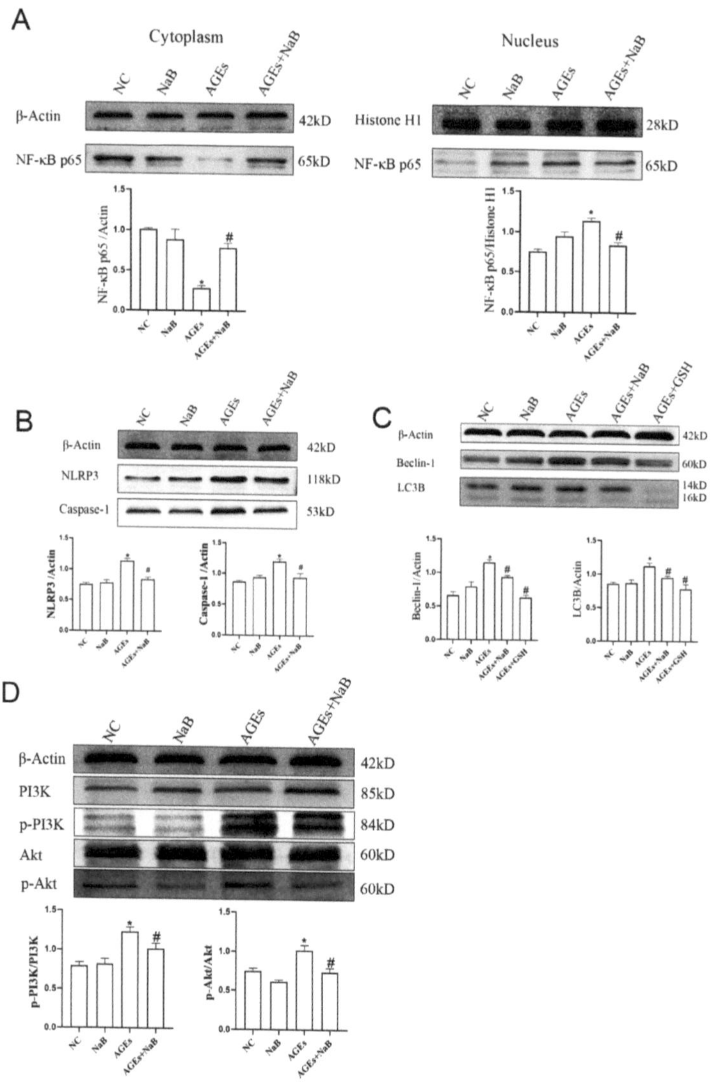

Figure 3. The expression of autophagy-related proteins and inflammation of THP-1 cells (n = 3/group). (**A**): The distribution of NF-κB p65 protein in nucleus and cytoplasm of THP-1 cells. (**B**): The expressions of NLRP3 and Caspase-1 associated with inflammation of THP-1 cell. (**C**): The expressions of Beclin-1 and LC3B associated with autophagy of THP-1 cell. (**D**): The expressions of PI3K, p-PI3K, Akt, p-Akt in THP-1 cells. NC: normal control; NaB: 400 μmol/L NaB; AGEs: 400 μg/mL AGEs; AGEs+NaB: pre-treated with 400 μmol/L of NaB for 24 h, and then treated with 400 μg/mL of AGEs; AGEs+GSH: pre-treated with 1 mmol/L of GSH for 1 h, and then treated with 400 μg/mL of AGEs; *: compared with the NC group, $p < 0.05$; #: compared with the AGEs group, $p < 0.05$.

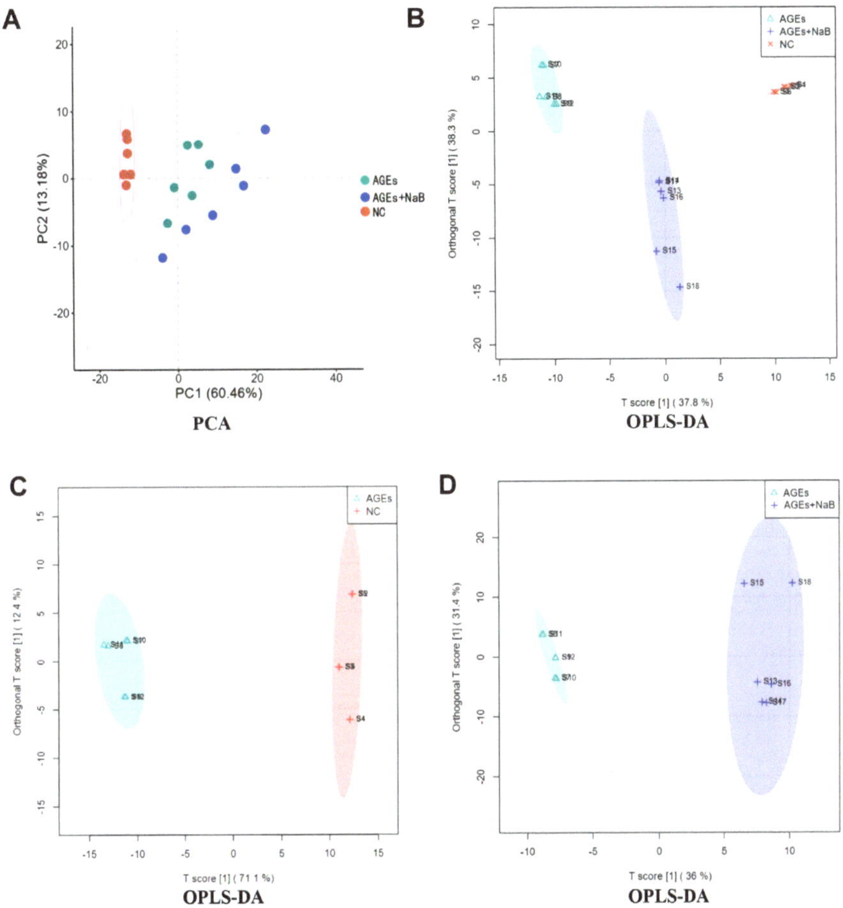

Figure 4. The PCA (**A**) and OPLS-DA (**B–D**) score plot. The NC group (red), the AGEs group (green) and the AGEs+NaB group (blue) (n = 6/group).

Table 1. Trend of differential metabolites of THP-1 cells using t-test (p < 0.05) and OPLS-DA analysis (VIP > 1).

Compounds	Pathway	m/z	Retention Time (min)	VIP	p-Value	Change
Between AGEs group and NC group						
L-Glutamic acid	D-glutamine and D-glutamate metabolism	344.06	1.11	1.4	0.04	↓
Phosphatidylcholine (PC) (18:3(9Z,12Z,15Z)/18:0)	Glycerophospholipid metabolism	806.57	23.34	1.03	0.04	↓
Biotinyl-5′-AMP	Biotin metabolism	298.57	1.82	1.06	0.03	↑
Oxidized glutathione	Glutathione metabolism	613.16	1.82	1.11	0.02	↑
Ceramide (d18:1/12:0)	Glycerophospholipid metabolism	512.5	21.51	1.15	0.01	↓
Prostaglandin F1a (PGF1a)	Glycerophospholipid metabolism	730.54	21.89	1.06	0.04	↑
N-Methyltryptamine	Tryptophan metabolism	387.19	5.36	1.06	0.02	↑
Elaidic acid	Fatty acid biosynthesis	356.35	5.52	1.09	0.01	↓
Androsterone	Steroid hormone biosynthesis	291.23	7.77	1.15	0	↑
Palmitic acid	Fatty acid biosynthesis	272.26	4.88	1.13	0	↑
13S-hydroxyoctadecadienoic acid	Biosynthesis of unsaturated fatty acids	314.27	4.92	1.18	0	↑

Table 1. *Cont.*

Compounds	Pathway	*m/z*	Retention Time (min)	VIP	*p*-Value	Change
	Between AGEs+NaB group and AGEs group					
UDP-N-acetylmuraminate	D-glutamine and D-glutamate metabolism	608.09	0.97	1.49	0.02	↓
Sphingomyelin (SM) (d18:1/16:0)	Sphingolipid metabolism	703.57	23.35	1.61	0.02	↓
Biotinyl-5′-AMP	Biotin metabolism	298.57	1.82	1.59	0	↓
Phosphatidyl ethanolamine (PE)	Glycerophospholipid metabolism	814.63	23.33	1.47	0.03	↑
Thromboxane B2 (TXB2)	Glycerophospholipid metabolism	771.51	23.34	1.54	0.01	↓
L-Glutamic acid	D-glutamine and D-glutamate metabolism	344.06	1.11	1.4	0.04	↑
Undecanoic acid	Fatty acid biosynthesis	390.36	11.81	1.39	0.04	↑
Prostaglandin F1a (PGF1a)	glycerophospholipid metabolism	365.27	9.16	1.47	0.02	↑
phosphatidylcholine (PC)	glycerophospholipid metabolism	482.36	6.96	1.59	0.01	↑

Note: ↑ indicates increase, ↓ indicates decrease, between AGEs group and NC group vs. NC group, between AGEs group and AGEs+NaB group vs. AGEs group.

2.5. NaB Affects Cellular Metabolic Pathway

The identified differential metabolites were input into the MetaboAnalyst database to construct KEGG analysis of the metabolic pathway, and set the critical value of the influence value of the metabolic pathway as 0.1. If the value was higher than this, it was regarded as a potential target pathway. AGEs treatment could significantly affect multiple metabolic pathways in THP-1 cells, including the D-glutamine and D-glutamate metabolism, biotin metabolism, sphingolipid metabolism, arginine biosynthesis, alanine, aspartate, and glutamate metabolism (Figure 5). The results of the metabolomics study showed that NaB had an excellent therapeutic effect on AGEs-induced metabolic disorders in THP-1 cells.

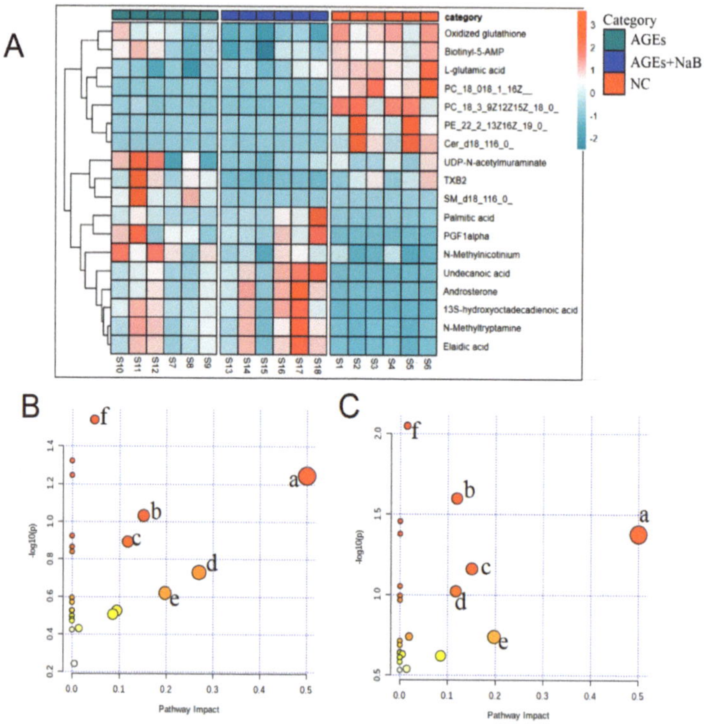

Figure 5. The cluster heat map and metabolic pathway analysis (n = 6/group). (**A**): Cluster heat map for the NC group (red), the AGEs group (green), the AGEs+NaB group (blue). (**B**): The metabolic

pathway analysis of THP-1 cell metabolites in the NC and AGEs groups. a: D-glutamine and D-glutamate metabolism, b: biotin metabolism, c: arginine biosynthesis, d: sphingolipid metabolism, e: slanine, aspartate and glutamate metabolism, f: glutathione metabolism. (**C**): The metabolic pathway analysis of THP-1 cell metabolites in the AGEs and AGEs+NaB groups. a: D-glutamine and D-glutamate metabolism, b: glycerophospholipid metabolism, c: biotin metabolism, d: arginine biosynthesis metabolism, e: alanine, aspartate and glutamate metabolism, f: sphingolipid metabolism.

3. Discussion

Gut microbiota plays an important role in the development of T2DM [25]. In fact, the damage of the intestinal mucosal barrier and the long-term chronic inflammation in the body caused by the release of intestinal bacterial endotoxin into the blood through the damaged mucosa can aggravate the insulin resistance and T2DM [26,27]. SCFAs are the products of anaerobic fermentation of undigested carbohydrates in the colon. The concentrations of SCFAs in the intestine were as high as 70–140 mmol/L, and they were easily absorbed into the blood, which can affect the immune and inflammatory response of tissues through the peripheral circulation system [28]. SCFAs can exert anti-inflammatory effects by inhibiting the recruitment and activation of white blood cells and the secretion of various pro-inflammatory cytokines, enhancing the cells' phagocytic ability. SCFAs are a potential regulator of the "gut-kidney axis", which plays an anti-inflammatory function in the kidney, among which NaB has the most significant anti-inflammatory effect [29]. Studies have shown that dietary supplements with increased dietary fiber intake, and NaB supplementation can prevent insulin resistance and improve diabetic kidney injury in patients with diabetes, but the specific mechanism needs to be further studied [30].

Diabetic nephropathy is one of the T2DM complications, mainly manifested as glomerular filtration and ultrastructure damage, glomerular basement membrane thickening, and mesangial cell proliferation [31]. In a state of hyperglycemia, protein combines with excess glucose to form a large number of AGEs that are not easily cleared. AGEs can damage renal structure and function by direct cross-linking and binding to macromolecular substances such as proteins, lipids, and nucleic acids [32,33]. In addition, AGEs can interact with the RAGE on the surface of monocytes-macrophages, microglia, and renal innate cells to trigger oxidative stress and produce a large number of ROS [34]. These ROS can increase the concentration of cellular lipid peroxide MDA and lead to the activation of the inflammatory response, which promotes macrophage recruitment and aggravating kidney injury [35]. The results of this study showed that the levels of ROS and MDA were significantly increased in the AGEs treatment group, confirming that AGEs can lead to high levels of oxidative stress response in macrophages. When treated with NaB, the productions of MDA and SOD were significantly decreased. These results suggest that NaB may improve diabetes nephropathy by interfering with inflammatory response.

It is true that inflammation is a protective reaction made by the body, but excessive inflammatory reactions can be harmful. A large number of inflammatory factors produced by the body are the important causes of diabetic kidney injury. Protein kinase B (Akt) is a serine/threonine-specific protein kinase that plays an important biological function in glucose metabolism related to cell apoptosis, survival, and proliferation. Previous studies showed that the hyperglycemic state was able to cause Akt phosphorylation and activate the PI3K/Akt pathway [36]. Meanwhile, the activated Akt can up-regulate IKB-α phosphorylation and NF-κB signaling activation. NF-κB signaling is central to the inflammatory response, activating the NLRP3 inflammasome and subsequently increasing the expression of pro-inflammatory cytokines by macrophages [37]. In this study, we found that the expressions of pro-inflammatory cytokines of IL-1β and TNF-α in the AGEs group were significantly increased, while the expressions of IL-1β and TNF-α were suppressed when treated with NaB.

Autophagy is a special protective phenomenon for mammalian eukaryotic cells to maintain their internal environment. Autophagy can wrap damaged organelles and excessive biological macromolecules in specific membrane structures and send them to lysosomes for degradation and reuse [38]. Autophagy is regulated by a variety of intracellular stimulatory signals, such as oxidative stress, inflammation, and nutritional status. When cells were stimulated by an external environment such as starvation, hypoxia, and oxidative stress, autophagy would be enhanced to maintain the stability of internal environment [39]. LC3B is a hallmark protein on the autophagosome membrane, and the concentration of Beclin-1 reflects the degree of autophagy. It was shown that ROS could induce cellular autophagy [40], and we found the AGEs could also induce the autophagy in THP-1 cells. Our hypothesis was that AGEs induce autophagy as a result of increased intracellular oxidative stress. It was known that moderate autophagy could inhibit inflammation and the expressions of pro-inflammatory cytokines [41]. However, excessive autophagy would cause a large amount of normal cell death and organ function loss [42]. In addition, our results showed that AGEs can also up-regulate the expressions of p-PI3K, p-Akt, NLRP3, and Caspase-1, and enhance the entry of NF-κB p65 protein into the nucleus. However, NaB can reverse the effect of AGEs on THP-1 cells. The results of this study suggested that NaB might inhibit autophagy by inhibiting the intracellular PI3K/Akt pathway.

Glycerol phospholipids are the main substances in cell membrane phospholipids, accounting for about 60% of lipid molecules [43]. Both phosphatidylcholine and phosphatidylthanolamine are metabolites of glycerol phospholipids, which are mainly involved in glycerol phospholipid metabolism [44]. Phospholipids such as Lyso-phosphatidylcholine, phosphatidylcholine, and sphingomyelin have turned out to be associated with insulin resistance and diabetes [45]. A high concentration of AGEs in diabetic patients can activate various protein kinases and increase lipoprotein-associated phospholipase A2 activity. As a result, the degradation of phospholipid compounds in the body is accelerated, leading to a decrease in phospholipid levels in the body. Increasing the number of SCFAs-producing bacteria in the intestine can increase the concentration of phosphatidylcholine precursor and phosphatidylcholine [46]. In this study, the concentration of phosphatidylcholine in THP-1 cells was significantly increased after pre-incubation with NaB, which may be related to glycerol and phospholipid metabolism.

Glutamic acid is the most abundant amino acid in the human body, which provides energy for cell metabolism. It is also the hub of the transformation of purine, pyrimidine, amino sugar, and other amino acids in the body [47,48]. In addition to this, glutamine also participates in the immune response [49,50]. Glutamine could enhance macrophage phagocytosis and immune response by providing ATP for macrophages. Inhibition of glutamine metabolism in macrophages can lead to the transformation of macrophages into M1 type and up-regulate the secretion of pro-inflammatory cytokines, while glutamine supplementation can promote the polarization of macrophages into M2 type and play an anti-inflammatory role [51]. Uridine diphospho-N-acetylglucosamine (UDP-N-acetylglucosamine) is a donor substrate for nucleoplasmic proteins modified with serine and threonine residues by N-acetylglucosamine (O-GlcNAc), which is the final product of the hexosamine biosynthesis pathway. The activation of the hexosamine pathway leads to the production of uridine dipho-sphate N-acetylglucosamine (UDP-GlcNAc) from glucose and glutamine, in which acetyl-CoA and UTP are consumed [52], and O-GlCNAc acetylation promotes inflammation [53]. In this study, the intervention with NaB had a callback effect on glutamine and reduced the concentration of UDP-GlcNAc. After the treatment of NaB, the concentration of glutamine increased significantly while the concentration of UDP-N-acetylmuraminate was decreased. The results indicated that NaB suppressed AGEs-induced inflammation via up-regulating glutamine concentration and glutamic acid metabolism, while down-regulating the concentration of UDP-N-acetylglucosamine. In addition, the relationship between the metabolic changes of biotin and biotinyladenylate and autophagy and inflammation also needs further study.

The present study has several limitations that need to be addressed in additional studies. (1) The beneficial effect of sodium butyrate on AGEs-induced diabetes rodent models has not been investigated, which requires further research. (2) According to our results, the oxidized GSH was reduced in AGEs-treated cells compared with NC cells, and there is a belief that it is related to the self-protective effect of cells at the early stage, but the exact mechanism remains to be determined. (3) Pharmacological inhibition of vacuolar-type H+-ATPase by bafilomycin A1 can regulate autophagy, and further investigation is needed to determine whether bafilomycin A1 affects sodium butyrate's autophagy inhibition effect.

4. Materials and Methods

4.1. Cell Culture and Treatment

THP-1 cells were purchased from Shanghai Cell Bank and cultured in RPMI 1640 medium (Biological Industries, Kibbutz, Israel) containing 10% fetal bovine serum (Biological Industries, Kibbutz, Israel) at 37 °C and 5% CO_2. THP-1 cells were stably cultured to the fourth generation and then seeded into 6-well plates with 1×10^6 cells per well. In this experiment, the cells were divided into 4 groups, including the normal control group (NC group), the sodium butyrate group (NaB group), the advanced glycation end products group (AGEs group), the AGEs+NaB group (AGEs+NaB group). In this experiment, THP-1 cells were induced to macrophage formation with the concentration of 100 ng/mL phorbol-12-myristate-13-acetate (PMA). The PMA was added into the cell culture medium. After 24 h, the NC, NaB, and AGEs groups underwent follow-up treatment, respectively, while the AGEs+NaB group was added with PMA and NaB at the beginning of the experiment, and then added with AGEs 24 h later. The NaB and the AGEs+NaB groups were pre-treated with 400 μmol/L of NaB for 24 h, and then the AGEs+NaB group was treated with 400 μg/mL of AGEs for another 24 h. The AGEs groups were only treated with 400 μg/mL of AGEs for 24 h. The cells in the NC group were cultured normally without any treatment. The cells of each group were collected at the end of the experiment for follow-up study.

4.2. Real-Time Quantitative PCR (qRT-PCR)

Total RNA of THP-1 cells was extracted by Trizol method (Vazyme Biotechnology, Nanjing, China) and reverse transcribed into cDNA (Vazyme Biotechnology, China) for qRT-PCR detection (CFX96, Bio-Rad, Hercules, CA, USA). The total PCR reaction system included 10 μL SYBR Green Master premix (Vazyme Biotechnology, China), 0.4 μL (10 μmol/L) of upper and lower primers, and 2 μL cDNA template. The reaction process included initial denaturation at 95 °C for 5 min, denaturation at 95 °C for 3 s, annealing at 58 °C for the 20 s, and extension at 72 °C for 30 s, with a total of 40 cycles. Using GAPDH as an internal reference, the relative gene expression was calculated by formula 2-ΔΔCt. The PCR primers' sequences are shown in supplementary Table S1.

4.3. Western Blotting Analysis

Total cellular proteins were extracted and performed as follows. Collected all the cells and washed twice with precooled phosphate buffer. The proteins were extracted by RIPA lysate buffer (Beyotime Biotechnology, Shanghai, China) and boiled with 5 × SDS-PAGE loading buffer for 10 min. The nucleus and cytoplasmic proteins were isolated strictly according to the instructions (Beyotime Biotechnology, China). All samples were subjected to SDS-PAGE electrophoresis, transferred to PVDF membranes (0.45 μm, Thermo Fisher Scientific, Waltham, MA, USA), then blocked with 5% skim milk for 1 h at room temperature. The rabbit antibodies including NLRP3 (BA3677, 1:250, Boster, Wuhan, China), Caspase-1 (Catalog No.: A16792, 1:1000, Abclone, Wuhan, China), Beclin-1 (A7353, 1:500, Boster, China), LC3B (Catalog No.: A19665, 1:500, Abclone, China), NF-κB p65 (Catalog No.: A10609, 1:500, Abclone, China), Histone H1 (Catalog No.: A4342, 1:500, Abclone, China), PI3K (Cat.#:AF6241, 1:1000, Affinity, China), p-PI3K (Cat.#:AF3242, 1:1000, Affinity, China), Akt (WL0003b, 1:1000, Wanlei, China), p-Akt (WLP001a, 1:1000, Wanlei, China) and β-actin (Catalog No: BA2305, 1:5000, Boster, China) were diluted with TBST buffer and incubated at 4 °C overnight. The PVDF membranes were washed 3 times in TBST buffer and incubated with the goat anti-rabbit IgG-HRP (1:10000, Boster, China) for 1 h. The ECL chemiluminescence kit (Millipore, Boston, MA, USA) was used, and the Image J software was used to analyze and calculate the expression of target proteins. All experiments were repeated 3 times.

4.4. The Cell Oxidative Stress Related Factors Detection

The DCFH-DA fluorescent probe kit (Beyotime Biotechnology, Shanghai, China) was used in this experiment to detect the intracellular ROS concentration. The superoxide dismutase (SOD) and malondialdehyde (MDA) kits were produced by Jiancheng Bioengineering Institute (Nanjing, China). The collected THP-1 cells were crushed in a mortar without nuclease, centrifuged at 16,400× *g* for 10 min, and the precipitation was discarded. The supernatant was used to detect the activities of SOD and MDA. In ROS detection, the DCFH-DA probe was diluted with serum-free RPMI 1640 medium at 1:1000. All the cells were suspended in the diluted DCFH-DA probe suspension and incubated for 20 min at 37 °C and mixed every 5 min to make sure the probe was in full contact with the cell. Finally, the serum-free medium was used to wash the cells 3 times to remove the DCFH-DA probe that did not bind to the cells, and suspended with sterile PBS buffer. According to the kit instructions, intracellular ROS was detected by Flow Cytometer (BD Biosciences, Franklin Lake, NJ, USA). In addition, the glutathione (GSH) group (AGEs+GSH group) was set in this experiment. GSH is a commonly used ROS clearance agent. In this study, the final concentration of GSH was 1 mM and the acting time was 1 h. DCFH-DA fluorescent probe kit was used to detect fluorescent removal of ROS induced by AGEs with GSH. All experiments were repeated 3 times.

4.5. Cell Collection and Preparation

Collected all the cells and added 2 mL of precooled 80% (*v*/*v*) methanol, placed at −80 °C for 20 min, then incubated at −20 °C for 1 h after ultrasonic treatment for 10 min. Then centrifuged at 4 °C at 16,400× *g* for 5 min. Collected the supernatant and dried in a freeze dryer. The lyophilized cell extract was redissolved with 200 µL 50% acetonitrile aqueous solution and fully dissolved by ultrasonic treatment for 10 min. After centrifugation at 4 °C and 16,400× *g* for 15 min, the supernatant was taken for subsequent analysis. The quality control sample consisted of 10 µL of each sample.

4.6. Chromatography and Mass Spectrometry Conditions

The chromatographic experiment was performed on an ACQUITY UPLC BEH C18 column (Waters, Milford, MA, USA, 100 mm × 2.1 mm, 1.7 µm) in the ultra-high-performance liquid chromatography (UPLC) system (Waters, USA). The column temperature was maintained at 30 °C. The flow rate was set at 0.2 mL/min. The sample injection volume was 5 µL. Solvent A was water mixed with 0.1% formic acid, and solvent B was acetonitrile. The gradient elution programs were: 0–2 min, 1–1%B; 2–8 min, 1–35%B; 8–11 min, 35–40%B; 11–14 min, 40–60%B; 14–17 min, 60–99%B; 17–18 min, 99–99%B; 18–19 min, 99–1%B; 19–20 min, 1–1%B. Electrospray ionization source (ESI) was used in high-resolution quadrupole mass spectrometer. Used ESI electrospray ionization method, and positive and negative ion mode was used for acquisition. The flow and temperature of desolvation gas were 700 L/h and 320 °C, respectively. The flow rate of the cone back flush gas was 50 L/h. The capillary voltage was 3 kV; cone hole voltage was 40 V; and source deviation voltage was 80 V. The mass scan range was 50~1500 Da.

4.7. Multivariate Data Processing and Data Processing

Peak recognition, peak matching, peak alignment, and normalization were performed on the raw data UPLC-MS/MS data by using Progenesis QI 2.3 (Waters, USA). The preprocessing results generated a data matrix including retention time (R/T), mass to charge ratio (*m*/*z*), and peak intensity. All these data were processed by peak area normalization and 80% rule, and then imported into the software for unsupervised component analysis (principal component analysis PCA) and supervised component analysis (partial least-squared discriminant analysis OPLS-DA) that supplied the intuitive diversity of the different groups. The variable importance in the projection (VIP) score reflects the contribution of the analyzed variables to the OPLS-DA model. Potential differential metabolites were screened according to the criteria of VIP > 1 and $p < 0.05$. The identification and confirmation of these differential metabolites were performed in the human metabolome database (http://www.hmdb.ca/, accessed on 10 September 2022). Based on the KEGG database, the metabolic pathways associated with potential differential metabolites were identified.

4.8. Statistical Analysis

SPSS 22.0 statistical software was used for data analysis, and data were expressed as mean ± standard deviation. Independent sample *t*-test was used for comparison between two groups, and one-way analysis of variance was used for comparison between multiple groups. Statistical graphs were generated by GraphPad Prism 8 software, $p < 0.05$ was considered statistically significant.

5. Conclusions

Our results indicated that NaB could inhibit AGEs-induced macrophage oxidative stress and inhibit AGEs-induced inflammation via a PI3K-dependent autophagy pathway. Further untargeted metabolomics analysis showed that NaB may inhibit AGEs-induced macrophage inflammation via altering cell metabolisms, which include D-glutamine and D-glutamate metabolism; glycerophospholipid metabolism; biotin metabolism; arginine biosynthesis metabolism; alanine, aspartate, and glutamate metabolism; and sphingolipid metabolism. Our results show a protective effect of butyrate to limit early molecular events underlying inflammation, suggesting a new idea for the prevention and treatment of diabetic complications, such as diabetic nephropathy.

Supplementary Materials: The following supporting information can be downloaded at: https://www.mdpi.com/article/10.3390/molecules27248715/s1, Table S1: The primers used in this experiment.

Author Contributions: Conceptualization, D.W. and L.W.; methodology, M.Y., X.L. and C.S.; software, J.H.; validation, M.L. and Z.Q.; formal analysis, D.W.; investigation, C.S., J.T. and Y.L.; resources, D.W.; data curation, L.W.; writing—original draft preparation, M.Y. and L.W.; writing—review and editing, D.W., L.W.; visualization, J.H.; supervision, D.W. and L.W.; funding acquisition, X.L. and L.W. All authors have read and agreed to the published version of the manuscript.

Funding: This work was supported by National Natural Science Foundation of China (grant number 82000475).

Institutional Review Board Statement: Not applicable.

Informed Consent Statement: Not applicable.

Data Availability Statement: The data presented in this study are available on request from the corresponding authors.

Conflicts of Interest: The authors declare no conflict of interest.

Sample Availability: Samples of the compounds are available from the authors.

References

1. Calle, P.; Hotter, G. Macrophage Phenotype and fibrosis in diabetic nephropathy. *Int. J. Mol. Sci.* **2020**, *21*, 2806. [CrossRef] [PubMed]
2. Jha, J.C.; Banal, C.; Chow, B.S.; Cooper, M.E.; Jandeleit-Dahm, K. Diabetes and kidney disease: Role of oxidative stress. *Antioxid. Redox Signal.* **2016**, *25*, 657–684. [CrossRef] [PubMed]
3. Weng, J.P.; Bi, Y. Epidemiological status of chronic diabetic complications in China. *Chin. Med. J.* **2015**, *128*, 3267–3269. [CrossRef]
4. Papachristoforou, E.; Lambadiari, V.; Maratou, E.; Makrilakis, K. Association of glycemic indices (Hyperglycemia, glucose variability, and hypoglycemia) with oxidative stress and diabetic complications. *J. Diabetes Res.* **2020**, *2020*, 7489795. [CrossRef]
5. Singh, A.; Kukreti, R.; Saso, L.; Kukreti, S. Mechanistic insight into oxidative stress-triggered signaling pathways and type 2 diabetes. *Molecules* **2022**, *27*, 950. [CrossRef]

6. Perrone, A.; Giovino, A.; Benny, J.; Martinelli, F. Advanced glycation end products (AGEs): Biochemistry, signaling, analytical methods, and epigenetic effects. *Oxid. Med. Cell. Longev.* **2020**, *2020*, 3818196. [CrossRef]
7. Byun, K.; Yoo, Y.; Son, M.; Lee, J.; Jeong, G.B.; Park, Y.M.; Salekdeh, G.H.; Lee, B. Advanced glycation end-products produced systemically and by macrophages: A common contributor to inflammation and degenerative diseases. *Pharmacol. Ther.* **2017**, *177*, 44–55. [CrossRef]
8. Chow, F.Y.; Nikolic-Paterson, D.J.; Atkins, R.C.; Tesch, G.H. Macrophages in streptozotocin-induced diabetic nephropathy: Potential role in renal fibrosis. *Nephrol. Dial. Transplant.* **2004**, *19*, 2987–2996. [CrossRef] [PubMed]
9. Li, Y.J.; Chen, X.C.; Kwan, T.K.; Loh, Y.K.; Singer, J.; Liu, Y.Z.; Ma, J.; Tan, J.; Macia, L.; Mackay, C.R.; et al. Dietary fiber protects against diabetic nephropathy through Short–Chain Fatty Acid–Mediated Activation of G Protein–Coupled receptors GPR43 and GPR109A. *J. Am. Soc. Nephrol.* **2020**, *31*, 1267–1281. [CrossRef] [PubMed]
10. Huang, W.; Man, Y.; Gao, C.; Zhou, L.; Gu, J.; Xu, H.; Wan, Q.; Long, Y.; Chai, L.; Xu, Y.; et al. Short-chain fatty acids ameliorate diabetic nephropathy via GPR43-mediated inhibition of oxidative stress and NF-κB signaling. *Oxid. Med. Cell. Longev.* **2020**, *2020*, 4074832. [CrossRef]
11. Holscher, H.D. Dietary fiber and prebiotics and the gastrointestinal microbiota. *Gut. Microbes* **2017**, *8*, 172–184. [CrossRef] [PubMed]
12. Silva, J.P.B.; Navegantes-Lima, K.C.; de Oliveira, A.L.B.; Rodrigues, D.V.S.; Gaspar, S.L.F.; Monteiro, V.V.S.; Moura, D.P.; Monteiro, M.C. Protective mechanisms of butyrate on inflammatory bowel disease. *Curr. Pharm. Des.* **2018**, *24*, 4154–4166. [CrossRef]
13. Yip, W.; Hughes, M.R.; Li, Y.; Cait, A.; Hirst, M.; Mohn, W.W.; McNagny, K.M. Butyrate shapes immune cell fate and function in allergic asthma. *Front. Immunol.* **2021**, *12*, 628453. [CrossRef] [PubMed]
14. Hui, W.; Yu, D.; Cao, Z.; Zhao, X. Butyrate inhibit collagen-induced arthritis via Treg/IL-10/Th17 axis. *Int. Immunopharmacol.* **2019**, *68*, 226–233. [CrossRef] [PubMed]
15. Tan, A.W.; Li, R.H.; Ueda, Y.; Stern, J.A.; Hussain, M.; Haginoya, S.; Sharpe, A.N.; Gunther-Harrington, C.T.; Epstein, S.E.; Nguyen, N. Platelet priming and activation in naturally occurring thermal burn injuries and wildfire smoke exposure is associated with intracardiac thrombosis and spontaneous echocardiographic contrast in feline survivors. *Front. Vet. Sci.* **2022**, *9*, 892377. [CrossRef] [PubMed]
16. Koch, E.; Nakhoul, R.; Nakhoul, F.; Nakhoul, N. Autophagy in diabetic nephropathy: A review. *Int. Urol. Nephrol.* **2020**, *52*, 1705–1712. [CrossRef]
17. Liu, N.; Shi, Y.; Zhuang, S. Autophagy in chronic kidney diseases. *Kidney Dis.* **2016**, *2*, 37–45. [CrossRef]
18. Zhuo, X.; Wu, Y.; Yang, Y.; Gao, L.; Qiao, X.; Chen, T. Knockdown of LSD1 meliorates Ox-LDL-stimulated NLRP3 activation and inflammation by promoting autophagy via SESN2-mesiated PI3K/Akt/mTOR signaling pathway. *Life Sci.* **2019**, *233*, 116696. [CrossRef]
19. Lou, D.; Zhang, X.; Jiang, C.; Zhang, F.; Xu, C.; Fang, S.; Shang, X.; Zhang, J.; Yin, Z. 3beta,23-dihydroxy-12-ene-28-ursolic acid isolated from cyclocarya paliurus alleviates NLRP3 inflammasome-mediated gout via PI3K-AKT-mTOR- dependent autophagy. *Evid. Based Complement. Altern. Med.* **2022**, *2022*, 5541232. [CrossRef]
20. Bhansali, S.; Bhansali, A.; Dutta, P.; Walia, R.; Dhawan, V. Metformin upregulates mitophagy in patients with T2DM: A randomized placebo-controlled study. *J. Cell. Mol. Med.* **2020**, *24*, 2832–2846. [CrossRef]
21. Shi, C.S.; Shenderov, K.; Huang, N.N.; Kabat, J.; Abu-Asab, M.; Fitzgerald, K.A.; Sher, A.; Kehrl, J.H. Activation of autophagy by inflammatory signals limits IL-1beta production by targeting ubiquitinated inflammasomes for destruction. *Nat. Immunol.* **2012**, *13*, 255–263. [CrossRef] [PubMed]
22. Isaka, Y.; Takabatake, Y.; Takahashi, A.; Saitoh, T.; Yoshimori, T. Hyperuricemia-induced inflammasome and kidney diseases. *Nephrol. Dial. Transplant.* **2016**, *31*, 890–896. [CrossRef] [PubMed]
23. Choe, J.Y.; Jung, H.; Park, K.Y.; Kim, S.K. Enhanced p62 expression through impaired proteasomal degradation is involved in caspase-1 activation in monosodium urate crystal-induced interleukin-1β expression. *Rheumatology* **2014**, *53*, 1043–1053. [CrossRef]
24. Zhang, W.Q.; Zhao, T.T.; Gui, D.K.; Gao, C.L.; Gu, J.L.; Gan, W.J.; Huang, W.; Xu, Y.; Zhou, H.; Chen, W.-N.; et al. Sodium butyrate improves liver glycogen metabolism in type 2 diabetes mellitus. *J. Agric. Food Chem.* **2019**, *677*, 694–705. [CrossRef] [PubMed]
25. Scheithauer, T.P.M.; Rampanelli, E.; Nieuwdorp, M.; Vallance, B.A.; Verchere, C.B.; Van Raalte, D.H.; Herrema, H. Gut microbiota as a trigger for metabolic inflammation in obesity and type 2 diabetes. *Front. Immunol.* **2020**, *11*, 571731.
26. Tilg, H.; Zmora, N.; Adolph, T.E.; Elinav, E. The intestinal microbiota fuelling metabolic inflammation. *Nat. Rev. Immunol.* **2020**, *20*, 40–54. [CrossRef] [PubMed]
27. Ma, Q.; Li, Y.; Li, P.; Wang, M.; Wang, J.; Tang, Z.; Wang, T.; Luo, L.; Wang, C.; Zhao, B. Research progress in the relationship between type 2 diabetes mellitus and intestinal flora. *Biomed. Pharmacother.* **2019**, *117*, 109138. [CrossRef]
28. Koh, A.; De, V.F.; Kovatcheva-Datchary, P.; Backhed, F. From dietary fiber to host physiology: Short-chain fatty acids as key bacterial metabolites. *Cell* **2016**, *165*, 1332–1345. [CrossRef]
29. Li, L.Z.; Tao, S.B.; Ma, L.; Fu, P. Roles of short-chain fatty acids in kidney diseases. *Chin. Med. J.* **2019**, *132*, 1228–1232. [CrossRef]
30. Huang, W.; Guo, H.L.; Deng, X.; Zhu, T.T.; Xiong, J.F.; Xu, Y.H.; Xu, Y. Short-Chain Fatty Acids Inhibit oxidative stress and inflammation in mesangial cells induced by high glucose and lipopolysaccharide. *Exp. Clin. Endocrinol. Diabetes* **2017**, *125*, 98–105. [CrossRef]

31. Giralt-López, A.; Molina-Van den, B.M.; Vergara, A.; García-Carro, C.; Seron, D.; Jacobs-Cachá, C.; Soler, M.J. Revisiting experimental models of diabetic nephropathy. *Int. J. Mol. Sci.* **2020**, *21*, 3587. [CrossRef]
32. Schmidt, A.M. Diabetes mellitus and cardiovascular disease. *Arterioscler. Thromb. Vasc. Biol.* **2019**, *39*, 558–568. [CrossRef] [PubMed]
33. Fedintsev, A.; Moskalev, A. Stochastic non-enzymatic modification of long-lived macromolecules—A missing hallmark of aging. *Ageing Res. Rev.* **2020**, *62*, 101097. [CrossRef]
34. Itakura, M.; Yamaguchi, K.; Kitazawa, R.; Lim, S.Y.; Anan, Y.; Yoshitake, J.; Shibata, T.; Negishi, L.; Sugawa, H.; Nagai, R.; et al. Histone functions as a cell-surface receptor for AGEs. *Nat. Commun.* **2022**, *13*, 2974. [CrossRef] [PubMed]
35. Hu, Q.; Qu, C.; Xiao, X.; Zhang, W.; Jiang, Y.; Wu, Z.; Song, D.; Peng, X.; Ma, X.; Zhao, Y. Flavonoids on diabetic nephropathy: Advances and therapeutic opportunities. *Chin. Med.* **2021**, *16*, 74. [CrossRef]
36. Pan, Q.Z.; Li, K.; Yang, Z.D.; Gao, M.; Shi, J.H.; Ren, S.P.; Zhao, G.Q. Dexmedetomidine attenuates high glucose-induced HK-2 epithelial-mesenchymal transition by inhibiting AKT and ERK. *Biomed. Environ. Sci.* **2020**, *33*, 323–330.
37. Liu, B.; Piao, X.; Niu, W.; Zhang, Q.; Ma, C.; Wu, T.; Gu, Q.; Cui, T.; Li, S. Kuijieyuan decoction improved intestinal barrier injury of ulcerative colitis by affecting TLR4-dependent PI3K/AKT/NF-κB oxidative and inflammatory signaling and gut microbiota. *Front. Pharmacol.* **2020**, *11*, 1036. [CrossRef]
38. Mizushima, N.; Levine, B. Autophagy in human diseases. *N. Engl. J. Med.* **2020**, *383*, 1564–1576. [CrossRef] [PubMed]
39. Li, B.; Zhou, P.; Xu, K.; Chen, T.; Jiao, J.; Wei, H.; Yang, X.; Xu, W.; Wan, W.; Xiao, J. Metformin induces cell cycle arrest, apoptosis and autophagy through ROS/JNK signaling pathway in human osteosarcoma. *Int. J. Biol. Sci.* **2020**, *16*, 74–84. [CrossRef]
40. Yang, X.; Song, X.; Li, Z.; Liu, N.; Yan, Y.; Liu, B. Crosstalk between extracellular vesicles and autophagy in cardiovascular pathophysiology. *Pharmacol. Res.* **2021**, *172*, 105628. [CrossRef]
41. White, E.; Lattime, E.C.; Guo, J.Y. Autophagy regulates stress responses, metabolism, and anticancer immunity. *Trends Cancer* **2021**, *7*, 778–789. [CrossRef]
42. Deretic, V. Autophagy in inflammation, infection, and immunometabolism. *Immunity* **2021**, *54*, 437–453. [CrossRef] [PubMed]
43. Wu, S.; Mao, C.; Kondiparthi, L.; Poyurovsky, M.V.; Olszewski, K.; Gan, B. A ferroptosis defense mechanism mediated by glycerol-3-phosphate dehydrogenase 2 in mitochondria. *Proc. Natl. Acad. Sci. USA* **2022**, *119*, e2121987119. [CrossRef] [PubMed]
44. Ouahoud, S.; Fiet, M.D.; Martínez-Montañés, F.; Ejsing, C.S.; Kuss, O.; Roden, M.; Markgraf, D.F. Lipid droplet consumption is functionally coupled to vacuole homeostasis independent of lipophagy. *J. Cell. Sci.* **2018**, *131*, jcs213876. [CrossRef] [PubMed]
45. Al-Sulaiti, H.; Diboun, I.; Agha, M.V.; Mohamed, F.F.S.; Atkin, S.; Dömling, A.S.; Elrayess, M.A.; Mazloum, N.A. Metabolic signature of obesity-associated insulin resistance and type 2 diabetes. *J. Transl. Med.* **2019**, *17*, 348. [CrossRef] [PubMed]
46. Liebisch, G.; Plagge, J.; Höring, M.; Seeliger, C.; Ecker, J. The effect of gut microbiota on the intestinal lipidome of mice. *Int. J. Med. Microbiol.* **2021**, *311*, 151488. [CrossRef] [PubMed]
47. Peplinska-Miaskowska, J.; Wichowicz, H.; Smolenski, R.T.; Jablonska, P.; Kaska, L. Comparison of plasma nucleotide metabolites and amino acids pattern in patients with binge eating disorder and obesity. *Nucleosides Nucleotides Nucleic Acids* **2021**, *40*, 32–42. [CrossRef]
48. Stine, Z.E.; Schug, Z.T.; Salvino, J.M.; Dang, C.V. Targeting cancer metabolism in the era of precision oncology. *Nat. Rev. Drug Discov.* **2022**, *21*, 141–162. [CrossRef]
49. Huang, M.; Xiong, D.; Pan, J.; Zhang, Q.; Sei, S.; Shoemaker, R.H.; Lubet, R.A.; Montuenga, L.M.; Wang, Y.; Slusher, B.S.; et al. Targeting glutamine metabolism to enhance immunoprevention of EGFR-Driven lung cancer. *Adv. Sci.* **2022**, *9*, e2105885. [CrossRef]
50. Shariatpanahi, Z.V.; Eslamian, G.; Ardehali, S.H.; Baghestani, A.R. Effects of early enteral glutamine supplementation on intestinal permeability in critically ill patients. *Indian J. Crit. Care Med.* **2019**, *23*, 356–362. [CrossRef] [PubMed]
51. Ren, W.; Xia, Y.; Chen, S.; Wu, G.; Bazer, F.W.; Zhou, B.; Tan, B.; Zhu, G.; Deng, J.; Yin, Y. Glutamine metabolism in macrophages: A novel target for obesity/type 2 diabetes. *Adv. Nutr.* **2019**, *10*, 321–330. [CrossRef] [PubMed]
52. Bolanle, I.O.; Palmer, T.M. Targeting protein O-GlcNAcylation, a link between type 2 diabetes mellitus and inflammatory disease. *Cells* **2022**, *11*, 705. [CrossRef] [PubMed]
53. Petrus, P.; Lecoutre, S.; Dollet, L.; Wiel, C.; Sulen, A.; Gao, H.; Tavira, B.; Laurencikiene, J.; Rooyackers, O.; Checa, A.; et al. Glutamine links obesity to inflammation in human white adipose tissue. *Cell. Metab.* **2020**, *31*, 375–390. [CrossRef] [PubMed]

MDPI AG
Grosspeteranlage 5
4052 Basel
Switzerland
Tel.: +41 61 683 77 34

Molecules Editorial Office
E-mail: molecules@mdpi.com
www.mdpi.com/journal/molecules

www.ingramcontent.com/pod-product-compliance
Lightning Source LLC
Chambersburg PA
CBHW041413210926
51850CB00013B/3258